Oxford Series in Ecology and Evolut

Edited by Paul H. Harvey, Robert M. May, H.

MW00844284

The Comparative Method in Evolutionary Biology
Paul H. Harvey and Mark D. Pagel
The Cause of Molecular Evolution
John H. Gillespie
Dunnock Behaviour and Social Evolution
N. B. Davies
Natural Selection: Domains, Levels, and Challenges
George C. Williams
Behaviour and Social Evolution of Wasps: The Communal Aggregation Hypothesis
Yosiaki Itô
Life History Invariants: Some Explorations of Symmetry in Evolutionary Ecology
Eric L. Charnov
Quantitative Ecology and the Brown Trout
J. M. Elliott
Sexual Selection and the Barn Swallow
Anders Pape Møller
Ecology and Evolution in Anoxic Worlds
Tom Fenchel and Bland J. Finlay
Anolis Lizards of the Caribbean: Ecology, Evolution and Plate Tectonics
Jonathan Roughgarden
From Individual Behaviour to Population Ecology
William J. Sutherland
Evolution of Social Insect Colonies: Sex Allocation and Kin Selection
Ross H. Crozier and Pekka Pamilo
Biological Invasions: Theory and Practice
Nanako Shigesada and Kohkichi Kawasaki
Cooperation Among Animals: An Evolutionary Perspective
Lee Alan Dugatkin
Natural Hybridization and Evolution
Michael L. Arnold
Evolution of Sibling Rivalry
Douglas Mock and Geoffrey Parker
Asymmetry, Developmental Stability, and Evolution
Anders Pape Møller and John P. Swaddle
Metapopulation Ecology
Ilkka Hanski
Dynamic State Variable Models in Ecology: Methods and Applications
Colin W. Clark and Marc Mangel
The Origin, Expansion, and Demise of Plant Species
Donald A. Levin
The Spatial and Temporal Dynamics of Host-Parasitoid Interactions
Michael P. Hassell
The Ecology of Adaptive Radiation
Dolph Schluter

Parasites and the Behavior of Animals
Janice Moore
Evolutionary Ecology of Birds
Peter Bennett and Ian Owens
The Role of Chromosomal Change in Plant Evolution
Donald A. Levin
Living in Groups
Jens Krause and Graeme Ruxton
Stochastic Population Dynamics in Ecology and Conservation
Russell Lande, Steiner Engen and Bernt-Erik Sæther
The Structure and Dynamics of Geographic Ranges
Kevin J. Gaston
Animal Signals
John Maynard Smith and David Harper
Evolutionary Ecology: The Trinidadian Guppy
Anne E. Magurran
Infectious Diseases and Primates Socioecology
Charles L. Nunn and Sonia M. Altizer
Computational Molecular Evolution
Ziheng Yang
The Evolution and Emergence of RNA Viruses
Edward C. Holmes
Aboveground–Belowground Linkages: Biotic Interactions, Ecosystem Processes, and Global Change
Richard D. Bardgett and David A. Wardle
Principles of Social Evolution
Andrew F. G. Bourke

Principles of Social Evolution

ANDREW F. G. BOURKE

School of Biological Sciences, University of East Anglia, UK

OXFORD
UNIVERSITY PRESS

OXFORD

UNIVERSITY PRESS

Great Clarendon Street, Oxford OX2 6DP

Oxford University Press is a department of the University of Oxford.
It furthers the University's objective of excellence in research, scholarship,
and education by publishing worldwide in

Oxford New York

Auckland Cape Town Dar es Salaam Hong Kong Karachi
Kuala Lumpur Madrid Melbourne Mexico City Nairobi
New Delhi Shanghai Taipei Toronto

With offices in

Argentina Austria Brazil Chile Czech Republic France Greece
Guatemala Hungary Italy Japan Poland Portugal Singapore
South Korea Switzerland Thailand Turkey Ukraine Vietnam

Oxford is a registered trade mark of Oxford University Press
in the UK and in certain other countries

Published in the United States
by Oxford University Press Inc., New York

First published 2011
Reprinted 2011 (twice)

British Library Cataloguing in Publication Data

Data available

Library of Congress Cataloging in Publication Data

Data available

Typeset by SPI Publisher Services, Pondicherry, India
Printed in Great Britain
on acid-free paper by
CPI Group (UK) Ltd, Croydon, CR0 4YY

ISBN 978–0–19–923115–7 (Hbk.)
       978–0–19–923116–4 (Pbk.)

3 5 7 9 10 8 6 4

To Tracey, Thomas, and William

Preface and Acknowledgements

I set out the aims of this book in the first chapter, so here I confine myself to recording my warm thanks to the large number of people who have been instrumental in one way or another in its production.

To start with, I thank the editors, Paul Harvey and Robert May, for inviting me to write a book for the Oxford Series in Ecology and Evolution. Ian Sherman at Oxford University Press saw the synopsis through its early stages, and I thank as well the anonymous assessors whose useful comments he solicited. Helen Eaton from Oxford University Press guided the book through the writing and production, and I greatly appreciate the support and input provided by her and all the production team.

I am grateful to my colleagues in the School of Biological Sciences at the University of East Anglia for their help. Tracey Chapman, Matt Gage, and Andy Johnston commented on draft chapters, and David S. Richardson and Doug Yu shared their expertise on, respectively, social vertebrates and mutualisms. For feedback on draft chapters, I likewise thank the past and present members of my research group, especially Edd Almond, Christiana Faria, Lucy Field, Jacob Holland, Tim Huggins, and Lorenzo Zanette. Edd Almond also helped in the preparation of some of the figures. This book, though not a textbook, was written at the same time as I taught an annual undergraduate module on 'Social Evolution'. I am indebted to three cohorts of UEA students for providing me with the opportunity to teach biology's most exciting topic, and for their many contributions, some made unwittingly as I tried out various ways of explaining what I wanted to say. However, humour being such a capricious thing, I have included even the jokes that seemed to amuse only me. A special mention goes to the Class of 2010 for their comments on a late draft of Chapter 1.

Of colleagues further afield, I thank Steve Frank, Andy Gardner, Eric Lucas, Tom Wenseleers, and Stuart West for discussions on intricate conceptual or empirical matters, or for allowing me to see unpublished material, or both.

Other colleagues kindly gave permission for their figures to be reproduced, or sent high-resolution versions of their figures for this purpose. My thanks go to them, and I similarly thank the relevant publishers for permission to reproduce copyright material. Details of sources for such material, with acknowledgements, are given in the corresponding table and figure legends.

My final thanks go to my family. I am profoundly grateful to my wife and colleague Tracey Chapman for her constant personal and professional support, and in particular the truly selfless gift of time. And I thank our sons, Thomas and William, for sharing with us the many pleasures, both small and large, of family life.

Contents

Preface and Acknowledgements   vii

1   **An expanded view of social evolution**   1

1.1   The biological hierarchy, the evolution of individuality,
and the major evolutionary transitions   2
1.2   Strengths of the 'major transitions view' of evolution
and aims of this book   4
1.3   Defining major evolutionary transitions and their
component stages   6
1.4   Inclusive fitness theory and the evolution of cooperation   21
1.5   Challenges remaining in the study of social evolution   23
1.6   Summary   26

2   **A primer in inclusive fitness theory**   28

2.1   Hamilton's rule and relatedness   28
     Social actions   28
     Relatedness   31
     Hamilton's rule   32
2.2   The effect of levels of relatedness on evolvable types
of social action   34
     Cooperation (narrow sense)   38
     Altruism   38
     Selfishness   40
     Spite   41
2.3   Social conflict and the tragedy of the commons   41
     Examples of kin-selected conflict   42
     Intragenomic and intergenomic conflict   43
     The tragedy of the commons   45
2.4   Assumptions of inclusive fitness theory   47
     The scale of social behaviour relative to dispersal   47

	Causes of relatedness, interests of other loci, green-beard genes, and consequences for social evolution	49
	Facultative gene expression	52
	Genes for social actions in nature	52
2.5	The value of inclusive fitness theory	57
	Relationship of inclusive fitness theory with multilevel selection theory	57
	Evidence for inclusive fitness theory	59
	Recent critiques of inclusive fitness theory	63
2.6	Summary	71

3 The major transitions in light of inclusive fitness theory — 74

3.1	Egalitarian versus fraternal major transitions	74
	Interactions within species	74
	Interactions between species	75
	Shared genes versus shared reproductive fate	78
3.2	Conflict resolution	79
	Self-limitation	80
	Coercion	84
3.3	Life cycles and the major transitions	86
	Unitary propagule (bottleneck present)	88
	Group propagule (bottleneck absent)	92
3.4	Summary	93

4 Social group formation — 95

4.1	Pathways of social group formation	95
	Pathways of social group formation among non-relatives	95
	Pathways of social group formation among relatives	96
4.2	Genetic factors in social group formation	101
	Genetic factors in social group formation among non-relatives	101
	Genetic factors in social group formation among relatives	106
4.3	Ecological factors in social group formation	110
	Ecological factors in social group formation among non-relatives	110
	Ecological factors in social group formation among relatives	113
4.4	Synergistic factors in social group formation	121
	Synergistic factors in social group formation among non-relatives	121
	Synergistic factors in social group formation among relatives	122
4.5	Hamilton's rule and social group formation	123
4.6	Summary	127

5	**Social group maintenance**	**129**
5.1	Limitation of exploitation: principles and processes	129
5.2	Limitation of exploitation from outside	130
	Recognition of self versus non-self in social groups of non-relatives	130
	Recognition of self versus non-self in social groups of relatives	131
	Recognition systems are imperfect	134
	Some forms of social group defence against external exploitation select for genetic variation within groups	136
5.3	Limitation of exploitation from inside: self-limitation through negative frequency-dependence	137
	Social bacteria	138
	Cytoplasmic male sterility	138
	Social insects	140
5.4	Limitation of exploitation from inside: self-limitation through excessive costs to the group	140
	Non-transmissible cancers in multicellular organisms	141
	Transmissible cancers in multicellular organisms	142
	Selfish reproduction within eusocial societies	145
	Worker social parasites of intraspecific origin in eusocial societies	146
	Limitation of exploitation by excessive costs in interspecific mutualisms	147
5.5	Limitation of exploitation from inside: limitation by others through coercion	148
	Enforced uniparental inheritance of mitochondria	148
	Enforced fairness in meiosis	150
	Enforced suppression of cytoplasmic male sterility and other forms of sex ratio distortion	152
	Coercion in eusocial societies	153
	Enforced fairness in interspecific mutualisms	158
5.6	Predicting the outcome of the limitation of exploitation	159
5.7	Summary	161
6	**Social group transformation**	**162**
6.1	The size-complexity hypothesis for social group transformation	162
6.2	Simple versus complex social groups	164
	Simplicity and complexity in social groups	164
	Evidence for size-associated syndromes of simplicity and complexity in multicellular organisms	168
	Evidence for size-associated syndromes of simplicity and complexity in eusocial societies	170

Number of independent evolutions of complexity
in social groups 172
Complexity, sexual reproduction, and genetic variation 175
6.3 External drivers leading to greater size in social groups 176
Short-term ecological drivers of greater size in social groups 176
Long-term evolutionary drivers of greater size in social groups 177
6.4 Effect of increasing size of the social group on group complexity 179
Reproductive division of labour within multicellular organisms 179
Non-reproductive division of labour within multicellular organisms 183
Reproductive division of labour within eusocial societies 184
Non-reproductive division of labour within eusocial societies 190
6.5 Self-reinforcing social evolution in social group transformation 190
Positive feedback favouring large group size
in multicellular organisms 190
Positive feedback favouring large group size in eusocial societies 191
6.6 The size-complexity hypothesis: conclusions 193
6.7 Summary 195

7 Synthesis and conclusions 198

7.1 The principles of social evolution: a summing-up 198
7.2 Open questions in the study of social evolution 201
7.3 The next major transition 201
7.4 Summary 205

References 206

Author Index 245

Subject Index 253

Taxonomic Index 263

I

An expanded view of social evolution

You are a scientifically curious protozoan living in a pond in Earth's very distant past, one and a half billion years ago. Having an inventive and adventurous nature, you build yourself a time machine and travel to the future. Your own world is inhabited almost entirely by single-celled creatures (unicells). Some are eukaryotes like yourself, defined by their possessing, among other things, intracellular organelles such as mitochondria or chloroplasts. The rest are prokaryotes (the Bacteria and Archaea), which lack organelles. Only the prokaryotes form multicellular structures, and these are threads, films, or mats that fail to grow very large, cannot move, and lack much in the way of internal complexity.

Arriving at the present day, you don a very small protective suit equipped with sensors and instruments with which to study your surroundings. With a shock, you become aware that the planet is now inhabited by creatures utterly alien to your previous experience. They are huge, many are mobile, and all possess a complex external and internal anatomy. Once your cilia have stopped trembling, you deploy your scientific instruments and begin to study them. With mounting astonishment you realize that each of these organisms is in fact a colony of billions of cells. These cells are like yourself but are physically stuck together by organic glue and diversified into dozens of bizarre forms. Most are sterile, having grown incapable of dividing in order to serve the organism as a whole. In short, you find that your descendants risk seeing their individuality extinguished through becoming tiny, subordinated parts of organisms of a wholly new kind, which have themselves diversified into a multitude of forms. We know these monsters as tigers, trees, toadstools, and their like, or multicellular animals, plants, and fungi.

Thunderstruck by this glimpse of the future, you flee to your own era. Then, perhaps seeking reassurance in an Edenic past, you turn the dial of your time machine in the opposite direction and travel backwards in time for another billion and a half years. Once at your destination, you rediscover your scientific curiosity and again begin to study the fauna of the time. This previous world turns out to be inhabited exclusively by prokaryotes. But your studies deliver another resounding intellectual jolt. Pondering on the lack of eukaryotes, studying the results from your field genome-sequencing kit, and carefully observing bacterial anatomy and behaviour, you realize that you yourself are a colonial organism. You and all other eukaryotic unicells arose from a symbiotic fusion of your ancestors with bacteria. Your formerly distinct bacterial partners live on as your mitochondria or chloroplasts, having propagated themselves inside your forebears in the unbroken line that leads to yourself.

In sum, you are an amalgam of simpler, more ancient cells whose descendants face a future as enslaved subunits of massive, multicellular conglomerates. You think about the processes that brought about such an extraordinary pair of transformations. You wonder what the next one will be.

1.1 The biological hierarchy, the evolution of individuality, and the major evolutionary transitions

Stripped of its elements of obvious fantasy and granted some allowance for uncertainty in the dates, the tale with which this chapter opens reflects science fact. The history of life on Earth has often been portrayed as the succession of different taxonomic groups of organisms, with, for example, the Age of Fishes leading to the Age of Reptiles, and the Age of Reptiles in turn being followed by the Age of Mammals. This is the view of traditional textbooks and lives on in present-day television documentaries, which expertly combine the most astounding twenty-first century photography and computer animation with the stodgiest kind of nineteenth-century zoology. Such a view is discredited by its unfounded assumption that ecological dominance is a one-dimensional quality readily attributable to single taxa. It also fails to see beyond the superficial differences between separate taxa to the more fundamental properties that they have in common. As Knoll (2003) nicely put it, 'Such catalogs of received wisdom can be memorized, but there isn't a lot to think about'. Another view sees the history of life as a succession of technical innovations, such as the origin of photosynthesis, of land-dwelling, of flight, and so on. This view is more defensible because it offers one way of explaining changes in the complexity of life over time. But it again fails to provide any sort of underlying, unifying framework with which to explain the logic of such changes.

An altogether different view of life's history overcomes these deficiencies by focusing on the hierarchical organization of the units of life. Genes occur in cells and cells fuse to form other cells. Cells may occur within multicellular organisms, and multicellular organisms occur, at least in some cases, in societies. Since units at each level also occur independently, and each level requires the presence of the lower ones, it follows that in the history of life there were distinct events in which genes grouped into cells, cells grouped to become a different type of cell, cells further grouped into multicellular organisms, and multicellular organisms grouped into societies. Fundamentally, the history of life has been the history of the grouping of biological units into higher-level units and the subsequent consolidation of the new higher-level units into integrated collectives, with this process, once started, having been repeated several times to generate the biological hierarchy we observe today.

This hierarchical view of life's fundamental evolutionary history arose in the late-nineteenth century and early twentieth century (Mackie 1986; Buss 1987). For example, in a 1924 text, A. Dendy wrote that 'evolution consists to a very large extent, if not mainly, in the progressive merging of individualities of a lower order in others of a higher order' (quoted in Mackie 1986). The construction of analogies between different levels of organization, especially between multicellular organisms and societies, also has a long history (Wheeler 1911; Wilson 1971). However, perhaps because, in their time, these early ideas lacked a clear theoretical underpinning, they failed to achieve widespread influence in the evolutionary synthesis of the mid-twentieth century. Similarly, in the latter half of the century, they suffered in the reaction against the naive group selectionism with which they were frequently linked (Williams 1966).

The modern form of the hierarchical view was foreshadowed by Bonner (1974), who considered 'a number of major steps in evolution where one leaps from one level of complexity to the next' via a 'compounding of units on a previous level'. Along with the origin of cellular life, the steps that Bonner (1974) highlighted were the origin of eukaryotes, the origin of multicellular organisms, and the origin of social organisms. The modern hierarchical view gathered momentum in the late-1980s with the works of Buss (1987) on the evolution of individuality and Maynard Smith (1988) on evolutionary progress and levels of selection. Buss's (1987) book, *The evolution of individuality*, presented the first treatment of the modern hierarchical view in a single, dedicated volume. Evolution of individuality refers to the evolution of individuals, in the sense of stable, integrated collectives, via the grouping together of formerly independent units. (I discuss the use of 'individual' in this sense more fully in Section 1.3.) Buss (1987) focused on the evolution of individuality in multicellular organisms. Conceptually, he placed his exposition in the framework of multilevel selection theory but clouded some issues by unnecessarily setting multilevel selection theory at odds with other theories of selection such as Hamilton's (1963, 1964) inclusive fitness theory. A comprehensive synthesis of the hierarchical view of evolution, spanning many levels and adopting a cogent approach to selection theory, came with the publication of Maynard Smith and Szathmáry's (1995) landmark book on the major transitions in evolution, a concept with which the idea of the evolution of individuality is closely intertwined. The term 'major transition in evolution' was proposed by Maynard Smith and Szathmáry (1995) to describe the evolution of each successive level of life's hierarchy. As we have seen, Bonner (1974) had already written of 'major steps in evolution' in this context. He also used the term 'transition point' to describe each step. Likewise, both Buss (1987) and Maynard Smith (1988) used 'transition' for the same purpose.

The concepts of the evolution of individuality and major transitions are themselves underpinned by a key insight whose roots stretch back to the early 1960s, being based in the gene's-eye (or 'selfish gene') view of natural selection and, by implication, in inclusive fitness theory (Hamilton 1963; Dawkins 1976, 1982). This insight is that the individuality emerging at each major evolutionary transition is a contingent state. Specifically, it is contingent upon the absence or suppression of within-individual conflict (e.g. Leigh 1977, 1991; Alexander and Borgia 1978; Eberhard 1980; Cosmides and Tooby 1981; Dawkins 1982; Buss 1987; Maynard Smith 1988; Bourke and Franks 1995; Maynard Smith and Szathmáry 1995). For, if the level of internal conflict is too great, the higher level of organization either fails to emerge or is unstable and collapses. The challenge has been to understand what kinds of process contribute to the stable evolution of each new level in the hierarchy of major transitions (Buss 1987; Maynard Smith 1988; Leigh 1991; Maynard Smith and Szathmáry 1995).

1.2 Strengths of the 'major transitions view' of evolution and aims of this book

The 'major transitions view' of evolution offers a more profound and scientifically satisfying vision of life's fundamental evolutionary history than its rivals for several reasons. First, it provides an evolutionary explanation for the biological hierarchy itself (Maynard Smith and Szathmáry 1995; Okasha 2006). It does this by viewing the hierarchy as the cumulative product of selection on organisms to form higher-level collectives. In the case of the level of the multicellular organism in particular, this is a substantial insight. Multicellular organisms are such a salient feature of life that it is easy to take their existence for granted. But the hierarchical view, aided by a gene's-eye interpretation of natural selection, demonstrates that the multicellular organism cannot be taken as a 'given'. It is not a form of biological organization that living matter must inevitably adopt, but an evolved construct in its own right, and as such its existence requires explanation (Dawkins 1976, 1982). As Buss (1987) stated, 'Individuality is a derived character'. This is one reason why a dogged insistence that the essence of Darwinian natural selection is selection at the level of the individual (e.g. Mayr 1997), where 'individual' has its traditional meaning of a single unicellular or multicellular organism, is incomplete. It leaves open what selective process brought about such organisms in the first place. Second, the major transitions view helps explain the increase in the complexity of living things over evolutionary time (Bonner 1974; Maynard Smith and Szathmáry 1995). It does this in two ways. One arises simply because a nested hierarchy is necessarily more complex than each of its lower-level constituents (Wieser 1997; McShea and Changizi 2003). The other comes about because each major transition creates conditions for the evolution of mechanisms for excluding would-be exploiters and for reducing internal conflict, which themselves add to the overall complexity of the new level (Chapters 5, 6). In the same way, to use an analogy from Dawkins (1982), computer systems in a world with malicious hackers are more complex than they would be in a solely well-intentioned world.

The major transitions view does not, however, propose that an increase in life's complexity over time is inevitable. Instead, each step in a major transition depends only on the selective conditions prevailing at the time of its occurrence. It is entirely possible that, within any one lineage, the right conditions never occur, with the result that neither does the next transition. In short, there need be no directional bias driving individuals to associate into new and higher levels of organization, and indeed the fossil record provides little evidence for such bias (McShea 1996, 2001; Marcot and McShea 2007). Hence another strength of the major transitions view of why complex life evolves is that it is consistent with the tenet that evolution by natural selection is not inherently progressive (Maynard Smith 1988; Maynard Smith and Szathmáry 1995).

Lastly, the major transition view of life's history has immense potential to provide a unified explanation for the evolution of a huge range of biological systems. The field of study that equips us to investigate the conditions for a stable transition to a new level of individuality is social evolution. To put this another way, the problem of how individuality arises and is maintained is the problem of the evolution of cooperation. (Here, I use cooperation broadly speaking, to include both altruism and narrow-sense cooperation as defined in Section 2.1.) Genes must cooperate to form a genome within a cell, cells must cooperate to form a multicellular organism, and multicellular organisms must cooperate to form a society. In a phrase that Lachmann et al. (2003) applied to organisms, each hierarchical level is 'composed of layers upon layers of cooperation'. As previously mentioned, if the selfish interests of the formerly independent units cannot be sufficiently subordinated to the collective interest of the higher-level entity, such an entity will be unstable and the transition to a new level of individuality will fail.

The working hypothesis of this book is that, as proposed by previous authors (Leigh 1991; Maynard Smith and Szathmáry 1995; Queller 2000), common principles of social evolution apply at each step in the evolution of individuality, regardless of the taxa involved and regardless of the level within the hierarchy of organization. This is an important and grand vision that conceptually unifies an immense range of superficially different phenomena in diverse taxa under the banner of social evolution (Leigh 1995; Queller 1997; Frank 2007a). For example, it implies not only that the evolution and behaviour of different forms of animal society can be considered in a single conceptual framework (e.g. Wilson 1971, 1975), but also that the same framework can be used to accommodate intracellular phenomena, such as the behaviour of selfish genetic elements (Burt and Trivers 2006). Reciprocally, applying a hierarchical understanding to life's evolutionary history has led to an expanded view of social evolution. Social evolution has grown outwards from the study of the beehive and the baboon troop to embrace the entire sweep of biological organization. It claims as its subject matter not just the evolution of social systems narrowly defined, but the evolution of all forms of stable biological grouping, from genomes and eukaryotic unicells to multicellular organisms, animal societies, and interspecific mutualisms.

My overall aim in this book is to present a fresh synthesis of this expanded view of social evolution. Specifically, I seek to articulate its theoretical basis and its principles as fully as possible, to investigate the extent to which its principles are applicable across different taxa and hierarchical levels, and to highlight new or little-appreciated principles, in addition to those already recognized. It is worth spelling out why these are valuable goals. The basic theory underpinning social evolution, namely Hamilton's (1963, 1964) inclusive fitness theory (kin selection theory), has been in place for many years and, despite lingering disagreements (Section 2.5), provides well-understood theoretical tools with which to conduct evolutionary analyses of sociality. In addition, the application of inclusive fitness theory to the various hierarchical levels of life has already advanced considerably (Maynard Smith and Szathmáry 1995; Keller 1999; Michod 2000; Queller 2000; Frank 2003; Korb and Heinze 2004; Ratnieks and Wenseleers 2008; Boomsma 2009). However, no single work has sought to integrate the main findings of these and related studies. I aim to place inclusive fitness theory centre-stage in the analysis of the major transitions as a whole. In addition, the expansion of the field of social evolution and of inclusive fitness theory to cover their new domains is still not widely appreciated beyond a fairly small group of enthusiasts, even among evolutionary biologists and behavioural ecologists. Many still regard the business of social evolution as the beehive and baboon troop alone. I seek to show them otherwise. With notable exceptions (e.g. Queller and Strassmann 2009), even those actively researching evolutionary processes at the different hierarchical levels (for example, the evolution of multicellularity or the evolution of eusociality), though plainly aware of the parallels between them, tend to occupy separate intellectual worlds. By offering a new look at the issues, I hope to increase the level of interchange between these worlds.

Another motivation for the present work concerns my title's invocation of 'principles' of social evolution. In the last decade or so, the fundamental theory of social evolution has undergone a rich mathematical flowering, with several authors having devised powerful new ways of algebraically expressing, validating, and extending its basic conclusions (e.g. Frank 1998; Rousset 2004; Grafen 2006; Lehmann and Keller 2006a; Gardner et al. 2007a). Likewise, the empirical study of social phenomena—at the level of genomes, organisms, or societies—has proceeded at a tremendous rate, leading to a large body of data from which generalizations can be drawn. Between mathematical foundations and empirical generalizations lie what I regard as the principles of social evolution. These are conceptual fusions of theory and data that embody broad truths about social evolution without necessarily being universally applicable or easily expressible in formal terms. Examples include the principles of conflict resolution via self-limitation or coercion (Sections 3.2, 5.1).

Although principles of social evolution in this sense have not been ignored (e.g. Bourke and Franks 1995; Frank 2003; Ratnieks et al. 2006), they have not been as widely articulated as their importance merits. Even Maynard Smith and Szathmáry (1995) themselves, while the first to consider each of the major transitions within a consistent evolutionary framework, did not always explicitly enunciate the common principles at work across the various hierarchical levels that they considered (Leigh 1995; Queller 1997). By concentrating on principles of social evolution, this book therefore aims to fill a gap in our current approach. In the process, it seeks to bring the field up-to-date by incorporating advances made from the mid-1990s onwards, such as the important conceptual and theoretical work of Frank (1995, 1998), Queller (2000), and Wenseleers and Ratnieks and colleagues (e.g. Wenseleers et al. 2003, 2004a, 2004b), and, on the empirical side, achievements such as the elucidation of the phylogeny of multicellular evolution in the volvocine algae by Herron and Michod (2008). Finally, I aspire not just to document the field, but to advance it. I aim to do this by, for example, suggesting new ways of classifying its phenomena (Section 1.3) and investigating those issues, such as social group transformation (see below), where our understanding needs to be improved (Section 1.5).

1.3 Defining major evolutionary transitions and their component stages

Maynard Smith and Szathmáry (1995, pp. 3–4) defined a major transition in evolution as a transition 'in the way in which genetic information is transmitted between generations' such that 'entities that were capable of independent replication before the transition can replicate only as part of a larger whole after it'. This broad definition allowed them to classify as major evolutionary transitions a wide range of processes (Table 1.1). However, not all of these processes (e.g. the evolution of human language) involved the evolution of individuality in the current sense. In a significant development, Queller (1997, 2000) further subdivided the major transitions into 'egalitarian' and 'fraternal' transitions (Table 1.2). This essentially grouped the major transitions according to whether they involved a union of unrelated entities (egalitarian) or of related entities (fraternal). Queller's (1997, 2000) proposal was a critical step because, as well as suggesting clusters of traits common to each type of transition (Table 1.2), it linked the class to which a given transition belongs directly to the formal type of social behaviour that it embodies, as defined within inclusive

Table 1.1   The major transitions in evolution as defined by Maynard Smith and Szathmáry (1995). Reproduced from Maynard Smith and Szathmáry (1995) with kind permission of E. Szathmáry

Replicating molecules	⇒	Populations of molecules in compartments
Independent replicators	⇒	Chromosomes
RNA as gene and enzyme	⇒	DNA + protein (genetic code)
Prokaryotes	⇒	Eukaryotes
Asexual clones	⇒	Sexual populations
Protists	⇒	Animals, plants, fungi (cell differentiation)
Solitary individuals	⇒	Colonies (non-reproductive castes)
Primate societies	⇒	Human societies (language)

Table 1.2   Two kinds of major evolutionary transition as defined by Queller (1997, 2000). Reproduced from Queller (2000) with kind permission of the author and Royal Society Publishing

	Egalitarian	Fraternal
Examples of cooperative alliances forged	Different molecules in compartments; genes in chromosomes; nucleus and organelles in cells; individuals in sexual unions	Same molecules in compartments; same organelles in cells; cells in individuals; individuals in colonies
Units	Unlike, non-fungible [non-interchangeable]	Like, fungible [interchangeable]
Reproductive division of labour	No	Yes
Control of conflicts	Fairness in reproduction; mutual dependence	Kinship
Initial advantage	Division of labour; combination of functions	Economies of scale; later division of labour
Means of increase in complexity	Symbiosis	Epigenesis
Greatest hurdle	Control of conflicts	Initial advantage

fitness theory (Section 3.1). Put briefly, a union of unrelated partners can involve cooperation (in its narrow sense) but not altruism, since altruism cannot evolve when interacting partners are unrelated. Conversely, a union of related partners can involve either cooperation or altruism, since positive relatedness permits the evolution of both of these forms of social behaviour. These distinctions are important because they help explain why egalitarian transitions are characterized by partners that, like other cooperators, do not sacrifice their ability to reproduce (hence their 'egalitarianism'). Examples include organelles within eukaryotic cells and the male and female halves of the genome in sexual organisms. By contrast, fraternal transitions are often characterized by the evolution of non-reproductive altruists. Here, relevant examples are non-reproductive cells within multicellular organisms and workers within social insect colonies (Queller 1997, 2000).

Note that some transitions may involve both egalitarian and fraternal components, and that whether a particular transition was egalitarian or fraternal may be unknown (Queller 2000). Specifically, a very early transition, the grouping of originally separate replicating molecules into genomes held within compartments (cells), may have involved the grouping of copies of the same molecules (fraternal transition), as well as the grouping of unlike ones (egalitarian transition). The very antiquity of this transition makes it unlikely that one could ever be certain about how it occurred (Queller 2000). Nonetheless, because of the direct link of the egalitarian–fraternal distinction with the forms of social evolution permissible under different regimes of relatedness, this distinction remains highly useful.

For the purposes of the present book, I therefore adopt a narrower definition of a major evolutionary transition than that developed by Maynard Smith and Szathmáry (1995), while embracing the distinction between egalitarian and fraternal transitions introduced by Queller (1997, 2000). I limit my use of the term major evolutionary transition to refer specifically to those transitions that involve the evolution of individuality, or at least that create groupings that are candidates for being regarded as individuals. Various definitions of 'individual' have been proposed (e.g. Buss 1987; Michod 2000). By 'individual' in this book I mean some stable, physically discrete entity that is composed of interdependent parts acting in a coordinated manner to achieve common goals and is typified by the very property of lacking a high degree of within-individual conflict (e.g. Dawkins 1982, 1990; Queller 1997, 2000). 'Physically discrete' here means that the parts of the individual are either physically joined to one another or tend to remain in close proximity. An individual in this sense could be the product of the evolution of individuality at any of the hierarchical levels of life. For example, it could be a single cell, a multicellular organism, or a eusocial society.

Two points need making about this definition of an individual. First, it need not be a concern if the definition is not precise, i.e. if in a given case it is hard to decide if some entity qualifies as an individual or not. Evolution is a process of continuous change, so it is not surprising if its products have blurry edges or a mosaic of features. This does not stop definitions in evolutionary biology being useful. Second, this definition widens the meaning of individual beyond that in ordinary use in biology. Some authors would prefer instead to widen the meaning of the term 'organism' for the same purpose, i.e. to use organism to mean everything here termed an individual (Queller and Strassmann 2009; West and Kiers 2009). The problem being addressed would then be the evolution of 'organismality' (Queller and Strassmann 2009). Still others would like to reintroduce the term 'superorganism' to refer to higher-level groupings of organisms, in particular colonies of eusocial insects (e.g. Hölldobler and Wilson 2009). I have stuck with the more established expanded use of individual (and will employ organism in its traditional sense), but do not imply a dogmatic preference for this semantic choice. The terminology may settle in time and meanwhile the key point is that all agree that there is a special property found in the natural world, here termed individuality, that requires an evolutionary explanation.

Under the above definition of individual, I identify six major transitions involving the evolution of individuality (Table 1.3). All involve groupings of previously separate entities (e.g. genes, cells, organisms, species) to form higher-level, stable collectives (e.g. genomes, multicellular organisms, eusocial societies, interspecific mutualisms). They include some transitions that are egalitarian (evolution of eukaryotic cells, zygotes, and interspecific mutualisms), some that are fraternal (evolution of multicellularity and eusocial societies), and one (evolution of genomes) that, as mentioned above, is likely to have involved both egalitarian and fraternal elements. Several are known to have occurred more than once independently and others may have done so (Box 1.1; Figs 1.1, 1.2; Table 1.4). For example, it is not widely appreciated that the transition to multicellularity and the transition to eusocial societies each occurred multiple times independently, at least 25 times in the case of the evolution of multicellularity and at least 24 times in the case of the evolution of eusociality (and many more, depending on how broadly eusociality is defined; Box 1.1). In addition, the stunning achievements of palaeontology and molecular phylogenetics now mean that several particular instances of the major transitions can be dated, at least approximately (Box 1.1; Table 1.4). Hence, it is estimated that the first cell arose around 3500 mya (million years ago), the first eukaryotic cell around 2000 mya, the first multicellular eukaryotes around 1200 mya, and the first unmistakably eusocial societies of multicellular organisms around 150 mya (Box 1.1).

I do not claim that the six cases defined above represent the only way of classifying the major transitions that involve the evolution of individuality. There are always likely to be several ways in which complex phenomena can be classified. For the present way, I again claim the merit of usefulness. Indeed, there are several ways in which my classification is an oversimplification. For example, the grouping of self-replicating molecules (genes) into genomes must itself have involved a number of key, intermediate steps (Maynard Smith and Szathmáry 1995). The origin of sexual reproduction in eukaryotes is foreshadowed in genetic exchange between prokaryotes (Maynard Smith et al. 1991; Narra and Ochman 2006). As well as involving the formation of zygotes, sexual reproduction in eukaryotes creates populations of interbreeding organisms sharing a common gene pool. The creation of such populations, though each is not an individual in the sense employed here, nonetheless has substantial evolutionary consequences. In fact both Leigh (1995) and Queller (1997) pointed out that several aspects of the evolution of sex seem to

Table 1.3 Six major evolutionary transitions leading to the evolution of individuality, with examples of phenomena proposed to fall under, or influence, the three stages of each transition. This scheme compresses transitions 1 and 2 of Maynard Smith and Szathmáry (1995) into one, and omits their transitions 3 and 8 (Table 1.1). It adds the evolution of interspecific mutualisms as a separate transition. This was clearly considered as such by Maynard Smith and Szathmáry (1995), without being included in their original table. The order of listing of each transition is not necessarily chronological and a given transition may have occurred more than once

Major evolutionary transition	Examples of phenomena at each stage		
	Social group formation	Social group maintenance	Social group transformation
1. Separate replicators (genes) ⇒ cell enclosing genome	Origin of compartmentalized genomes	Control of selfish DNA	Evolution of large, complex genomes
2. Separate unicells ⇒ symbiotic unicell[1]	Origin of eukaryotic cells	Control of organellar reproduction	Evolution of hybrid genomes through transfer of genes from organellar to nuclear genome
3. Asexual unicells ⇒ sexual unicell	Origin of zygotes	Control of meiotic drive	Evolution of obligate sexual reproduction
4. Unicells ⇒ multicellular organism	Origin of multicellular organisms	Control of selfish cell lineages (cancers)	Evolution of segregated, early-diverging germline
5. Multicellular organisms ⇒ eusocial society	Origin of societies	Control of conflict with dominance, punishment, or policing	Evolution of dimorphic reproductive and non-reproductive castes
6. Separate species ⇒ interspecific mutualism	Origin of interspecific mutualisms	Control of cheating (sanctions)	Evolution of physically conjoined social partners

[1] Strictly speaking, one should classify the evolution of the eukaryotic cell as an instance of transition 6 (evolution of interspecific mutualisms). But, because of its antiquity and unique importance, I have followed precedent in treating it separately.

sit outside the major transitions framework. However, I include the evolution of sex despite this because sufficient aspects fall within the framework to make including it potentially illuminating (Bourke 2009). Overall, I ignore some complexities because I believe there are large gains to our understanding to be had from concentrating on broad similarities between evolutionary phenomena rather than on their detailed differences. But I acknowledge that these gains may come at the cost of overlooking some interesting and important processes.

Box 1.1 Timing and frequency of the six major transitions involving the evolution of individuality.

This box presents details of six major transitions leading to the evolution of individuality that have taken place over life's history (Table 1.3). Without aiming to be comprehensive, it seeks to document when each transition occurred, how many times independently it occurred, and the identity of the taxa involved. Note that the dates given, and some of the phylogenetic and taxonomic conclusions, are not always agreed upon by researchers (e.g. Cavalier-Smith 2006). This is because palaeontological and molecular dating methods involve various sources of uncertainty and do not always concur. Similarly, different molecular reconstructions of phylogeny do not necessarily produce the same evolutionary tree. Such uncertainties increase the earlier the events or phylogenetic splits under consideration took place. In addition, early occurrences of any one major transition may be unknown to researchers because they have left no trace in fossil or extant taxa, making it hard definitively to decide whether transitions that appear to have occurred just once are truly unique (Vermeij 2006). Therefore, while the information in this box is intended to reflect the current state of knowledge, many details and even some broader patterns are provisional and may prove incorrect in light of future research, as many in the field stress (e.g. Baldauf 2003; Ruiz-Trillo et al. 2007).

1. Separate replicators (genes) ⇒ cell enclosing genome

The transition leading to the first cells is the one that is, for obvious reasons, most veiled in mystery. Life originated when the first self-replicating molecules arose (Cairns-Smith 1985; Dawkins 1986, 2004). Most sources (Wilson 1992; Ridley 2000; Knoll 2003) put the origin of life at 3800–4000 million years ago (mya). Between the origin of life and the first cell, many important events must have occurred; for example, a switch from RNA to DNA as the primary replicator and the origin of the genetic code (Maynard Smith and Szathmáry 1995). Cells may have originated independently more than once (with unsuccessful designs going extinct) but, because all known cells share essentially the same genetic code, this is a case where there is no evidence that there was more than one origin (Dawkins 2004). The first cells were prokaryotes, as represented today by the Bacteria and the Archaea (Tudge 2000). The first prokaryotic cells (of unspecified type) appear in the fossil record *c.* 3500 mya (Schopf 2006). The origin of the first cells must also have seen the origin of asexual reproduction in the form of cell division.

2. Separate unicells ⇒ symbiotic unicell

This transition involved the formation of unicellular eukaryotes from the symbiotic fusion of two cells. Eukaryotic cells are much larger than prokaryotic ones, with extant eukaryotic cells typically being 1000 times larger by volume (Bonner 1974). Eukaryotic cells also differ in possessing a nucleus, an internal cytoskeleton, and organelles, i.e. mitochondria or plastids such as chloroplasts (Tudge 2000; Knoll 2003). (Henceforth, I use 'chloroplast' to signify any plastid, since the minor types of plastid are believed to be derived from chloroplasts (Tudge 2000).) The evidence is overwhelming that, ancestrally, mitochondria are symbiotic proteobacteria and chloroplasts are symbiotic cyanobacteria (Margulis 1970; Knoll 2003); for example, these organelles retain their own genomes and, in molecular phylogenies, they respectively group among the proteobacteria and cyanobacteria (Rand et al. 2004; Williams et al. 2007). The deduction is that they were acquired when another cell permanently incorporated them within its cytoplasm (endosymbiosis). The acquisition of the proteobacteria that became mitochondria is thought to have occurred before the

Box 1.1 (*Cont.*)

acquisition of the cyanobacteria that became chloroplasts (Knoll 1992; Bhattacharya et al. 2007). The dates of these events are particularly uncertain, with large dating discrepancies among and within studies based on fossils, on 'fossil molecules' (molecules found in rocks and believed to characterize particular taxa), and on molecular phylogenies. Fossil eukaryotes of unclear affinities occur in rocks formed between 1900 and 1300 mya (Knoll 1992; Knoll et al. 2006). This indicates that the two endosymbiotic events occurred before, or during, this interval, with suggestions that fossil molecules mark a much earlier date now judged to be unreliable (Brocks et al. 1999; Rasmussen et al. 2008a). Using molecular phylogenies, Hedges et al. (2004) dated the origin of mitochondria at 1800–2300 mya. Molecular dates for the origin of chloroplasts vary from 1600 mya (Yoon et al. 2004) and 1500–1600 mya (Hedges et al. 2004) to 825–1162 mya (Douzery et al. 2004).

The mitochondrial endosymbiosis is believed to have occurred once only (Gray et al. 1999). One view is that a prokaryotic cell (probably an Archean) acquired eukaryotic features (nucleus, cytoskeleton, etc.) and then symbiotically incorporated a bacterial cell, but it is also possible that two prokaryotic cells fused, with the compound cell then acquiring the rest of its eukaryotic features (Tudge 2000; Knoll 2003; Embley and Martin 2006; Zimmer 2009a). The acquisition of the cyanobacteria that became chloroplasts (by a eukaryotic cell that already contained mitochondria) probably also occurred once (Rodríguez-Ezpeleta et al. 2005; Bhattacharya et al. 2007; Burki et al. 2008; Archibald 2009), at least primarily. Remarkably, however, some photosynthetic eukaryotic lineages (e.g. cryptophyte algae) secondarily acquired chloroplasts, by symbiotically incorporating other eukaryotic, algal unicells that themselves already harboured chloroplasts (Archibald and Keeling 2002; Knoll 2003; Bhattacharya et al. 2007). Such secondary endosymbioses are reckoned to have happened at least three times independently (Archibald 2009). In addition, the unicellular eukaryote *Paulinella chromatophora* appears to be in the early stages of acquiring a photosynthetic cyanobacterial symbiont in a process that, when complete, would represent a primary endosymbiosis independent of, and far more recent than, the one that originally gave rise to chloroplasts (Archibald 2006; Rodríguez-Ezpeleta and Philippe 2006; Nakayama and Ishida 2009).

3. Asexual unicells ⇒ sexual unicell

Although bacteria transfer genetic material to one another (Maynard Smith et al. 1991; Narra and Ochman 2006), sexual reproduction involving the formation of haploid gametes (meiosis) from a diploid cell, and the fusion of gametes to form a zygote, occurs only in eukaryotes. Sexual reproduction in this sense could therefore not have preceded the origin of eukaryotes. Sex appears almost universal in living eukaryotes, with asexual taxa representing secondary losses (Zimmer 2009b), but there is little direct evidence in the fossil record of the origin of eukaryotic sex (Bell 1982; Schopf 1994; Knoll 2003). However, by mapping mode of reproduction onto a phylogeny of the eukaryotes, Dacks and Roger (1999) concluded that (a) sexual reproduction with meiosis is ancestral to all eukaryotes (implying that it also arose once, in a unicellular eukaryote) and (b) initially sexual reproduction was facultative, not obligate. This explains why, for example, most extant unicellular eukaryotes have facultative sex (Dacks and Roger 1999), with those apparently lacking meiosis representing cases of cryptic sexual reproduction or secondary loss (Dacks and Roger 1999; Ramesh et al. 2005). From the morphology of its spores, the multicellular red alga *Bangiomorpha* (from 1200 mya) was reported by Butterfield (2000) to represent 'the oldest reported occurrence [of sexual reproduction] in the fossil record'. If so, the

Box 1.1 (*Cont.*)

origin of eukaryotic sex can be placed between the origin of eukaryotes (from above, roughly 2000 mya) and 1200 mya.

4. Unicells ⇒ multicellular organism

Grosberg and Strathmann (2007) estimated that, in the prokaryotes and the eukaryotes combined, multicellularity evolved from unicellular ancestors at least 25 times independently. In the prokaryotes, multicellularity is always simple, whereas in eukaryotes multicellularity may be simple or complex (Ridley 2000). Examples of multicellular prokaryotes include *Anabaena* (Cyanobacteria) that forms multicellular filaments (Kaiser 2001; Golden and Yoon 2003), *Bacillus subtilis* (Firmicutes), in which wild strains but not laboratory strains form multicellular fruiting bodies (Branda et al. 2001), Myxobacteria (Proteobacteria) that form multicellular foraging swarms and fruiting bodies (Shimkets 1990, 1999; Velicer and Vos 2009), *Streptomyces* (Actinobacteria) that forms multicellular filaments and mycelia (Flärdh and Buttner 2009), and unclassified magnetotactic bacteria occurring in spheres of 20 to 45 cells (Keim et al. 2004). Fossils of multicellular prokaryotes of various morphologies are found in rocks dating far into the deep past (Schopf 2006; Butterfield 2009). In particular, fossils of bacterial filaments occur in some of the oldest rocks, namely those formed between 3500 and 2500 mya (Schopf 2006). Myxobacteria are believed, from molecular phylogenies, to have arisen more recently, between 1000 and 565 mya (Shimkets 1990, 1999). Very ancient rocks harbour evidence of microbial mats or biofilms as well as bacterial filaments (Tice 2008). However, present-day bacterial biofilms, to the extent they comprise mixtures of different strains and species, may be best viewed as communities of competitors, implying that there is no straightforward parallel between biofilm formation and multicellularity (Nadell et al. 2009).

In the eukaryotes alone (Fig. 1.1), it is estimated that multicellularity has evolved at least 16 times independently (Bonner 1998; Carroll 2001; King 2004; Grosberg and Strathmann 2007). Examples of simple multicellular eukaryotes include species of volvocine algae, which lie within the Chlorophyceae, a division of the green algae within the Kingdom Plantae (Tudge 2000; Herron and Michod 2008); the cellular slime moulds or social amoebae such as *Dictyostelium*, which lie within the Amoebozoa (Bonner 2003a; Dawkins 2004; Schaap et al. 2006); and the Choanoflagellates, which are aquatic flagellates representing the sister group of the animals (Dawkins 2004; Hedges et al. 2004; King 2004; Carr et al. 2008). Complex multicellularity occurs in five eukaryote lineages (Tudge 2000; Grosberg and Strathmann 2007), which include all the organisms one typically calls to mind when thinking of 'animals' or 'plants'. These are the animals strictly speaking (Kingdom Animalia), the plants strictly speaking (Kingdom Plantae), fungi (Kingdom Fungi), red algae (Kingdom Rhodophyta), and brown algae (Kingdom Phaeophyta). Note, therefore, that some taxa (e.g. Plantae), include both simple and complex multicellular forms.

Dates for the evolution of multicellularity in these various lineages are uncertain. In the cellular slime moulds, multicellularity is thought on indirect evidence to have arisen less than 490 mya (Bonner 2009). All of them have a multicellular fruiting body, suggesting a single origin (Schaap et al. 2006). A molecular phylogenetic analysis suggests that multicellularity in the volvocine algae also arose once, but more recently, i.e. an estimated 234 mya (Herron and Michod 2008; Herron et al. 2009). Because all animals (strictly speaking) are multicellular, the date of the split of animals and Choanoflagellates provides a rough date for the evolution of multicellularity in animals. The molecular phylogeny of Douzery et al. (2004) dates this split at 761–957 mya, whereas that of Hedges et al. (2004) dates

Box 1.1 (*Cont.*)

it considerably earlier, at 1450 mya. Love et al. (2009) reported fossil molecules indicative of sponges in rocks from more than 635 mya as the earliest evidence of animals in the fossil record. As mentioned earlier, a multicellular red alga, *Bangiomorpha*, occurs in rocks formed 1200 mya (Butterfield 2000, 2009; Knoll 2003).

5. Multicellular organisms ⇒ eusocial society

Eusociality has traditionally been defined as occurring when a group of multicellular organisms exhibits a reproductive division of labour (i.e. reproduction is concentrated in one or a few members of the group), cooperative brood care, and an overlap of parental and offspring generations (Wilson 1971). As several authors have pointed out, a distinction occurs within eusocial societies between those without an irreversible worker (helper) caste and those with an irreversible worker caste (e.g. Crespi and Yanega 1995; Boomsma 2007). Here, to be irreversibly a worker means, as an adult, to have a morphology suited for working and to be incapable of changing (e.g. by moulting) into a reproductive form. In the terminology of the present book, societies without irreversible workers are simple eusocial societies and those with them are complex eusocial societies (Sections 1.3, 6.2). The former also correspond closely to the kinds of vertebrate societies traditionally termed 'cooperative breeders' (e.g. Emlen 1991), in which some members are helpers but are not irreversibly committed to a helper role (Crespi and Yanega 1995). Some authors have argued that the term 'eusocial' should effectively be confined to complex eusocial societies (Crespi and Yanega 1995), while others have advocated a more general definition (Sherman et al. 1995). In this book I take the latter approach, using a broad, loose definition of eusociality that includes as eusocial any society with an appreciable reproductive division of labour, irreversible or not. This is because any level of reproductive division of labour implies a degree of altruism and altruism is a social trait of particular interest (Section 1.4). Under this definition, eusociality has arisen many times independently, at least 24 times according to current evidence (Table 1.4). It occurs principally in insects (Fig. 1.2), but is also found in crustaceans (social shrimps) and at least one taxon of mammals (the mole-rats). The earliest estimated date for an origin of eusociality in any taxon is 170 mya, i.e. in the mid-Jurassic (Table 1.4).

The origin of eusociality does not, therefore, seem to be as rare an evolutionary event as has sometimes been suggested (e.g. Wilson and Hölldobler 2005), although the number of origins of complex eusociality is a different matter (Section 6.2). Furthermore, as already implied, the number of independent origins of eusociality rises considerably depending on how widely one casts the eusocial net. First, Agnarsson et al. (2006, 2007) have estimated that there have been 18–19 independent origins of sociality in spiders (e.g. *Anelosimus*). However, social spiders are not usually considered eusocial because, although they exhibit cooperative brood care, they lack a clear-cut reproductive division of labour (Avilés 1997; Whitehouse and Lubin 2005).

Second, on the logic that simple eusocial societies and cooperatively-breeding vertebrates are at the same grade of sociality, the latter need including in the estimation of the number of eusocial origins. Cooperative breeding in vertebrates has itself arisen many times independently, though how many times is hard to specify because of uncertainties in phylogenies, in determining grades of sociality, and in distinguishing ancestral non-sociality from secondary loss of sociality (Solomon and French 1997; Cockburn 1998; Bennett and Owens 2002; Ligon and Burt 2004; Russell 2004). For example, depending on the methods used, Ligon and Burt (2004) estimated that across birds there

Box 1.1 *(Cont.)*

have been 28 independent gains of cooperative breeding and 20 losses, or 38 gains and 12 losses.

Third, as highlighted by Wilson (1975), the so-called colonial marine invertebrates also represent the transition from solitary multicellular ancestors to a sophisticated form of sociality, since some species exhibit a reproductive division of labour (and hence are eusocial from the present standpoint). The colonial marine invertebrates include members of the phyla or subphyla Cnidaria (e.g. siphonophores, corals, and colonial sea anemones), Urochordata (e.g. sessile colonial tunicates and pelagic colonial tunicates such as Salps), and Bryozoa. All consist of colonies of polyps or zooids, with a single polyp or zooid being homologous to the solitary multicellular ancestor. Not all exhibit a reproductive division of labour. But, in many cases, some polyps or zooids within the colony specialize on reproductive functions and others specialize on non-reproductive functions, such as defence, feeding, or (in mobile species) locomotion. They may also share resources and coordinate their behaviour (Mackie 1986; Harvell 1994; Ayre and Grosberg 2005; Dunn and Wagner 2006; Dunn 2009). In essence, these cases represent eusocial societies whose members happen to be physically stuck together (Fig. 1.3). The number of origins of coloniality in these organisms seems poorly known, let alone the number of origins of a reproductive division of labour (McShea and Changizi 2003). However, the Bryozoa, for example, form a monophyletic group with the earliest fossil representatives (which are colonial) being found in the Lower Ordovician, 440–490 mya (Fuchs et al. 2009).

6. Separate species ⇒ interspecific mutualism

Mutualistic relationships between separate organisms of different species span a huge range of interactions, varying in the taxa involved and their degree of interdependence, some being facultative and opportunistic, others obligate and extremely intimate (Sachs et al. 2004). The partners may be a unicellular organism and a multicellular organism, or two multicellular organisms. Hence this transition has happened many times, but how many is hard to specify. Researchers have examined the number of origins of interspecific mutualisms, and their relative dates of occurrence, in just a few cases. One is the lichens. Lichens are intimate associations between fungi (mainly Ascomycota) and photosynthetic algae or cyanobacteria (Lutzoni and Miadlikowska 2009). In Ascomycota, phylogenetic evidence suggests that lichen formation has occurred three to five times independently (Lutzoni and Miadlikowska 2009). When these events happened is unclear, but associations between fungi and cyanobacteria resembling lichens have been found in the fossil record from around 600 and 400 mya (Taylor et al. 1995; Yuan et al. 2005; Karatygin et al. 2009).

Having adopted the foregoing definition of a major evolutionary transition, I propose that each transition can be usefully divided into three principal stages. These are (a) *social group formation*, (b) *social group maintenance*, and (c) *social group transformation* (Fig. 1.4). I employ the term 'social group' in these labels, despite its teetering on the edge of tautology, to mean a stable group of any entities that cooperate

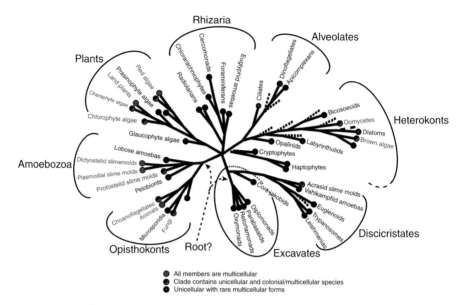

Fig. 1.1 The phylogenetic distribution of multicellularity among eukaryotes. Reproduced from King (2004) with kind permission of the author.

Table 1.4 Animal taxa in which there has been an independent transition from solitary living to eusociality (defined broadly), i.e. from multicellular organisms to a society. Taxa are divided into orders of insects and other groups. In ants, two studies yielded two date ranges for the origin of eusociality. Other compilations of the number of independent origins of eusociality include those of Grimaldi and Engel (2005), Crozier (2008), and Hughes et al. (2008)

Taxon	Number of independent origins of eusociality	Approximate date of origin of eusociality	References
Blattodea Termites[1]	1	140–145 mya	No. origins: Thorne (1997); Inward et al. (2007b) Date: Grimaldi and Engel (2005)
Coleoptera Ambrosia beetle (*Austroplatypus incompertus*)	1	No data	No. origins: Kent and Simpson (1992); Smith et al. (2009)
Hemiptera Aphids (Pemphigidae, Hormaphididae)	6–9	No data	No. origins: Stern (1994, 1998); Stern and Foster (1996)

Table 1.4 *Cont.*

Taxon	Number of independent origins of eusociality	Approximate date of origin of eusociality	References
Hymenoptera			
Ants (Formicidae)	1	115–135 mya; 140–170 mya	No. origins and dates: Brady et al. (2006a); Moreau et al. (2006)
Bees (Allodapines)	1	>40 mya	No. origins and date: Chenoweth et al. (2007); Schwarz et al. (2007)
Bees (Corbiculate social bees: Apini, Bombini, Meliponini)	1–2	>65 mya (i.e. Late Cretaceous)	No. origins: Cameron and Mardulyn (2001); Thompson and Oldroyd (2004) Date: Grimaldi and Engel (2005)
Bees (Halictidae)	3	20–22 mya	No. origins and date: Brady et al. (2006b)
Wasps (Social Vespidae: Polistinae, Stenogastrinae, Vespinae)	2	c. 100 mya (i.e. mid-Cretaceous)	No. origins: Hines et al. (2007); Hunt (2007) Date: Grimaldi and Engel (2005)
Wasps (Pemphredoninae: *Microstigmus* spp.)	1	No data	No. origins: Hughes et al. (2008)
Wasps (Polyembryonic parasitoid wasps, Encyrtidae)	≥1	No data	No. origins: Cruz (1981); Giron et al. (2007)
Thysanoptera			
Thrips (Phlaeothripinae)	1	No data	No. origins: Chapman et al. (2008)
Other groups			
Shrimps (*Synalpheus* spp)	3	<7 mya	No. origins: Duffy et al. (2000); Duffy (2007) Date: Morrison et al. (2004); Duffy and Macdonald (2010)
Mole-rats (Bathyergidae)	2	<40–48 mya	No. origins: Faulkes et al. (1997) Date: O'Riain and Faulkes (2008)

mya, millions of years ago

[1] Termites represent a clade within the cockroaches (Order Blattodea) and hence can no longer be classified in their own order, Isoptera (Inward et al. 2007a).

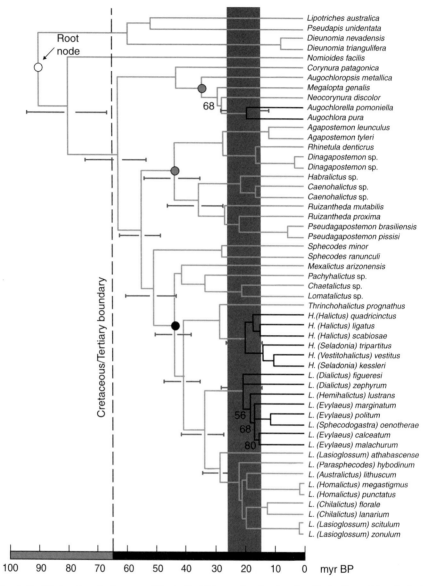

Fig. 1.2 Molecular phylogeny of the halictid bees, dated using fossils (indicated by circles), illustrating three independent origins of eusociality (black branches). Two of these cases also contain species that have secondarily reverted to solitary living (*Halictus quadricinctus*, *Lasioglossum calceatum*, and *L. lustrans*). The shaded vertical bar represents a period of climatic warming. Myr BP, millions of years before present. Reproduced from Brady et al. (2006b) with kind permission of the authors and Royal Society Publishing.

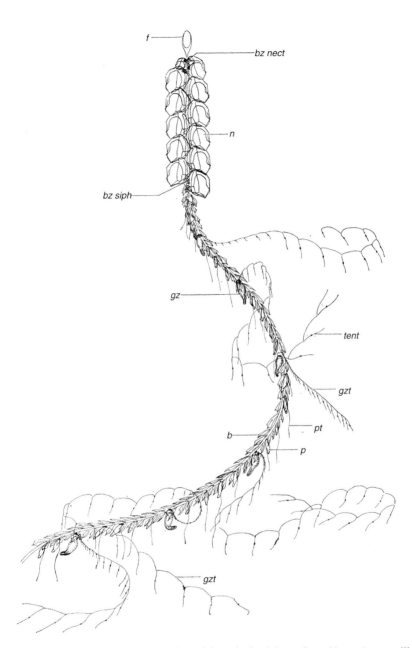

Fig. 1.3 Mackie's (1964) classic drawing of the pelagic siphonophore *Nanomia cara*, illustrating eusociality in a colonial marine invertebrate. The zooids are highly differentiated and include a float (pneumatophore, f), forms for swimming (nectophores, n), forms for eating (gastrozooids, gz) from which tentacles (gzt) project for capture of plankton as prey, and forms for excretion or defence (palpons, p). Mature colonies also bear reproductive zooids (gonozooids). The entire colony varies in size with number of zooids but may be 25 cm long. Drawing by George Mackie reproduced from Mackie (1964) with kind permission of the artist/author and Royal Society Publishing.

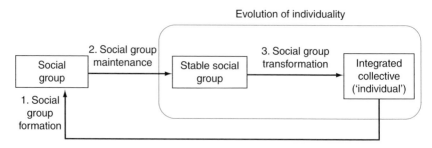

Fig. I.4 Stages involved in a major evolutionary transition. See Section 1.3 for further details.

in ways that make the group a potential candidate for consideration as an individual. Note, therefore, that by this definition a multicellular organism and an animal society each qualifies as a form of social group (of cells and animals, respectively). I avoid 'group' alone to distinguish social groups from temporary aggregations, such as flocks, herds, or shoals, whose evolution and ecology have been well treated by Krause and Ruxton (2002). 'Social group' is also a term already employed in the literature in senses essentially the same as that used here (Pamilo 1989; Pacala et al. 1996; Wilson and Wilson 2007; Gardner and Grafen 2009).

Social group formation refers to the processes involved in the origin of social living, i.e. the initial spread of genes for social behaviour through a population (Table 1.3). Social group maintenance refers to the processes that are involved in the stable persistence of social groups once they have originated (Table 1.3). Both social group formation and social group maintenance are ways of describing what have effectively been the main foci in the field of social evolution to date. For example, in the study of social insects, researchers have extensively considered these processes under the headings of the origin and maintenance of eusociality (Bourke and Franks 1995; Crozier and Pamilo 1996). Social group transformation is hereby defined as the set of processes that transforms a stable social group into an obligate collective with a high degree of interdependence of its parts and sufficient overall integration to be considered an individual and candidate for participation, as a subunit, in the transition to the next hierarchical level (Table 1.3). Put another way, social group transformation is the stage in a major transition at which the new level of individuality emerges (Fig. 1.4). Such a process almost has to occur before the next major transition can take place, because potential subunits that have not themselves achieved sufficient individuality are unlikely to be able to band together to form a new kind of social group.

An example of social group transformation is the change, over evolutionary time, from simple to complex multicellularity. Simple multicellular organisms are those with a weakly-defined germline (lineage of gamete- or spore-producing cells) and few somatic cell types (i.e. few cell types among those cells making up the non-reproductive part of the body, or soma). Complex multicellular organisms are those with a strongly-defined germline and many somatic cell types (Ridley 2000; King

2004; Grosberg and Strathmann 2007). Another example of social group transformation is the evolutionary change from simple to complex eusociality in insects. Simple eusocial societies are those with behaviourally distinct but morphologically alike members that are potentially equally capable of reproduction or working. Complex eusocial societies are those with reproductive forms (queens or kings, collectively 'reproductives') and non-reproductive or semi-reproductive forms (workers) that are morphologically distinct and, as adults, irreversibly committed to their social roles (Alexander et al. 1991; Bourke 1999). Social group transformation has not been ignored as a concept (e.g. Queller 1997; Bourke 1999; McShea and Changizi 2003; Michod 2007; Boomsma 2009), but it has rarely been explicitly defined and it has not received the degree of attention paid to the formation and maintenance of social groups (Chapter 6). All three stages of a major transition—social group formation, maintenance, and transformation—overlap to some extent. The borders between them are not absolute, and, for example, some factors that assist maintenance may also assist transformation. Nonetheless, to pick out processes most important at each stage, it remains useful to subdivide each major transition in this way.

1.4 Inclusive fitness theory and the evolution of cooperation

As mentioned earlier, the evolution of individuality and the major evolutionary transitions both invoke the problem of the evolution of cooperation (in its broad sense). In this book I take the view that a fundamental problem posed by the evolution of cooperation has been solved by Hamilton's (1964) inclusive fitness theory. Hamilton's theory addresses the evolution of four types of social behaviour, namely cooperation (narrow sense), altruism, selfishness, and spite. These represent all possible types of social behaviour as formally defined by the nature of the costs and benefits experienced by social partners (Section 2.1). Though Hamilton (1964) did a great service in formalizing the concept of selfishness, selfishness had always been easy to explain with natural selection theory. Indeed, the expectation of selfishness has frequently been taken to epitomize the theory. Spite as a separate form of social behaviour was not recognized before Hamilton's (1964) work. Therefore, historically, it was the evolution of cooperation that seemed to pose a fundamental problem for Darwinism. This was the puzzle of how self-sacrificial behaviour could evolve in a world where Darwinian natural selection appears to favour organisms that selfishly maximize their reproduction. Hamilton's (1964) theory solved it through the insight, later expanded by Dawkins (1976), that cooperation at any one level, such as the organismal level, has its basis in selfishness at the level of genes. When an organism exhibits self-sacrificial behaviour, it does so because a gene for self-sacrifice is levering its way into the next generation either through enhancing the future reproduction of its current bearer (direct benefit, as in narrow-sense cooperation) or through promoting aid to relatives that bear the same gene (indirect or kin-selected benefit, as in altruism; Section 2.1). Hence, from the viewpoint of the focal gene, there is selection for maximizing transmission to future generations, just as Darwinian natural selection leads us to expect. It follows that, despite occasional claims to the contrary, the evolution of cooperation is no longer quite the fundamental challenge to natural selection theory that it once seemed.

Nonetheless, the evolution of cooperation invokes another general problem, which is the problem of how cooperation remains stable in the face of cheating. In social groups built on cooperation, what prevents selfish individuals prospering by taking the benefits generated within the group while producing none of their own? Unchecked, this would lead to cooperation being swamped by an ever-rising level of selfish cheating. What controls cheating is essentially the problem of social group maintenance (Chapters 3, 5). In brief, when social partners are related, inclusive fitness theory shows that relatedness itself acts as a brake on cheating because too high a level of cheating harms group members sharing one's own genes (Frank 1995). In addition, regardless of whether social partners are related or unrelated, various types of coercion, such as punishment and policing, have been invoked as repressors of cheating (e.g. Clutton-Brock and Parker 1995; Frank 2003; Ratnieks and Wenseleers 2008). A general finding is that the evolution of coercion itself depends on conditions specified by inclusive fitness theory (Section 3.2). In the special case in which social partners are unrelated and social benefits are exchanged (reciprocity), another set of models has been developed that seeks to find conditions for the stability of cooperation (Trivers 1971; Axelrod and Hamilton 1981; Hammerstein 2003a). Historically, these models have often been perceived as being separate from inclusive fitness theory, but they may usefully be regarded as falling within its remit (e.g. Lehmann and Keller 2006a). This is because they deal with a specific case of Hamilton's (1964) four-way classification of social behaviours at one point on the continuum of possible relatedness levels (namely narrow-sense cooperation between non-relatives).

Overall, therefore, Hamilton's (1964) inclusive fitness theory has proved to have remarkable generality and versatility. As we will see, it also musters considerable explanatory firepower. On top of this, it can be derived robustly from first principles of the genetical theory of natural selection (Grafen 1985, 2006; Frank 1998). It is its combination of breadth, fit to data, and profundity that, as several previous authors have argued (e.g. Trivers 1985; Bourke and Franks 1995; Lehmann and Keller 2006a; West et al. 2007a), renders Hamilton's (1964) theory the foremost candidate for a general theory of social evolution.

Like any large and active field of investigation, the theoretical study of social evolution is not free from disagreements and unresolved issues (e.g. Taylor and Nowak 2007; West et al. 2007a). Paradoxically, while the potential richness of inclusive fitness theory as a general theory of social evolution is still underappreciated, the theory is sometimes perceived as an entrenched orthodoxy. A tendency therefore exists for iconoclastically-minded theoreticians to derive models of cooperation in novel ways and then announce them to be fundamental additions to existing theory (e.g. Killingback et al. 2006; Nowak 2006; Ohtsuki et al. 2006; Traulsen and Nowak 2006). It is healthy for orthodoxies to be continually challenged by new theories and new data. However, to date, these models have proved to fall short of true novelty, as other authors have shown that their results are capable of being derived from inclusive fitness theory (e.g. Grafen 2007a, 2007b; Lehmann et al. 2007a, 2007b; West et al. 2007a). Indeed, inclusive fitness theory has a long history of successfully assimilating apparent challenges and alternatives (Grafen 1984; Queller 1992; Lehmann and Keller 2006a). This is not surprising when one considers its deep foundations in the theory of natural selection. Although it is premature to declare a consensus, a substantial body of opinion therefore holds that claims of fundamental extensions to inclusive fitness theory will have to be radically innovative, as well as robust, to be accepted as such (e.g. Lehmann and Keller 2006a; West et al. 2007a). For all these reasons, Hamilton's (1964) inclusive fitness theory will underpin the conceptual reasoning employed throughout this book.

1.5 Challenges remaining in the study of social evolution

In the previous section I argued that Hamilton's (1964) theory of inclusive fitness is a general theory of social evolution that has solved the historical problem presented by the evolution of cooperation. Inclusive fitness theory has also gone far in addressing the subsequent problem of what controls cheating in stable social groups. This raises an obvious question: what are the outstanding problems in the study of social evolution? At the ultimate (evolutionary) level, and as far as the expanded view of social evolution espoused in this book is concerned, I suggest that two key issues remain unresolved and underexplored.

The first key unresolved issue is: *to what extent do common principles operate at each stage of a major transition across all levels in the biological hierarchy, and what is the empirical evidence for their operation?* The stages in question are those that I have previously defined (Section 1.3), namely social group formation, social group maintenance, and social group transformation. For example, do unicellular organisms group together to form multicellular organisms for demonstrably the same reasons as multicellular organisms group together to form societies? This is an issue worth exploring because, although it is almost axiomatic in the field that the major transitions are explicable via shared principles, neither the various principles themselves, nor the evidence for their operation, have been fully and explicitly set out (Section 1.2).

The second key unresolved issue is: *what are the main integrative processes that occur, within lineages, between one major transition and the next? In other words, how does social group transformation occur?* This issue is worth investigating, because emphasis in the field to date has been on how social groups form and are maintained, rather than on how an existing social group consolidates to become an individual (Section 1.3). Proposed examples of phenomena that fall within the scope of both key issues are presented in Table 1.3.

After setting out the conceptual background provided by inclusive fitness theory in more detail (Chapter 2), I first examine how the theory illuminates general aspects of the major transitions (Chapter 3). I then consider the pair of issues above in a series of chapters on each of the three stages in a major evolutionary transition (Chapters 4–6). I end with a brief synthesis and set of conclusions (Chapter 7). A large number of previous books and review articles have addressed the topic of this book, at least in part (e.g. Bonner 1974, 1988, 1998; Trivers 1985; Buss 1987; Maynard Smith and Szathmáry 1995; Dugatkin 1997; Frank 1998, 2003; Keller 1999; Michod 2000, 2007; Queller 2000; Ridley 2000; Hammerstein 2003a; Korb and Heinze 2004; Sachs et al. 2004; Travisano and Velicer 2004; Lehmann and Keller 2006a; Grosberg and Strathmann 2007; West et al. 2007a; Boomsma 2009; Queller and Strassmann 2009). The present work draws heavily on the insights of these and other authors, to whose achievements I am indebted. However, having provided the background, I seek to extend their treatments by concentrating on the two key, unresolved issues described above.

It might be helpful also to set out what this book does not attempt to do. I do not aim to produce a textbook, monograph of personal research, or advanced theoretical exposition, excellent examples of all of which are included among the previous works. Nor do I seek to present a taxon-based analysis (the division of the material in the present book will instead be thematic) or an approach based on climbing up through the biological hierarchy from genes to societies (instead, each chapter draws on examples and concepts across different levels of the hierarchy). The field of social evolution, even as traditionally defined, has grown immense. Therefore, for tractability, I do not attempt a comprehensive review or pretend to be exhaustive, but instead concentrate on widely applicable concepts and illustrative examples in a succinct synthesis. Where few data exist, I avoid trying to fill the gap with speculation alone. This means that my treatment is not balanced but is heavily biased towards the higher-level transitions, especially the transitions from unicellular to multicellular organisms and from multicellular organisms to eusocial societies. (The astute reader may also detect a bias for examples drawn from the eusocial Hymenoptera.) Because these two transitions each happened many times independently and, comparatively speaking, are among the transitions less remote in evolutionary time, there is more concrete information on them. For this reason, and to maintain the focus on comparisons across different transitions, I largely omit the very earliest transition (genes to genomes) and its substages. An additional justification for this neglect is that this transition was theoretically very thoroughly dealt with by Maynard Smith and Szathmáry (1995) and Szathmáry (2006). I note, however, that the compartmentation inherent in models for the evolution of the cell imposes a strong degree of kin structure on the system, so permitting inclusive fitness effects to operate (Michod 1983; Szathmáry and Demeter 1987; Frank 1994a; Maynard Smith and Szathmáry 1995; Queller 1997).

Despite my focus on the later transitions, I also do not cover one recent example, namely the appearance of complex social behaviours in humans. In general, we humans show a great deal of cooperative behaviour towards non-relatives and there is also evidence that we punish social cheats in an altruistic manner, i.e. at a cost to ourselves. To this extent, our social behaviour is unusually puzzling (Colman 2006; Sigmund 2007). Humans also occur in an unusually complex social hierarchy, involving families, ethnic groupings, cities, nations, and federations of nations. Stearns (2007) has presented a thought-provoking essay on how the major transitions viewpoint might be used to understand the complex hierarchy of post-agricultural human society. One reason why humans stick out among living organisms in terms of social evolution is undoubtedly our possession of language and cultural inheritance (Maynard Smith and Szathmáry 1995; Fehr and Fischbacher 2003; Richerson et al. 2003; Boyd and Richerson 2009). Although these attributes are not confined to humans, they are obviously most highly developed in us, rendering many of the puzzles posed by our social behaviour largely specific to ourselves. Humans apart, reciprocated interactions within species between non-relatives seem rare in nature (Hammerstein 2003b; Clutton-Brock 2009a). For these reasons, as well as omitting a detailed consideration of human social behaviour, I pass over much of the very sizeable theoretical literature on models for stability in reciprocated social interactions between unrelated members of the same species.

Although I emphasize shared principles of social evolution, I do not wish to imply that other approaches to investigating the major transitions in evolution are not worthwhile or that other outstanding issues in social evolution are unimportant. For example, large-scale environmental changes in Earth's history are likely to have facilitated some of the major transitions, especially the earlier ones. Specifically, a rise in atmospheric oxygen levels early in the Proterozoic Eon (2400 to 2200 mya) is thought to have been associated with the evolution of aerobic bacteria and hence of eukaryotic cells containing mitochondria (Knoll 2003; Embley and Martin 2006; Sessions et al. 2009). Likewise, a rise in marine oxygen levels late in the Proterozoic Eon (1000 to 550 mya), by permitting large body size, may have facilitated the evolution of complex multicellular organisms (Knoll 2003; Payne et al. 2009). In addition, either independently of rises in global oxygen levels or not, the distant past may also have seen one or more worldwide glaciations, an idea known as the 'Snowball Earth' hypothesis (Hoffman et al. 1998; Kopp et al. 2005). If such events occurred, they would undoubtedly have affected the environmental conditions, and perhaps also the selective conditions, under which major transitions occurred (Boyle and Lenton 2006). However, I do not further consider such phenomena, interesting and important as they may be, because I wish to focus on ultimate rather than the proximate (non-evolutionary) drivers of the major transitions, since ultimate drivers were required for each major transition to occur irrespective of the nature of the prevailing large-scale environmental conditions.

To give another example of an important topic that deserves mention but that this book does not fully cover, the social complexity that occurs at each new major transition might often involve an element of self-organization. Self-organization refers to the surprisingly large degree of higher-level order that arises when many subunits interact together while obeying a few, relatively simple local rules. There is ample evidence that self-organization plays a major part in both the development of multicellular organisms and the organization of behaviour within animal societies (Seeley 1995; Camazine et al. 2001; Boomsma and Franks 2006). Self-organization must interact with natural selection through the local rules that each subunit follows being subject to genetic variation, so allowing selection to pick between adaptive and non-adaptive patterns of higher-level order (Bourke and Franks 1995; Camazine et al. 2001). But in the sense that self-organization alone does not therefore necessarily produce adaptive order, it is a proximate factor in social evolution. Again, then, I do not treat it any depth, although later I briefly explore how self-organization might operate as a factor reinforcing increases in group size (Section 6.5).

Finally, each major transition entails a series of genetic and genomic changes to meet the requirements of life at the next hierarchical level. These include changes in genome size and in the regulation of gene expression, along with both the co-option of pre-existing genes and the evolution of novel genes (Ridley 2000; West-Eberhard 2003; Lynch 2007). In particular, complex multicellularity required the evolution of novel genes and proteins underlying the regulation of gene expression to permit the development of the organism from a zygote to an adult body and to generate polymorphism among cells and tissues (Carroll 2001; Kaiser 2001; Szathmáry and Wolpert 2003; King 2004; Rokas 2008). At the level of societies, as molecular genetics and genomics progress in the power of their techniques (e.g. Honeybee Genome Sequencing Consortium 2006), researchers are also making rapid progress in uncovering the molecular-genetic basis of sociality (e.g. Robinson et al. 2005, 2008; Toth et al. 2007). However, although in the next chapter I present summary evidence for a genetic basis to social behaviour (Section 2.4), I do not attempt a detailed review of the genetic and genomic mechanisms involved in social evolution. Once more, this is to maintain a focus on evolutionary principles. In sum, my essential goal is to explore how common principles of social evolution shape biological organization across different taxa at all hierarchical levels.

1.6 Summary

1. At a fundamental level, the history of life has involved the repeated grouping together of biological units into higher-level units followed by the consolidation of each new higher-level unit into an integrated collective. This process has generated the stable biological hierarchy (e.g. genes in cells, cells in organisms, organisms in societies) observed today. Each transition to a new hierarchical level of organization is termed a major transition in evolution.

2. The 'major transitions view' of life provides an evolutionary explanation for the biological hierarchy itself and for increases in complexity of living things in terms of the selective conditions prevailing at the time of each transition. Because a successful transition requires that previously independent entities unite and then that their selfish interests become subordinated to the collective interest of the social group, each transition involves a process of social evolution. The major transitions view, therefore, represents a substantial expansion of the field of social evolution.

3. This book focuses on those major transitions that involve the evolution of individuality, i.e. the formation, from previously independent units, of integrated collectives qualified to be regarded as new individuals in their own right. These include genomes within cells, the eukaryotic cell, zygotes, multicellular organisms, eusocial societies, and interspecific mutualisms. Major transitions can be broadly classified into those that are 'egalitarian' (involving unrelated units) and those that are

'fraternal' (involving related units). It is proposed that every major transition has three principal stages: (a) social group formation, (b) social group maintenance, and (c) social group transformation. Social group formation refers to the initial spread of social behaviour and social group maintenance refers to processes that keep social groups stable. Social group transformation is defined as the process that transforms a stable social group into an individual that is a candidate for participation in the next major transition.

4. Hamilton's inclusive fitness theory (kin selection theory) provides a general theory of social evolution powerful and versatile enough to serve as the conceptual foundation for understanding the major transitions in evolution.

5. With this background, two underexplored issues in social evolution can be identified. First, to what extent do common principles operate at each stage of the major transitions, and what is the empirical evidence for their operation? Second, what are the principles underlying social group transformation? This book aims to investigate these issues in detail, with the premise that some of life's most fundamental evolutionary processes are processes of social evolution.

2

A primer in inclusive fitness theory

2.1 Hamilton's rule and relatedness

This chapter serves as an introduction to inclusive fitness theory, which underlies the entire approach to be taken in this book (Section 1.4). Later chapters then apply the theory to the major transitions, first to the transitions in general and then to their component stages. Since Hamilton (1963, 1964) initially proposed inclusive fitness theory, also known as kin selection theory, the scientific literature on it has become huge. Its history, as an idea within population genetics and selection theory, is complex (Grafen 1985). Recent reviews and theoretical expansions include those of Frank (1998), Rousset (2004), Grafen (2006), Lehmann and Keller (2006a), Gardner et al. (2007a), West et al. (2007a), Gardner and Foster (2008), and Wenseleers et al. (2010). This chapter takes an elementary approach, building on that presented by Bourke and Franks (1995) but informed by recent developments. My hope is that such an approach will allow the theory's basic logic to be grasped more easily, while also providing an introduction to its many ramifications.

Social actions

Hamilton (1964) proposed a four-way classification of social behaviour based on whether social partners gain or lose offspring through the performance of the behaviour. Say there are two interacting social partners. It will be assumed in this chapter that they are members of the same species, although this assumption will be revisited in the following chapter (Section 3.1). One partner (the actor) does something to the other (the recipient) with the result that, on average, each experiences a change in their expected offspring number (direct fitness), measured over the entire lifetime of the focal partner. Such a behaviour is called a social action (Fig. 2.1). For example, the actor might steal food from the recipient, and on average this might increase the actor's expected offspring number and decrease that of the recipient. Hamilton's (1964) four types of social action are cooperation (narrow sense), altruism, selfishness, and spite (Table 2.1). Cooperation (narrow sense) is defined as the social action in which both partners gain offspring through the performance of the action. Altruism is defined as the social action in which the actor (or altruist in this case) loses offspring and the recipient (or beneficiary) gains offspring. Selfishness is defined as the social action in which the actor gains offspring and the recipient loses. Hence food-stealing, in the example above, is an instance of selfishness. Finally, spite is defined as the social action in which both partners lose offspring (Table 2.1).

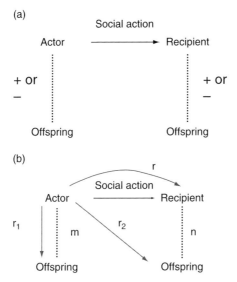

Fig. 2.1 (a) Diagram illustrating a social action, which occurs when an organism (the actor) performs a behaviour such that it either gains (+) or loses (–) in expected offspring number and a social partner (the recipient) also either gains or loses in expected offspring number. (b) The same social action, but with values added for relatedness and changes in offspring number: m = change in actor's offspring number, n = change in recipient's offspring number, r = relatedness of actor and recipient, r_1 = relatedness of actor and actor's offspring, r_2 = relatedness of actor and recipient's offspring.

Note that each of these terms means something close to its everyday meaning, but that, in defining social actions, biologists are concerned only with the *effects* of actions, namely their effects on expected offspring numbers. They are not concerned with what the actor intends or is feeling, if indeed it can intend or feel at all. Hence, for example, one can properly speak of bacterial altruism when using 'altruism' in its biological sense. Note too that the effects in question are effects at the level of the entity (e.g. cell or organism) performing the behaviour, which is why they are expressed in terms of offspring number. This is because it seems most natural to describe a social action at the level of whatever unit is carrying out the action. It follows that genes, unlike organisms, can never themselves show biological altruism. Put another way, all genes are selfish, in the sense that they are selected to maximize their rate of transmission to the next generation (Dawkins 1976), but genetic selfishness may manifest itself as either selfishness or altruism (or cooperation or spite) at the organismal level. Finally, note that narrow-sense cooperation and altruism are often pooled together under the informal catch-all term, 'cooperation'. This can cause some confusion, which is why West et al. (2007b) suggested the term 'mutual benefit' for what is here termed narrow-sense cooperation. To keep a terminology more like that used in most

Table 2.1   Four types of social action

		Effect on recipient's expected offspring number	
		Gains (+)	Loses (–)
Effect on actor's expected offspring number	Gains (+)	Cooperation (narrow sense)	Selfishness
	Loses (–)	Altruism	Spite

After Hamilton (1964).

of the literature, in this book I use 'narrow-sense cooperation' to mean the social action in which both partners gain offspring and 'broad-sense cooperation' to mean cooperative behaviour in general, i.e. altruism or narrow-sense cooperation. The reviews of Sachs et al. (2004), Lehmann and Keller (2006a, 2006b), Bergmüller et al. (2007a, 2007b), West et al. (2007a, 2007b), and Bshary and Bergmüller (2008) deal more widely with terminological and associated biological issues in classifying social phenomena.

Selfishness as defined above is pervasive within species in the natural world, being manifested in various forms of aggression (cannibalism, infanticide, territorial exclusion) or other exploitative interactions, such as social or brood parasitism (cuckoo-like behaviours). But cooperation in its broad sense is also pervasive, and indeed on the major transitions view of evolution it underlies the very hierarchical structure of life. Examples of narrow-sense cooperation within species include cooperative colony founding by ant queens and cooperative foraging by pelicans (Dugatkin 2002; Clutton-Brock 2009a). Altruism within species is found in the sterile somatic cells of multicellular organisms, which aid the reproduction of the germline cells. Altruism also occurs in the non-reproductive zooids of colonial marine invertebrates such as siphonophores (Box 1.1), which aid the reproduction of the reproductive zooids. And it occurs in the sterile or partly-sterile workers of social insects, which aid the reproduction of the queen or king. Spite, whereby the actor undergoes a loss of offspring in order to inflict such a loss on the recipient, is rarer in the natural world. However, some examples have been identified, such as the release of costly toxins by bacteria in order to kill off competing bacteria (Gardner and West 2006; Hawlena et al. 2010). Inclusive fitness theory describes the conditions required for a gene for any of the four types of social action to spread through a population, i.e. to undergo natural selection. The theory was prompted by Hamilton's (1964) desire to explain the evolution of altruism, since altruism seemed to contradict the Darwinian expectation that all organisms should seek to maximize their offspring output (Section 1.4). Nonetheless the theory is a very general one and applies not just to altruism but, as stated, to all four types of social action.

Relatedness

A key parameter in inclusive fitness theory is genetic relatedness. In the theory, relatedness means the probability that a gene at a locus (chromosomal position) in one organism is present at the corresponding locus in another organism. More strictly, it means the probability of sharing the focal gene relative to the average probability that the two organisms share the gene, which is set by the gene's average frequency in the population. Kinship (i.e. being members of the same family or extended family) is the usual reason for organisms sharing genes with above-average frequency, although it is not the only possible one. For this reason, relatedness can usually be understood as the chance of gene-sharing that is brought about by being kin, over and above the average chance. This is why inclusive fitness theory has become known as kin selection theory (Maynard Smith 1964), to the extent that, for most purposes, these two labels are synonymous. (Inclusive fitness theory covers social actions between non-relatives, as well as relatives. Relatedness is a continuous variable and social actions between non-relatives can be thought of as the special cases arising when relatedness takes a zero value. But, because it can be linguistically awkward to describe social actions between non-relatives as occurring through kin selection, in this book I use mainly the descriptor 'inclusive fitness' for the theory.) Relatedness also has a technical definition as a regression coefficient, calculated from the regression of the recipients' gene frequency on the actors' gene frequency. This is a formal way of expressing the informal definition just presented (Pamilo and Crozier 1982; Grafen 1985; Bourke and Franks 1995).

An important point implicit in these definitions is that a relatedness of zero does not mean that two organisms have no genes in common. It only means that their chance of gene-sharing at corresponding loci is no different from that of two random members of the population, who share genes with average frequency. To turn this around, random members of the population are, by definition, those whose relatedness is zero. This matters because selection only 'cares' about gene frequencies other than average gene-frequencies. Adding offspring that carry a focal gene with average frequency to a population leaves the population gene-frequency unchanged; so, of necessity, no selection can have occurred. By contrast, adding offspring (such as the offspring of relatives) that carry a focal gene with above-average frequency to a population causes a selective increase in the focal gene's frequency. This is why the definition of relatedness as a deviation from the population average gene-frequency is critical in inclusive fitness theory (Grafen 1985, 1991; Bourke and Franks 1995; McElreath and Boyd 2007).

Let us now consider some relatedness values (r) that occur between different types of kin. Clones, such as offspring produced by asexual reproduction, are by definition genetically identical organisms, so their relatedness is 1. In sexual, outbreeding diploids, such as ourselves, offspring are genetically half one parent and half the other parent. The chance of gene-sharing due to kinship between a focal parent and its offspring is, therefore, the average of its relatedness to its own genome (r to self equals 1) and its relatedness to the genome of its mate (r to mate equals zero under outbreeding, since the mate is a random member of the population). Hence parent–offspring relatedness equals:

$$(0.5 \times 1) + (0.5 \times 0) = 0.5.$$

Following straightforward Mendelian inheritance, siblings in diploids share genes in the maternal half of the genome with a probability of 0.5 and genes in the paternal half of the genome with a probability of 0.5. Therefore, relatedness between diploid siblings, i.e. the average chance of gene-sharing across both halves of the genome, is:

$$(0.5 \times 0.5) + (0.5 \times 0.5) = 0.5.$$

Hamilton (1964) famously pointed out that haplodiploidy, as found in the Hymenoptera, creates relatedness values different from those found among diploids. In haplodiploids, females develop from fertilized eggs and are diploid, and males develop from unfertilized eggs and are haploid. Female siblings (sisters), being diploid, still consist of a maternal half of the genome and a paternal half. Their chance of gene-sharing in the maternal half is the ordinary 0.5. But their chance of gene-sharing in the paternal half is elevated to 1, since all gametes of their haploid father share the same gene at any one locus. So, averaging across the two halves of the genome shows relatedness among haplodiploid sisters to be:

$$(0.5 \times 0.5) + (0.5 \times 1) = 0.75.$$

This is the well-known finding that full sisters in the social haplodiploids (ants, bees, and wasps) are related by three-quarters, which is higher than the usual diploid value of 0.5 (Hamilton 1964). Similar calculations can be used to generate relatedness values for any selected sets of kin in diploids and haplodiploids (Fig. 2.2; Table 2.2).

Hamilton's rule

Inclusive fitness theory is expressed algebraically as a mathematical inequality called Hamilton's rule, which specifies the circumstances under which a gene for each type of social action will spread by natural selection (Grafen 1985, 1991). By definition, the condition for the natural selection of any gene is that its phenotype should bring about a positive change in the gene's frequency. Hamilton's rule defines this condition in the case of genes for social actions. So, as has been stressed before (Dawkins 1979; Grafen 1985; Bourke and Franks 1995), inclusive fitness theory is a logical extension of the genetical theory of natural selection. Genes for social actions are not especially different from any other type of gene, and no unusual assumptions need to be made to derive the theory. For example, during the origin of altruism, a gene for altruism could be simply a gene whose bearers fail to disperse, provided non-dispersers are more likely to show helping behaviour. This is because such a gene would then be the one that makes the difference between altruism being shown and altruism not being shown (Dawkins 1979; Bourke and Franks 1995). Hamilton's rule finds that whether a social action

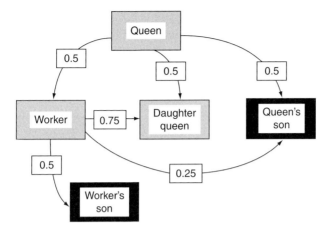

Fig. 2.2 Relatedness values in a colony of eusocial Hymenoptera headed by one, singly-mated queen, with workers being capable of producing male offspring asexually (arrhenotoky) (see also Table 2.2). Grey shading indicates females, black shading males. Workers are always female. The queen's female offspring are, depending on rearing environment, workers or new queens. The queen's relatedness to the son of a worker is queen–worker relatedness (0.5) × worker–son relatedness (0.5), i.e. 0.25. A worker's relatedness to the son of another worker is sister–sister relatedness (0.75) × worker–son relatedness (0.5), i.e. 0.375. If the queen mates multiply, say with k males, relatedness among her daughters becomes $0.5(0.5 + 1/k)$, so a worker's relatedness to the son of another worker then becomes $0.5(0.5 + 1/k) \times 0.5 = 0.25(0.5 + 1/k)$, which asymptotes to 0.125 as k rises (e.g. Bourke and Franks 1995).

Table 2.2 Relatedness values between different classes of kin in diploids and haplodiploids. The values assume monandry (females mate with one male each), monogyny (there is one breeding female, or queen, per family), and outbreeding (see also Fig. 2.2). Haplodiploid values are technically 'life-for-life relatedness' values, which are regression relatednesses weighted by relative sex-specific reproductive value. The distinction between these two ways of expressing relatedness has no consequences for the arguments in the present book Grafen (1986, 1991), Bourke and Franks (1995), Crozier and Pamilo (1996), and Bourke (1997) provide a full explanation, along with more extensive tables of relatedness values

Actor		Recipient	
		Female	Male
(a) Diploids			
Female or male	Offspring	0.5	0.5
	Sibling	0.5	0.5
(b) Haplodiploids			
Female	Offspring	0.5	0.5
	Sibling	0.75	0.25
Male	Offspring	1.0	0
	Sibling	0.5	0.5

undergoes selection is affected by the genetic relatedness between the social partners and by the changes in their expected offspring numbers occurring as a result of the social action.

A general form of Hamilton's rule can be derived quite simply from first principles of population genetics (Box 2.1). In words, Hamilton's rule states that a gene for a social action will undergo selection if the social action's effect on the gene's frequency via the actor's reproduction (measured as the change in the actor's offspring number), added to the social action's effect on the gene's frequency via the recipient's reproduction (measured as the change in the recipient's offspring number weighted by actor-recipient relatedness), is positive (Box 2.1, Equation 2.2). Note that if relatedness is zero, only the first of these terms remains (i.e. change in the actor's offspring number). So, as expected, one then recovers the condition for a gene with no systematic effect on relatives to undergo selection, namely that it increases expected offspring number. This is the condition required for 'ordinary' natural selection. Therefore, Hamilton (1964) demonstrated that to predict the course of selection on a social trait requires a measure of fitness that extends the classical fitness measure of personal offspring production to include an additional element based on the actor's social effects on the offspring production of others. This is why he called his theory inclusive fitness theory. The classical part of inclusive fitness, based on personal offspring production, is termed direct fitness and depends on the m term in Hamilton's rule (Box 2.1). The additional part, based on the actor's social effects on the offspring production of others, is termed indirect fitness and depends on the rn term in Hamilton's rule (Box 2.1). Together, therefore, direct and indirect fitness make up inclusive fitness.

Stepping back, one can begin to see that Hamilton's rule implies that organisms are selected to perform social actions as if they valued others according to how closely related they are, with the reproduction of others being valued more highly the greater the degree of relatedness (Grafen 1985, 1991). This insight is expanded upon in the following section.

2.2 The effect of levels of relatedness on evolvable types of social action

With some basic ideas of inclusive fitness theory now in place, this section explores the predictions of the theory by examining each type of social action in turn and asking what form Hamilton's rule adopts for a particular social action when relatedness varies. In other words, it seeks to determine the specific conditions for each type of social action. In terms of understanding the major transitions, determining these conditions is important because it helps explain the nature of the relatedness regimes that underlie the different major transitions (Section 3.1).

Algebraically, Hamilton's rule can be generically expressed as the statement that a gene for a social action spreads if:

$$m + rn > 0,$$

Box 2.1   A simple derivation of Hamilton's rule

There are many formal proofs of Hamilton's (1964) rule. Recent ones are in Frank (1998), Gardner et al. (2007a), McElreath and Boyd (2007), Gardner and Foster (2008), and Wenseleers et al. (2010). To keep the mathematics simple, but convey the essential logic, this box presents an informal derivation, taken largely from Bourke and Franks (1995) but with some minor changes in notation. An organism (the actor) bears a gene G for a social action. The effect of the performance of the action is that the actor's expected offspring number, measured over its entire lifetime, increases or decreases, and likewise the expected offspring number of a conspecific recipient also increases or decreases. Let the following terms be defined (Fig. 2.1):

m = the change in the actor's offspring number (hence m can be positive or negative);
n = the change in the recipient's offspring number (hence n can also be positive or negative);
r = the relatedness of the actor and the recipient;
r_1 = the relatedness of the actor and the actor's offspring;
r_2 = the relatedness of the actor and the recipient's offspring.

To make m and n more memorable, think of yourself as the actor and your neighbour as your social partner, so m is the change in offspring number experienced by *me* and n is the change in offspring number experienced by *neighbour*. The total change in gene frequency as a result of the social action being performed is the sum of the change through the actor's offspring and the change through the recipient's offspring. Change through the actor's offspring is the product of r_1 and m, since the action adds to the next generation (if m is positive) or takes away from the next generation (if m is negative) a total of m actor's offspring with a probability of r_1 of bearing the gene G (by the definition of relatedness). By the same logic, change through the recipient's offspring is the product of r_2 and n. Therefore, the total change in gene frequency is the sum of r_1m and r_2n.

The gene G (like any gene) will spread through the population if the total change in gene frequency brought about by its phenotype (here, the social action) is positive. In other words, the gene will spread if more gene copies are added to the next generation than are lost from it. So the condition for the gene to spread is:

$$r_1m + r_2n > 0. \qquad \text{(Equation 2.1)}.$$

This is Hamilton's rule in a very general form. It can be expressed in a more familiar form by dividing each side by r_1, so yielding:

$$m + (r_2/r_1)n > 0.$$

Next note that:

$$r_2 = r \times r_1,$$

since, as probabilities, relatednesses can be multiplied, and the actor's relatedness to the recipient's offspring (r_2) equals the product of the actor's relatedness to the recipient (r) and the recipient's relatedness to its offspring, which equals r_1 under the reasonable assumption that parent-offspring relatedness is the same for all parents in the population. From:

$$r_2 = r \times r_1$$

one obtains:

$$(r_2/r_1) = r,$$

where m is the change in the expected offspring number of the actor (over the actor's lifetime), n is the change in the expected offspring number of the recipient, and r is actor-recipient relatedness (Box 2.1). By definition, when the social action is cooperation (narrow sense), both m and n are positive (Table 2.1). So Hamilton's rule remains

$$m + rn > 0,$$

with m and n now referring to the positive values of the changes in offspring number of the actor and recipient, respectively. When the social action is altruism, m is negative and n is positive, so Hamilton's rule becomes:

$$-m + rn > 0.$$

When the social action is selfishness, m is positive and n is negative, so Hamilton's rule becomes:

$$m - rn > 0.$$

And lastly, when the social action is spite, both m and n are negative, so Hamilton's rule becomes:

$$-m - rn > 0$$

(Table 2.3).

Let us now examine what happens to these different versions of Hamilton's rule when relatedness varies. As a regression coefficient, relatedness can in principle take any value (Grafen 1985), but the range from $r < 0$ (negative relatedness) to $r = 1$ encompasses the likely set of values found in nature. At the lower end, $r < 0$ means that social partners are less genetically similar to one another than random members

Table 2.3 Evolvable social actions at different levels of relatedness and specific versions of Hamilton's rule for their evolution. m, n, r = positive values of change in actor's expected offspring number, change in recipient's expected offspring number, and relatedness, respectively (note that this usage of m, n, and r differs slightly from that in Box 2.1)

Relatedness	Social action	Hamilton's rule
(a) Relatedness takes any value		
	Cooperation	$m + rn > 0$
	Altruism	$-m + rn > 0$
	Selfishness	$m - rn > 0$
	Spite	$-m - rn > 0$
(b) Relatedness takes specified value		
$r < 0$	Cooperation	$m - rn > 0$
	Selfishness	$m + rn > 0$
	Spite	$-m + rn > 0$
$r = 0$	Cooperation	$m > 0$
	Selfishness	$m > 0$
$0 < r < 1$	Cooperation	$m + rn > 0$
	Altruism	$-m + rn > 0$
	Selfishness	$m - rn > 0$
$r = 1$	Cooperation	$m + n > 0$
	Altruism	$-m + n > 0$
	Selfishness	$m - n > 0$

of the population (Grafen 1985). At the upper end, $r = 1$ is the value found between members of the same clone.

We start by deriving the conditions for cooperation to evolve when relatedness varies across this range. When $r < 0$, one can see from the form of Hamilton's rule for cooperation given above (cooperation evolves if $m + rn > 0$) that Hamilton's rule becomes:

$$m - rn > 0,$$

with r now referring to the positive value of relatedness. When $r = 0$, the rule becomes simply $m > 0$. When $0 < r < 1$, the rule remains:

$$m + rn > 0.$$

And when $r = 1$, the rule becomes:

$$m + n > 0.$$

These different versions of Hamilton's rule for cooperation are summarized in Table 2.3. Now let us conduct the same exercise for altruism. When $r < 0$, Hamilton's rule becomes:

$$-m - rn > 0.$$

But this is mathematically impossible, since the sum of two negative values can never exceed zero. This means that altruism cannot evolve when relatedness is negative and so this case need not be considered further. When $r = 0$, the rule becomes $-m > 0$. This is also impossible (a negative number does not exceed zero), meaning that altruism cannot evolve when relatedness is zero, i.e. when the recipient is unrelated. So, again, this case can be discounted forthwith (although the general point that altruism cannot evolve between non-relatives is critical). When $0 < r < 1$, the rule remains:

$$-m + rn > 0.$$

And when $r = 1$, the rule becomes:

$$-m + n > 0.$$

Following the same reasoning with selfishness and spite, and again excluding impossible conditions, yields a complete set of Hamilton's rules for different levels of relatedness (Table 2.3). This seemingly dry algebraic exercise is rich in implications for the evolution of social behaviour, as we will now see by working through its main biological conclusions for each type of social action.

Cooperation (narrow sense)

Cooperation, it is evident (Table 2.3), can evolve at all levels of relatedness from negative values to 1. For example, when the actor and recipient are unrelated ($r = 0$), the condition for the evolution of cooperation is that $m > 0$. This means that cooperation can evolve between non-relatives simply if the actor experiences a net increase in its expected offspring number. It is the familiar finding that the evolution of cooperation between non-relatives requires that the actor, on average, gain a direct benefit over its lifetime (e.g. Lehmann and Keller 2006a; West et al. 2007a). However, relatedness between an actor and a recipient aids the evolution of cooperation, since the condition for the evolution of cooperation with $r > 0$ becomes:

$$m + rn > 0$$

(Table 2.3). This is an easier condition to fulfil than $m > 0$, since the fitness credit to the actor now includes the additional offspring of the recipient that cooperation brings about (weighted by relatedness).

Altruism

When relatedness is zero, as we saw above, Hamilton's rule for altruism becomes $-m > 0$, which cannot be satisfied. Biologically, an altruist donating aid to a non-relative undergoes a net loss of offspring over its lifetime (by the definition of altruism) and also fails to gain any fitness through the aid it donates, since the beneficiary's offspring bear a focal gene of the altruist with only average frequency (by the definition of zero relatedness). Adding such offspring to the next generation does not increase the focal gene's frequency, so altruism towards a non-relative does not evolve. In fact it would be selected against, since it would cause a decrease in the frequency of the focal gene (proportional to the $-m$ term). Strictly speaking, the conclusion is that altruism cannot evolve when relatedness between the actor and recipient is zero at the locus for social behaviour. This means altruism between non-kin cannot evolve, except in special cases where relatedness is positive at the locus for social behaviour but not at other loci (Section 2.4).

When relatedness between actor and recipient is positive, altruism can evolve according to Hamilton's rule, provided that:

$$-m + rn > 0$$

(Table 2.3), i.e. that $rn > m$. This is the canonical finding of inclusive fitness theory and means altruism can evolve by natural selection, but only when altruist and recipient are related. In addition, the gain to the altruist, measured as the number of extra offspring conferred on the beneficiary weighted by relatedness, must exceed the cost to it, measured as the number of young it loses. In other words, selection tolerates the loss of copies of the gene for altruism occurring via the altruist's reduced offspring output on condition that there is a more-than-compensating increase in the number of copies of the gene for altruism via the increased reproduction of the beneficiary. This way, the gene for altruism can experience a net gain despite the self-sacrifice of the actor. Moreover, for any relatedness level above zero, the evolution of altruism is facilitated by low cost to the actor and high benefit to the recipient (i.e. low m and high n), implying that altruism can evolve even when relatedness is low. So altruism does not require high relatedness to evolve, only relatedness greater than zero. Nonetheless, high levels of relatedness clearly facilitate the evolution of altruism, since they make it more likely that the fitness gain via the recipient's offspring (rn) will exceed the fitness loss via the actor's reduced offspring output (m).

It is also important to note that, from Hamilton's rule, altruism can evolve even if it is associated with the actor becoming totally sterile. That is, the fall in the actor's offspring number ($-m$) can be large enough to bring its reproductive output down to zero, yet Hamilton's rule can still be satisfied provided the indirect gain via the beneficiary exceeds this decrease ($rn > m$). Moreover, as an actor becomes less fecund, this condition becomes easier to fulfil. Say the number of offspring it could have alone is given by its maximum fecundity, N_1, and the number it could have as an altruist is given by N_2, where N_2 is less than N_1 (which must be true by the definition of altruism). Then the cost of altruism is:

$$m, = N_1 - N_2.$$

So, as the actor's maximum fecundity falls (N_1), so too must m fall, thereby allowing Hamilton's rule to be more easily satisfied.

To state Hamilton's rule for altruism in terms of the net gain in copy number experienced by the gene for altruism is to repeat the point that inclusive fitness theory explains altruism at the organismal level by selfishness at the gene level (Section 1.4). Pursuing this logic, we can see that, when relatives are the beneficiaries, altruists are aiding co-bearers of their own gene for altruism. Effectively altruism evolves when the gene for altruism directs aid at copies of itself (Dawkins 1976). This is particularly evident when altruists aid clones ($r = 1$). In this case, an actor is genetically indifferent over whether it reproduces or its clone does, since it values the clone-mate genetically as much as it values itself. So all that is then needed for altruism to evolve is that altruism be more efficient than personal reproduction, i.e. that $n > m$ (Table 2.3).

Selfishness

It is clear that selfishness, like cooperation, is capable of evolving at any level of relatedness (Table 2.3). When relatedness is zero, the condition for selfishness is simply $m > 0$, i.e. that the actor should experience a rise in offspring number (Table 2.3). This is straightforward natural selection for self-advantageous behaviour. Put another way, it is the condition for natural selection to occur for a 'non-social' trait in an unstructured population. 'Non-social' here means not that others are necessarily unaffected by selfishness, but only that the effects of selfishness fall randomly and not on any particular class of neighbours. Note that, as is intuitive, selfishness is more strongly selected for as the number of offspring it confers on the actor rises (greater m). When relatedness is positive, selfishness evolves if $m - rn > 0$ (Table 2.3). This condition shows that relatedness acts as a brake on the evolution of selfishness, since the left-hand side of this inequality falls as relatedness rises. In short, relatedness facilitates altruism and mitigates selfishness. In the case of selfishness, the reason is that causing a loss of offspring to a related recipient rebounds on the actor, since the offspring will bear the gene for selfishness with above-average frequency. By contrast, when relatedness is zero, selfish behaviour can evolve at any level of cost to others (the n term is absent from Hamilton's rule in this case). This is because a gene for selfishness has no evolutionary interest in its effects on conspecifics that bear it with only average frequency.

Note, however, that relatedness (short of clonality), although setting limits on the extent of selfishness, does not guarantee the absence of selfishness. There are many cases of fierce and even lethal competition between relatives, such as siblicide in ants (Sundström et al. 1996), birds (Mock and Parker 1997, 1998), and mammals (Wahaj et al. 2007). This may occur because an actor's inclusive-fitness interests are best served by its favouring itself or closer relatives over more distant classes of relatives in the group (Section 2.3), or because limited dispersal generates competition between kin and so alters the relevant form of Hamilton's rule (Section 2.4).

Spite

Spite is unique among the four types of social action in that it can only evolve when relatedness is negative (Table 2.3). Negative relatedness is required to turn the harm inflicted on recipients into a benefit that potentially outweighs the self-harm experienced by the actor. The fact that conditions for negative relatedness are comparatively restrictive (Grafen 1985) accounts for spite being the least frequent and least well-studied type of social behaviour, although, as previously mentioned, some examples have been suggested (Gardner and West 2006, 2009; Gardner et al. 2007b). In one example, the required element of negative relatedness was present because spiteful bacteria produced inhibitory toxins only when challenged with competitors of a slightly different strain (Hawlena et al. 2010).

Overall, therefore, this deconstruction of Hamilton's rule reinforces the point that inclusive fitness theory is a general theory of social evolution, applying to all four types of social action defined by Hamilton (1964). In fact, by incorporating selection on selfish traits that do not affect relatives, inclusive fitness theory also encompasses what has traditionally been regarded as the essential form of natural selection, i.e. 'individual-level' selection for self-advantage. It is now apparent, however, that this occurs as a special case within inclusive fitness theory. This point is not just an interesting but marginal curiosity. Inclusive fitness theory adds to our understanding of selection for self-advantage by showing that selfishness occurs at any level of cost to non-relatives. This is essentially why naive 'good for the species' views of natural selection break down. In addition, inclusive fitness theory shows that, when it affects relatives, selfishness is kept in check. This conclusion is important for understanding how cheating can be controlled within social groups (Section 3.2).

 The present exercise in deconstructing Hamilton's rule has also allowed us to deduce much about the conditions required for each type of social action. Specifically, it shows that each social action has an associated set of relatedness levels under which it can evolve (Table 2.3). Key findings are that cooperation can evolve when non-relatives interact but that the evolution of altruism always requires above-zero relatedness. Other authors have also stressed the general compass of inclusive fitness theory (e.g. Reeve 2001; Lehmann and Keller 2006a; West et al. 2007a). It is worth re-emphasizing here because, owing to the history of the theory, a sort of tunnel-vision has sometimes prevailed in evolutionary biology, whereby inclusive fitness theory has been regarded as applicable only to altruism between relatives. In addition, the theory needs to be considered in its complete form because, as such, it is highly relevant to the evolutionary basis of the various major transitions, as later chapters show.

2.3 Social conflict and the tragedy of the commons

The previous sections have used inclusive fitness theory to understand the conditions for the evolution of social actions from the standpoint of a single party, the actor.

A remarkable feature of the theory is that it also provides a framework for understanding what happens in social evolution when social groups harbour multiple actors with differing standpoints. The theory predicts that, when different parties in a social group are differentially related to group offspring, there is potential social conflict (Hamilton 1964; Trivers 1974; Ratnieks and Reeve 1992; Yamamura and Higashi 1992; Emlen 1995; Mock and Parker 1997). This is because in such cases the different parties will have different optima for their inclusive fitness, meaning that their inclusive fitnesses cannot simultaneously be maximized. In practice, a variety of processes can restore a degree of coincidence of fitness interest among social partners with less than clonal relatedness and resolve the conflicts that a lack of clonal relatedness engenders (Sections 3.2, 5.3–5.5). By contrast, in a clone, and only in a clone, all group members have a perfect coincidence of their inclusive fitness interests to start with (Hamilton 1964; Frank 1995; Gardner and Grafen 2009). Hence they lack potential social conflict. For these reasons, inclusive fitness theory represents the primary tool for analysing evolutionary conflicts of interest within social groups.

Examples of kin-selected conflict

Classic examples of kin-selected conflict include parent–offspring conflict (Trivers 1974), in which parent and offspring 'disagree' over the allocation of resources among offspring because parents value their offspring equally ($r = 0.5$), whereas each offspring values itself more highly ($r = 1$) than it values its siblings or half-siblings ($r \leq 0.5$). An example is parent–offspring conflict over provisioning of young in birds and mammals (Trivers 1974). Many types of kin-selected conflict have been identified in the eusocial Hymenoptera (Bourke and Franks 1995; Ratnieks et al. 2006). One well-known one is queen–worker conflict over the parentage of males (Trivers and Hare 1976; Ratnieks 1988). This arise because, if colonies are headed by one, singly-mated queen (Fig. 2.2; Table 2.2), the queen is more closely related to queen-produced males (her sons, $r = 0.5$) than to worker-produced males (her grandsons, $r = 0.25$), whereas non-reproductive workers (which form the mass of workers) are less closely related to queen-produced males (their brothers, $r = 0.25$) than to worker-produced males (their nephews, $r = 0.375$). So each party favours the rearing of a different set of male progeny. Another well-known type of kin-selected conflict in eusocial Hymenoptera is queen–worker conflict over the sex ratio. In general, the sex ratio that maximizes the inclusive fitness of any party depends on that party's relatedness asymmetry, which is defined as its relative relatedness to females and males (Trivers and Hare 1976; Boomsma and Grafen 1991; Bourke and Franks 1995; Crozier and Pamilo 1996). In a Hymenopteran colony headed by one, singly-mated queen, the queen's relatedness asymmetry equals her relatedness to new, daughter queens divided by her relatedness to sons, i.e. 0.5/0.5, and the workers' relatedness asymmetry equals their relatedness to sisters divided by their relatedness to brothers, i.e. 0.75/0.25 (Fig. 2.2; Table 2.2).

Hence the stable female:male sex ratio is 1:1 for the queen and 3:1 for the workers, wherein lies the conflict.

Note that, for any form of within-group conflict, the kin structure of the social group dictates that a potential conflict of evolutionary interests exists. However, as pointed out by Ratnieks and Reeve (1992), whether a potential conflict becomes actual, i.e. is expressed as overt behaviour, will depend on other factors. A number of such factors have now been identified (Ratnieks and Reeve 1992; Bourke and Franks 1995; Beekman and Ratnieks 2003; Beekman et al. 2003; Ratnieks et al. 2006; Helanterä and Ratnieks 2009). They can be conveniently expressed as a series of constraints. There may be genetic constraints (there is no genetic variation for conflict behaviour), constraints on power and opportunity (parties in potential conflict lack physical power or practical opportunities for self-interested manipulations), or informational constraints (information as to the optimal course of action may be lacking or inaccessible). As an example of an informational constraint, in colonies of the Honey bee (*Apis mellifera*), which are headed by a single multiply-mated queen, conflicts between patrilines of workers derived from different fathers appear absent because workers are unable to discriminate between different degrees of kin within the colony (Keller 1997; Châline et al. 2005). In other words, workers in this case cannot access the information that would allow each patriline to favour closer kin.

Intragenomic and intergenomic conflict

Inclusive fitness theory has, as in the examples above, traditionally been used to analyse the occurrence of conflict within family groups or eusocial societies. But its logic and tools can be applied to analyse evolutionary conflicts of interest at any level in the biological hierarchy. Take the intracellular, genomic level. Within cells, intragenomic conflicts occur between different sets of nuclear genes, while intergenomic conflicts occur between nuclear genes and genes in the genomes of organelles or other cytoplasmic elements (Eberhard 1980; Cosmides and Tooby 1981; Werren *et al.* 1988; Hurst et al. 1996; Haig 1997, 2004; Pomiankowski 1999; Hoekstra 2003; Burt and Trivers 2006; Engelstädter and Hurst 2009). These conflicts are readily understood using inclusive fitness analyses (Haig 1997), although not all researchers have expressed their analyses in this way.

An example of intragenomic conflict is the conflict between meiotic drive genes and other genes in the nuclear genome. In sexual organisms, meiosis is usually fair, meaning that all loci are represented in gametes and, within any one locus, each allele has the same chance (0.5) of entering a gamete. Meiotic drive occurs when a drive gene cheats by achieving more than 50% representation in the gametes, relative to its allele (Burt and Trivers 2006). This can be understood in terms of inclusive fitness theory in that the theory predicts potential conflict between alleles at the same locus in outbred sexual diploids because these alleles are unrelated (Keller and Reeve 1999; Wenseleers and Ratnieks 2001; Bourke 2009). Meiotic drive represents this potential conflict becoming an actual one. Meiotic drive genes frequently result in a reduced reproductive success of the genome as a whole (offset, for the drive genes, by their greater representation in offspring). So, given that genes at different loci in the genome are unrelated, inclusive fitness theory predicts that modifiers at other loci will be selected to suppress drive (Leigh 1977), and such modifiers are duly found (Section 5.5).

An example of intergenomic conflict is found in flowering plants with cytoplasmic male sterility. Cytoplasmic male sterility occurs when a hermaphroditic flowering plant (i.e. one with both female and male sexual structures) fails to produce pollen and therefore becomes 'male-sterile'. It is a common and widespread trait, being found in over 150 plant species from many families, including commercially important crop species such as Maize and Rice (Schnable and Wise 1998; Burt and Trivers 2006).

Why should a plant effectively castrate itself? The key to understanding this is that, in every case that has been examined, the gene for male sterility is borne in the mitochondrial genome and not in the nuclear genome. (Given the mitochondria reside in the cellular cytoplasm, this explains why plant geneticists termed the trait 'cytoplasmic' male sterility, though a better term in light of current knowledge would be mitochondrial male sterility.) Consider the evolutionary interests of genes in the mitochondrial genome. Mitochondria reproduce asexually and are passed on by a parent plant through its female gametes (ovules) but not its pollen. Therefore, relatedness of a mitochondrial gene to ovules is 1 but to pollen is effectively 0. Hence, from a mitochondrial gene's standpoint, the optimal allocation of resources between the plant's female and male functions is one-to-zero, or all-females. In short, as in traditional inclusive fitness analyses of sex ratio evolution, the mitochondrial gene's optimal sex allocation is dictated by its relatedness asymmetry (Bourke 2009). As a result, if it could influence how the plant directs resources into producing ovules versus pollen, the mitochondrial gene would favour only ovule production and no pollen production (Burt and Trivers 2006). This is evidently what happens, so accounting for cytoplasmic male sterility. Furthermore, as this interpretation predicts, male-sterile plants are indeed more productive of ovules and hence seeds (representing female function) than unaffected plants (Burt and Trivers 2006). The relatedness asymmetry of the nuclear genes remains at 1:1 (they are equally represented in ovules and pollen) and these genes are therefore in conflict with mitochondrial genes over the failure to produce pollen. So, as predicted, nuclear genes fight back to restore sex allocation to their preferred ratio (Section 5.5).

In fact, inclusive fitness considerations underlie cytoplasmic male sterility and similar cases to an even greater extent than is often realized (Hurst 1991; Pomiankowski 1999). Mitochondrial genes sitting in cells in, say, a leaf cannot reach out and affect what is taking place in the flowers. Instead what happens is that mitochondrial genes in the floral tissues giving rise to the plant's male sexual structures achieve male sterility by bringing about cell death within these tissues (Pomiankowski 1999; Burt and Trivers 2006). In so doing, the mitochondria in these cells destroy themselves. Let us apply Hamilton's rule for altruism to this case (using the form $rb - c > 0$ from Box 2.1). The cost to the self-destructing mitochondria (c) is nil, because they are not transmitted via pollen anyhow. Their relatedness (r) to the mitochondria in the ovules (their beneficiaries) approaches 1, since (barring mutations) mitochondria within a single organism are assumed clonal or nearly clonal (Section 5.5). Therefore, by Hamilton's rule, mitochondrial altruism is selected if $b > 0$, that is, if extra offspring are gained by the mitochondria in the rest of the plant. This condition is fulfilled, given that affected plants overproduce seeds, which of course contain copies of the parent plant's mitochondria. The conclusion is that cytoplasmic male sterility evolves because the mitochondria bearing the causative genes are kin-selected altruists and Hamilton's rule for their particular form of altruism is quantitatively fulfilled.

This analysis illustrates both the operation of social evolutionary processes in an unfamiliar context and the utility of inclusive fitness theory in understanding them. It also implies, as pointed out by Crespi (2001), that mitochondria sometimes act similarly to other kinds of social bacteria. Indeed, the conditions for mitochondrial self-sacrifice in cytoplasmic male sterility also apply to male-killing intracellular bacteria such as *Wolbachia* (Section 5.5), whose effects rely on bacteria in male embryos of the host sacrificing themselves to promote the fitness of their clonemates in the female embryos of the host (Hurst 1991; Pomiankowski 1999). Note that both mitochondria causing cytoplasmic male sterility and male-killing *Wolbachia* exhibit a fairly sophisticated conditionality. The particular mitochondria or bacteria causing the death of the tissues they reside in must be able to 'tell' that they are in the appropriate kind of tissue (i.e. floral tissues giving rise to the plant's male sexual structures or male embryos, respectively); otherwise they would mistakenly kill off tissues involved in female function. Hence the conditionality involved in kin-selected social behaviour (Section 2.4) is achievable even in relatively simple biological systems. Lastly, all cases of intra- and intergenomic conflict reinforce the point that even the most individualistic of all the levels of biological organization, the multicellular organism, is not free from internal conflicts of interest, so underscoring individuality's contingent nature (Alexander and Borgia 1978; Dawkins 1982). As a corollary, such cases exemplify the type of internal conflict that needs to be held in check if multicellular organisms are to retain their individuality.

The tragedy of the commons

Conflict within social groups at any level is often of the type in which there exists a 'public good' available to the group that any one group member would benefit from overexploiting but that would be depleted to the cost of the entire group if all group members behaved selfishly. An example comes from bacteria (*Pseudomonas fluorescens*), in which the public good is a mat formed from secreted polymers that keeps the bacteria floating close to aerial oxygen (Section 5.3). Selfish mutants benefit by saving themselves the costs of polymer production, but eventually this causes all cells in the mat to sink and die (Rainey and Rainey 2003). Another example comes from colonies of eusocial Hymenoptera, in which the public good is the workforce. Selfish female larvae benefit by developing as queens, so imposing a productivity cost on the colony through there being fewer workers (Bourke and Ratnieks 1999; Wenseleers et al. 2003; Wenseleers and Ratnieks 2004; Hughes and Boomsma 2008).

This type of conflict is encapsulated by the idea of the tragedy of the commons (Hardin 1968; Rankin et al. 2007a). In the analogy that Hardin made famous, this refers to an area of common ground on which each of a set of herdsmen may graze his cattle. Each herdsman gains a benefit by adding an extra animal to the common, which, as the owner of the animal, he alone profits from. This outweighs the cost to him of the extra animal reducing the total amount of forage on the common, since this cost is spread across all the herdsmen. Each herdsman is, therefore, as it were, selected to keep adding an extra animal to the common. Because all the herdsmen follow the same logic, the result is that the common becomes overgrazed and rendered useless. As Hardin (1968) put it, 'Therein is the tragedy . . . Freedom in a commons brings ruin to all'. Put another way, the tragedy of the commons arises because the economic or fitness optimum of a focal group member does not coincide with that of groupmates. Take any pair of groupmates. For every unit of profit yielded by the shared resource, the first would prefer to take it at the expense of the second, while the second would prefer to take it at the expense of the first. The result is that each keeps on taking until there is no longer enough resource to profit from.

The bearing of Hardin's (1968) analysis on pressing problems in human society, such as our overconsumption of natural resources or our overproduction of greenhouse gases, is painfully obvious. But for present purposes, the relevant issue is how, in general, the tragedy of the commons can be overcome. In human society, Hardin's (1968) proposed solution was coercion in one form or another. As an example he suggested taxation. This redistributes, ultimately by force, some fraction of personal profits back to the common pool, so redressing the balance between private and public interests. Another example might be a quota system that limits how much a single group member can draw from a shared resource. Hardin (1968) characterized such systems as forms of 'mutual coercion, mutually agreed upon by the majority ... affected'. The important point about these means of resolving the tragedy of the commons, as in other examples from human society, is that coercion occurs in a top-down manner. In other words, through at least a degree of foresight, rational planning, and agreement, human societies elect to be controlled by rules, such as tax laws or quotas, that limit runaway exploitation. Moreover, these systems are backed up by a centralized government, police force, and system of punishment for transgressors. Evolution cannot work this way. In natural systems the resolution of the tragedy of the commons, and of within-group conflict in general, has to evolve by natural selection in a bottom-up manner, without foresight, planning, conscious agreement, or centrally-managed government (Rankin et al. 2007a). This seems even more difficult to achieve. We will see how it might happen in later chapters (Sections 3.2, 5.3–5.5).

2.4 Assumptions of inclusive fitness theory

Now that we have examined inclusive fitness theory as it applies to both cooperation and conflict, and how each of these is affected by differing relatedness levels, we have reached a good point at which to take a closer look at some of the assumptions of the theory. Some assumptions of the theory are of a technical nature and others much more general. This section concentrates on the latter.

The scale of social behaviour relative to dispersal

The first assumption to be considered concerns the scale over which social actions are performed in relation to the scale of dispersal and competition. Take, as an example, Hamilton's rule as it applies to the evolution of altruism. The essential point is that the usual form of Hamilton's rule in this case ($rb - c > 0$; Box 2.1) assumes complete dispersal from their natal patch by the extra offspring (of the beneficiary) introduced into the population by the actor's altruism. If this assumption is not met, then some of the extra offspring will remain in competition with occupants of the natal patch, to which they are likely to be related. This additional competition can retard the evolution of altruism. Therefore, limited dispersal (population viscosity), while allowing relatedness to build up within a patch, can also generate competition between relatives, with the result that altruism may be counter-selected and even rendered incapable of evolving (West et al. 2002a; McElreath and Boyd 2007; Platt and Bever 2009).

The degree to which limited dispersal influences social actions can be modelled by deriving an extended version of Hamilton's rule (Box 2.2). Many classic cases of altruism in nature effectively involve an alternation of limited dispersal of altruists (leading to positive relatedness) with population-wide dispersal of the products of altruism, and the usual form of Hamilton's rule therefore applies (West et al. 2002a). For example, in a typical colony of social insects, the workers are non-dispersing offspring whose altruism brings extra reproductives into the world. These, as winged (alate) queens and males, then disperse from the natal colony to mate and found their own colonies elsewhere, in competition with the population as a whole. But in other systems, where relatives remain in competition with one another, one needs to consider the social effects of non-dispersal (Box 2.2). An example is the savage competition for mates between related male fig-wasps (West et al. 2001). In the context of the major transitions, the effects of limited dispersal are particularly important in explaining why some life cycles are more conducive to the evolution of altruism than others (Sections 3.3, 4.2).

Box 2.2   Hamilton's rule with limited dispersal

Limited dispersal (population viscosity) leads to competition between relatives, which can hinder the evolution of altruism (e.g. Taylor 1992; Wilson et al. 1992). This box demonstrates this point with a simple model. Many investigators have researched how limited dispersal affects social evolution, as reviewed by McElreath and Boyd (2007) and Platt and Bever (2009). The approach presented here is that of Grafen (1984), Queller (1994), and West et al. (2002a). We start with Hamilton's rule for altruism expressed (for consistency with the source papers) in its conventional notation, i.e. altruism evolves if:

$$rb - c > 0,$$

(Equation 2.3)

where r = altruist-beneficiary relatedness,
b = gain in offspring to the beneficiary,
c = loss in offspring to the altruist.

Say altruism occurs and so b offspring are added to the population and c are lost from it, leading to a net number of $(b - c)$ offspring being added to the population. Imagine some or all of these offspring fail to disperse. If so, they will compete (for example, by consuming resources) with other occupants of the natal patch. This could decrease the fitness gain of the altruist, if the altruist is related to those experiencing this additional competition. By contrast, if the products of altruism disperse completely, they effectively compete only in the population at large, i.e. with non-relatives of the altruist, so the altruist does not experience any decrease in fitness from this source. The loss of fitness to the altruist when there is limited dispersal is the product of the number of members of the population affected by the additional competition (let this equal d) and the altruist's relatedness to them (let this equal r_c). The loss of fitness therefore equals $r_c d$. Hamilton's rule for altruism then becomes:

$$rb - c - r_c d > 0.$$

(Equation 2.4)

Note that, as expected, if the altruist is not related to those affected by competition with the products of altruism, e.g. because there is complete dispersal, $r_c = 0$ and Equation 2.4 reverts to the original Hamilton's rule.

Say there is complete lack of dispersal and the total number of occupants of the local patch is fixed (this latter situation is referred to as local population regulation or inelastic population regulation). Then, for every occupant added to the patch, one must be lost. In this case:

$$d = (b - c).$$

Say also that the altruist's relatedness to the beneficiary is the same as the altruist's relatedness to those affected by the additional competition, which in the case of limited dispersal would be its patch-mates, i.e. that $r = r_c$. This would occur, for example, when patch sharing is used by the altruist as its cue governing who to select as a recipient. Then Equation 2.4 becomes:

$$rb - c - r(b - c) > 0,$$
$$\text{i.e. } -c + rc > 0.$$

(Equation 2.5)

The left-hand side of Equation 2.5 can never be positive, given that r does not exceed 1. Therefore, under conditions of complete non-dispersal, altruism cannot evolve (Grafen 1984; West et al. 2002a). What has happened is that the increased competition

Box 2.2 (*Cont.*)

from the products of altruism has negated the benefits of altruism on a one-to-one basis (the *rb* terms have cancelled out). All that is then left to the altruist is the cost of altruism (−*c*) and the fitness gain (*rc*) caused by the altruist's reduction in offspring output. The latter arises through the effective removal of some competitors from the patch but, because those affected by this reduced competition are patch-mates in general (and not solely the altruist's own offspring), it is never large enough to compensate for the cost of altruism.

The total number of occupants of the patch may not be fixed, and instead the patch population size may expand as it fills up with new members (perhaps because these new members can harvest extra resources). Such population elasticity will mitigate the negative influence of limited dispersal on the evolution of altruism, since additional occupants of the patch no longer displace existing ones, or not to the same extent (i.e. *d* in Equation 2.4 decreases). There are various other factors that can also reduce or nullify the effects of limited dispersal on the evolution of altruism (West et al. 2002a; Lehmann et al. 2008). Furthermore, relatedness can be redefined as the probability of gene-sharing relative to the patch average to make Equation 2.4 more like the original Hamilton's rule (Queller 1994; West et al. 2002a). But the main conclusion from this box is that that one needs to consider the scale over which altruism, dispersal, and competition occur when analysing social evolution (West et al. 2002a). Limited dispersal increases local relatedness but also brings like genes into competition, which hinders the evolution of altruism. Conversely, both elastic population regulation and an alternation of limited dispersal (increasing relatedness) with population-wide dispersal (promoting population-wide competition) facilitate the evolution of altruism.

Causes of relatedness, interests of other loci, green-beard genes, and consequences for social evolution

Understanding the next important assumption of inclusive fitness theory involves another look at what is meant by the relatedness term in Hamilton's rule. According to the theory, the critical relatedness is that at the locus for social behaviour. However, the evolutionary interests of loci in the genome apart from the locus for social behaviour require considering too. Kinship not only is universal in nature but also has the important property that it brings about, on average, the same level of relatedness between social partners across all loci in their genomes (that share a common inheritance system). Say an actor bears a gene for a social action and it directs the action towards a recipient, with their relatedness level being, through kinship, $r = 0.5$. From the way kinship affects relatedness, this means that r is 0.5 not just at the locus for the social action, but at every locus in the actor's and recipient's genomes. In turn, this means that, if Hamilton's rule is fulfilled for the gene for the social action, it will also be fulfilled for genes at all the other loci. Therefore, no genes will be selected to act as modifiers that prevent the social action taking place. The gain of the gene for the social action is also their gain

(Dawkins 1982). Hence kinship stands out among causes of relatedness because it brings about evolutionary 'agreement' across different loci in the genome with respect to the expression of social behaviour (Dawkins 1982; Grafen 1985; Bourke and Franks 1995).

Intragenomic conflict can arise if genes for social actions follow inheritance rules different from those of the majority of genes in the genome, even when relatives are aided. Examples are sex-linked meiotic drive genes, which follow different inheritance rules to those followed by the majority of genes, which are not sex-linked (Leigh 1977; Alexander and Borgia 1978; Dawkins 1982; Bourke and Franks 1995; Burt and Trivers 2006). In cases like this, Hamilton's rules for the genes for social actions and genes at loci in the rest of the genome would not be the same. The gene for the social action would, therefore, be an 'outlaw' (Alexander and Borgia 1978; Dawkins 1982) from the perspective of other loci, at which there would be selection for modifiers to suppress the social action. As mentioned in the previous section, such modifiers indeed exist in the case of meiotic drive.

Relatedness at the locus for a social action can arise in the total absence of kinship through the sharing of 'green-beard genes'. Green-beard genes are genes with three effects (Hamilton 1964; Dawkins 1976; Gardner and West 2009): they confer on their bearer an external label (the 'green beard'); they cause the bearer to recognize conspecifics that also bear the label (i.e. co-bearers of the green beard); and they cause the bearer to discriminate socially in favour of conspecifics with the label. A green-beard gene for altruism, therefore, channels aid towards co-bearers directly, instead of relying on the statistical probability of gene-sharing inherent in kinship to favour genetic co-bearers. Green-bearded social partners would, therefore, be related at the locus for social behaviour by $r = 1$, since each would have the green-beard gene, while, if they were not kin, being unrelated at all the other loci in their genomes. As we will see, cases are now known of real-life green-beard genes (Gardner and West 2009). However, the essential prediction from Hamilton's rule is upheld even in these cases, namely that the evolution of social actions depends on relatedness at the locus for social behaviour (Gardner and West 2009). Indeed, the green-beard concept originally arose as a thought experiment precisely to provide a means of fulfilling Hamilton's condition for altruism in the absence of the relatedness brought about by kinship (Hamilton 1964; Dawkins 1976).

Although green-beard genes occur in nature, they are expected to have unstable evolutionary dynamics. When a green-bearded altruist dispenses aid to a green-bearded beneficiary, genes at all loci elsewhere in the altruist's genome incur a cost. They lose out through the altruist's reduction in offspring output, and, being unrelated to genes at corresponding loci in the beneficiary, they fail to gain from the beneficiary's rise in offspring output. But genes other than the green-beard gene present in the genome of the green-bearded beneficiary gain a benefit, a free gift of altruism received. So the evolutionary dynamics of the situation depend on some detailed assumptions. Say altruism incurs a lifetime cost in direct fitness to green-bearded altruists, and other genes in a green-bearded altruist's genome can 'tell' they are in a donor of altruism and not a beneficiary. Then the green-beard gene and the set of all other genes in green-bearded altruists effectively have different Hamilton's rules, and the green-beard gene acts as an outlaw from the viewpoint of the other genes. These other genes should be selected to suppress the altruism. Genes other than the green-beard gene in the green-bearded beneficiary should do nothing, because they gain from the free gift.

Say, however, that any given green-beard bearer acts with equal frequency as both an altruist and a beneficiary over its lifetime. Then other genes in the genome in every green-beard bearer gain from the free gifts of altruism received, and the green-beard gene is not an outlaw. This argument was put forward by Ridley and Grafen (1981). But note that in this case the green-beard phenotype is not strictly altruistic. If everyone gains a net lifetime direct benefit, the social interaction is narrow-sense cooperation. So it is unsurprising that outlawry then vanishes.

Gardner and West (2009) argued that the green-beard gene is not an outlaw even in (implicitly) the first case considered above, in which green-beard altruists incur a lifetime cost in direct fitness and other genes in an altruist's genome can 'tell' they are in a donor of altruism. Their argument was that modifiers are selected to occur at both the green-beard locus itself and at other loci. This is most easily seen in the case where a modifier prises apart the components of the green-beard phenotype by suppressing aid-giving but allowing the green beard to remain. Such a modifier at a locus other than the green-beard locus would benefit from altruism received without paying its costs, and so would spread (Lehmann and Keller 2006a). Gardner and West's (2009) point is that this would also be true of the same modifier arising at the green-beard locus itself. Hence, they concluded, there is no interlocus conflict and green-beard genes are not outlaws. I do not dispute Gardner and West's (2009) analysis, but their denial of outlaw status to green-beard genes is a semantic point. If one defines an outlaw as any gene that, when under positive selection, incurs selection for modifiers at other loci (irrespective of whether the same modifier would prosper at the focal locus), a green-beard gene for (lifetime) altruism is still an outlaw. Such a definition is useful for present purposes because it highlights a key contrast between the status of a green-beard gene for altruism and a kin-selected gene for altruism. A green-beard gene for altruism, as an outlaw, provokes the evolution of modifiers that suppress altruism. A kin-selected gene for altruism, as a non-outlaw, does not. This is because, as shown earlier, if Hamilton's rule for a gene for altruism towards kin is satisfied, it must also be satisfied for genes at other loci in the genome (with the same inheritance system).

In sum, for these reasons it is very unlikely that green-beard genes lead to long-term stability in the evolution of altruism (Lehmann and Keller 2006a; Gardner and West 2009). Although kinship is not the only means of achieving relatedness between social partners, it is the most common. Moreover, when relatedness arises through kinship, but not otherwise, the evolution of altruism can proceed stably provided Hamilton's rule is satisfied.

Facultative gene expression

This assumption of inclusive fitness theory concerns the difference between carrying a gene and expressing it. The theory's logic is based on actors and recipients sharing a gene for a social action. In some cases the recipient will bear the gene for the social action and in others it will not, and the frequency with which it does is given by relatedness. In the cases when both the actor and the recipient bear the gene for the social action, one might legitimately wonder how they can exhibit different phenotypes, namely the actor phenotype and the recipient phenotype. The answer is that inclusive fitness theory assumes that genes for social actions can be borne by individuals but not expressed in them. For example, in the case of altruism, it assumes that both altruists and beneficiaries bear the gene for altruism but that only altruists express it (by behaving altruistically). In this way, the gene for altruism can aid copies of itself in other bodies, even though these bodies do not themselves exhibit altruism. In short, inclusive fitness theory assumes facultative, or conditional, gene expression (Charlesworth 1978; Bourke and Franks 1995; Queller and Strassmann 1998). This is not an assumption made just for convenience. The theory would not work without it. If all bearers of a gene for altruism obligately expressed the gene, no reproductive recipient would bear the gene and so altruism could not evolve (Bourke and Franks 1995).

The assumption of facultative expression of kin-selected behaviours is well substantiated. Cells within multicellular organisms that differentiate into various forms though derived from the same zygote (Bonner 1974; Buss 1987), and queen and worker castes in the eusocial Hymenoptera that develop from the same type of egg (Wheeler 1986), each testify to the repeated occurrence in nature of differences between social actors and recipients based on facultative gene expression. In this, social behaviour is not exceptional, since many other behavioural and morphological traits exhibit polymorphisms subject to facultative gene expression (West-Eberhard 2003).

Genes for social actions in nature

Finally, inclusive fitness theory, like any genetical theory of social evolution, of course assumes that genes for social actions are more than hypothetical constructs and in fact exist. This is indeed the case, as fully documented by some excellent recent reviews of genes that influence social behaviour across a diversity of taxa (Robinson et al. 2005, 2008; Donaldson and Young 2008; Smith et al. 2008; Gadau and Hunt 2009; Keller 2009). For illustrative purposes, I present selected examples in Table 2.4.

Two general points about the genetic basis of social behaviour are worth making. The first is that, contrary to the widespread intuition that something as complex as social behaviour should be under the control of many genes of small effect, examples of one or a few loci affecting social behaviour are well substantiated (Table 2.4). Some are in simple multicellular organisms, such as cellular

Table 2.4 Selected examples of genes influencing social behaviour

Organism	Name of locus	Social behaviour	References
Multicellular alga, *Volvox carteri*	*regA*	Expression of *regA* is associated with irreversible differentiation of somatic cells in the multicellular clone; so *regA* is a kin-selected gene for altruism.	Kirk (1998), Nedelcu and Michod (2006), Duncan et al. (2007), Nedelcu (2009)
Budding yeast, *Saccharomyces cerevisiae*	*FLO1*	*FLO1*-bearing cells clump (flocculate), protecting bearers from toxins; *FLO1* is a green-beard gene because cells with *FLO1* thereby aid other cells with *FLO1*; however, relatedness (via kinship) is presumably above-zero, as this occurred in haploid strains.	Smukalla et al. (2008)
Cellular slime mould, *Dictyostelium discoideum*	*csA* encoding protein gp80	*csA*-bearing cells stick together, gaining disproportionate representation in fruiting head; *csA* is a green-beard gene because cells with *csA* thereby aid other cells with *csA*; however, relatedness (via kinship) is also high in local groups of cells.	Queller et al. (2003), Gilbert et al. (2007)
Termite, *Cryptotermes secundus*	*Neofem2*	Expression of *Neofem2* in queen is required for workers to refrain from reproductive behaviour in queen's presence.	Korb et al. (2009)

Table 2.4 *Cont.*

Red imported fire ant, *Solenopsis invicta*	*Gp-9*, with alleles *B* and *b*	Complex of effects on workers' treatment of queens and queen phenotype, leading to social polymorphism involving single- and multiple-queen colonies; *b* is a green-beard gene because *Bb* workers execute *BB* queens and so indirectly favour other *b*-bearers.	Keller and Ross (1998), Krieger and Ross (2002), Gotzek and Ross (2007), Wang et al. (2008)
Cape honey bee, *Apis mellifera capensis*	*thelytoky*, with alleles + (dominant) and *th* (recessive)	Homozygous *th/th* workers produce female offspring asexually (thelytoky) and have a queen-like behavioural and physiological phenotype; wildtypes (+/+) and heterozygotes (+/*th*) produce male offspring asexually (arrhenotoky) and lack a queen-like behavioural and physiological phenotype	Lattorff et al. (2005, 2007)
Voles, *Microtus* spp.	*Avpr1a* encoding receptor V1aR	Increased expression of *V1aR* leads to increased affiliative behaviour and is associated with difference between polygamous and monogamous mating systems in some *Microtus* spp.	Lim et al. (2004), Fink et al. (2006), Young and Hammock (2007), Solomon et al. (2009)

slime moulds and yeast, where arguably the social behaviour that they influence is also simple. But others, such as genes affecting workers' reproductive behaviour in the thelytokous Cape honey bee *Apis mellifera capensis*, or workers' treatment of queens in the Red imported fire ant *Solenopsis invicta*, occur in complex multicellular organisms (Table 2.4). It does not follow from these examples that genetic influences on social behaviour always involve one or a few loci, since cases where they do so might be easier to detect. Inclusive fitness theory does not require genes for social actions to be confined to one or a few loci. Indeed, once kin-selected social behaviour is widespread, it would be expected that many genes might come to influence it because, given kinship creates uniform average relatedness values across loci, all loci would 'agree' with the behaviour being performed. But a new social behaviour presumably originates more easily if mutations are required at one or a few loci, and single-locus examples show that this can happen. It also does not follow that aspects of the environment are unimportant in how genes for social behaviours achieve their effects. In fact, social phenotypes may be heavily influenced by the social environment (Robinson et al. 2008; Keller 2009), an example being found in *S. invicta*, in which workers' behaviour and profiles of gene expression are influenced both by their genotype and by that of nestmates (Gotzek and Ross 2007; Wang et al. 2008).

The second general point concerns the fact that several of the better-understood cases of genes affecting social behaviour (Table 2.4) are green-beard genes, whose general properties were described earlier in this section. As an example, consider the gene *csA* in the cellular slime mould *Dictyostelium discoideum* (Table 2.4) The multicellular stage of *D. discoideum* consists of a fruiting body made up of a stalk of sterile cells supporting a fruiting head of reproductive spores. The gene *csA* has all three defining properties of a green-beard gene because it confers on cells bearing it an external label (an adhesive protein, gp80), it causes bearers to recognize co-bearers (cells sharing gp80 physically stick together at their gp80 sites), and it causes bearers to discriminate socially in favour of co-bearers. This last effect occurs through blocs of adhering gp80-bearing cells excluding cells lacking gp80 from the fruiting head (Queller et al. 2003). Haig (1996) predicted that genes encoding cell-adhesion molecules would be good candidates for green-beard genes, and the cases discovered since in *D. discoideum* and the yeast *Saccharomyces cerevisiae* confirm this insight in facultatively multicellular organisms (Table 2.4). One reason why green-beard genes are effective in such cases is that physically sticking together represents an easy-to-achieve means of being social for otherwise unicellular organisms, a topic to be returned to in a later chapter (Section 4.1). In *D. discoideum*, the *csA* locus must interact with kin-selected effects. This follows from the recent discovery of high relatedness within natural *D. discoideum* aggregates (Gilbert et al. 2007). In fact, as pointed out by Helanterä and Bargum (2007), known examples of green-beard genes tend to occur either as derived features within existing eusocial systems (e.g. the Red imported fire ant *Solenopsis invicta*) or in contexts where relatedness stemming from kinship is likely to be positive as in *D. discoideum* and *S. cerevisiae* (Table 2.4).

The reason for the rarity of green-beard genes outside these contexts appears to be that, as described above, they would have unstable evolutionary dynamics in the absence of kinship. In addition, should a green-beard gene increase to fixation, every member of the population would be green-bearded and no further social discrimination would occur. The existence of the green-beard phenomenon would then be hard to detect (Gardner and West 2009). Fixation might, however, be prevented if the gene were to suffer some cost as its frequency increased. For example, lethal recessivity would maintain polymorphism at a green-beard locus, since increased homozygosity would put a ceiling upon the focal gene's frequency. This is probably why, in the Red imported fire ant *Solenopsis invicta*, the effects of the green-beard allele *b*, which acts a lethal recessive (*bb* individuals fail to develop), are still manifest (Table 2.4). If *b* did not act as a lethal recessive, it would be expected to go to fixation and hence become undetectable (Keller and Ross 1998; Gardner and West 2009).

Despite the expected transitory nature of green-beard traits, their exact genetic basis may still be easier to establish than that of kin-selected traits. The reasons why concern the nature of kin-selected social behaviour. As discussed above, kin selection assumes facultative expression of genes. So when a kin-selected gene goes to fixation, the result is, in principle, a population that is genetically monomorphic at the focal locus but phenotypically polymorphic. In addition, recall again that kinship brings about the same average relatedness across all loci between social partners. Therefore, the gene for a kin-selected social action will, to some extent, be hidden within the genome, because there is nothing to make it stand out from all the other genes.

For these reasons, establishing the exact genetic basis of kin-selected traits remains a challenge (Bourke 2002, 2005), although at least one example is known from the multicellular alga *Volvox carteri* (Table 2.4). This gene was detected through being expressed only in somatic cells and, from sequence information, being found to occur in a non-social ancestor (Nedelcu and Michod 2006). Similarly, other kin-selected loci with a good chance of being detected include those that involve current polymorphisms of sociality and non-sociality, i.e. within facultatively social populations caught at the origin of social groups as in, for example, some halictid bees (Packer 1990). They might also include variants of existing social behaviours. Two examples occur in the Honey bee, the 'thelytoky' phenotype, whose genetic basis is known (Table 2.4), and the 'anarchic' phenotype. The latter is characterized by workers that lay more eggs than wildtype workers, with their eggs having some chemical immunity against being policed (Oldroyd et al. 1994; Martin et al. 2004). Its genetic basis is beginning to be established (Thompson et al. 2008). Nonetheless, although the detailed genetic characterization of kin-selected genes for social actions is at an early stage of investigation, ample evidence already exists that social behaviour is subject to genetic variation, just as in the case of other traits (Table 2.4). This way, a fundamental assumption of inclusive fitness theory is met.

2.5 The value of inclusive fitness theory

As stated at the beginning of this chapter, inclusive fitness theory has had a complex history since Hamilton (1963, 1964) first proposed it. Like many scientific ideas that make large claims about fundamental issues, it has provoked debate and disagreement. The theory's intimate connection with a reductionist, gene's-eye view of adaptive evolution cost it some support. In addition, at least in its early years, the nature and validity of its assumptions, as well as the status of Hamilton's rule as a robust theorem in population genetics, were not clear.

Several conceptual advances have combined to alleviate these problems. First, to the satisfaction of many, if not all, the centrality of a gene's-eye view of adaptive evolution, based on the concept that adaptations fundamentally serve the interests of genes because genes are replicators, was established (Dawkins 1976, 1982, 1986). Second, early criticisms of inclusive fitness theory were shown to arise from misunderstandings of its basis and claims (Hamilton 1972; Dawkins 1979; Grafen 1984). Third, the firm grounding of Hamilton's rule in established concepts in population genetics and selection theory was formally proven (Hamilton 1970; Grafen 1985; Frank 1998) and the relationship of inclusive fitness with classical definitions of fitness clarified (Grafen 1982; Queller 1996). Fourth, the equivalence was demonstrated between inclusive fitness models and intrademic group selection models for the evolution of altruism (Grafen 1984; Queller 1992; Reeve 2000a; West et al. 2007b), so overturning the long tradition of viewing these approaches as providing competing explanations for social evolution. Against this background, this section considers three related topics: the relationship of inclusive fitness theory with multilevel selection theory; the evidence for the theory; and recent critiques of the theory. The aim is to reinforce the case for the considerable value of the theory in understanding social evolution.

Relationship of inclusive fitness theory with multilevel selection theory

The complementarity of inclusive fitness theory and intrademic group selection is sometimes used to argue for a global equivalence of inclusive fitness theory with multilevel selection theory. The latter, also known as levels-of-selection or hierarchical selection theory, views natural selection as a process capable of occurring at any level in the biological hierarchy. (References to earlier works on multilevel selection were provided by Bourke and Franks (1995). More recent ones include those of Wilson (1997), Sober and Wilson (1998), Michod (2000), Gould (2002), Goodnight (2005), Okasha (2006), and Wilson and Wilson (2007).) The major transitions view of evolution seems especially suited to an interpretation based on multilevel selection, and, indeed, several of the latter's recent proponents have explicitly argued for its being so (e.g. Michod 2000; Okasha 2006; Wilson and Wilson 2007). In most respects, just as inclusive fitness theory and the appropriate form of group selection theory can be used with equal validity to analyse problems in social evolution, this is unproblematic and may be illuminating (e.g. Korb and Heinze 2004). But there remain important distinctions.

One such distinction concerns the issue of group-level adaptations. Multilevel selection theory equates levels of selection with levels of adaptation. For example, it regards mechanisms of conflict resolution (Section 3.2) as group-level adaptations, because they seem to serve the interests of the social group as a whole. However, taking an inclusive-fitness perspective, Gardner and Grafen (2009) argued that levels of selection and levels of adaptation are distinct. This is because, in their analysis, any social group (say an animal society) can only be considered as an entity that maximizes its fitness, and hence shows adaptations at its level, if within-group conflict is absent or completely suppressed. In theory this means only clonal societies can represent levels of adaptation at the group level. On this view, mechanisms of conflict resolution would be consequences, but not causes, of group-level adaptation (Gardner and Grafen 2009). In practice, though, Gardner and Grafen (2009) conceded that very low levels of within-group conflict lead to social groups with group-level adaptations.

Another way of putting this point is to assert that the inclusive fitness of a social group with less than clonal relatedness cannot be defined. Only the inclusive fitnesses of the different relatedness classes within it can be defined, and they will not necessarily coincide (Section 2.3). Hence a group with less than clonal relatedness has no inclusive fitness to maximize. This brings in an insight of Ratnieks and Reeve (1992). They cautioned against categorizing eusocial societies or similar entities as either 'superorganisms' (i.e. entities that exhibit group-level adaptations) or not. Instead, eusocial societies are 'superorganismal' in some contexts but not others (Bourke and Franks 1995). Something like the dance language of the Honey bee certainly looks like a group-level adaptation (Seeley 1997). But there is likely to be very little within-colony conflict over the organization of foraging by worker bees, since all parties are served by efficient foraging. By contrast, aspects of Honey bee reproduction, such as lethal fighting between young nestmate queens (Gilley and Tarpy 2005), look maladaptive at the colony level and this is explicable through within-colony conflicts of interest among the different relatedness classes finding expression in such behaviours. A eusocial society is, therefore, better regarded as a mixture of adaptive and maladaptive features at the colony level, depending on context (Ratnieks and Reeve 1992; Bourke and Franks 1995; Strassmann and Queller 2007; Queller and Strassmann 2009). A multicellular organism is likewise better regarded as a mixture of adaptive and maladaptive features at the organismal level (Dawkins 1982; Wilson and Sober 1989), since within-organism conflict occurs in the form of intragenomic and intergenomic conflict (Section 2.3). In other words, even a multicellular organism is organismal only in some contexts. The common factor is that adaptations at both the group level and the within-group level are adaptations of the genes underpinning them. Therefore, all debate over whether a eusocial colony is categorically a superorganism exhibiting group-level adaptations or not (Hölldobler and Wilson 2009), or over the exact point at which adaptation and fitness are transferred from one level in the biological hierarchy to the next (e.g. Michod 2000; Okasha 2006), can be circumvented by acknowledging that all adaptations, at any level, ultimately serve the self-interest of the genes responsible for them (Dawkins 1976, 1982).

This highlights another important distinction between inclusive fitness theory and multilevel selection theory, which is that inclusive fitness theory is an explicitly gene-centred view of natural selection. Consider the image of the Russian doll, which within multilevel selection theory has been used to describe the nested layering shown by the biological hierarchy (e.g. Buss 1999; Wilson and Wilson 2008). The snag with this attractive image is that a Russian doll is, as it were, Russian dolls all the way down. But, in the biological hierarchy, the innermost entity is not like the others. It is the gene, a replicator, while the other levels are, following Dawkins' (1982) analysis, not replicators but 'vehicles' for replicators. Put another way, cells are not the genes of organisms and organisms are not the genes of societies. Genes are the genes of organisms and societies. This matters given adaptations ultimately serve replicators, not vehicles. It is the replicators that travel down the generations, rising and falling in frequency. The vehicles arose after replicators and are their temporary constructs. Hence the replicators have primacy over vehicles both temporally and causally (Dawkins 1976, 1982). Adaptations may serve vehicles incidentally, when replicators share a coincidence of evolutionary interests. Again, this is why adaptations at the level of the vehicle may be said to occur when potential within-vehicle conflict is absent (Gardner and Grafen 2009). It is also the essence of the explanation, from inclusive fitness theory, for why major transitions occur.

In short, gene-centred selection theory can learn from multilevel selection theory the importance of hierarchical population structure in determining how gene selection proceeds. But multilevel selection theory should heed gene-centred selection theory's demonstration of the replicator-vehicle distinction (Dawkins 1982; Bourke and Franks 1995). A better image for the biological hierarchy would be the parcel in the children's game of 'pass the parcel'. Unwrap layer after layer of patterned paper, and what you find at the centre is not another piece of paper, but a gift—the *raison d'être* of the entire parcel.

Evidence for inclusive fitness theory

The best measure of the worth of any scientific theory is, of course, its fit to data, its ability successfully to explain what is known and make predictions about what is not known. On this measure, inclusive fitness theory has achieved a remarkable record of success. Scores of studies have supported the predictions of the theory across a huge range of taxa, including viruses (Turner and Chao 1999), microbes (Crespi 2001; West et al. 2007c), social insects (Bourke and Franks 1995; Crozier and Pamilo 1996; Ratnieks et al. 2001, 2006; Bourke 2005), and social vertebrates (Emlen 1991, 1997; Dickinson and Hatchwell 2004; Komdeur et al. 2008; Clutton-Brock 2009b; Hatchwell 2009). Even studies of social robots support the predictions of the theory (Floreano and Keller 2010). The benefits and costs of social traits are notoriously difficult to measure, which is why direct tests of Hamilton's rule are relatively rare. But some have been performed, for example in populations of bees and wasps in which females have the choice between helper behaviour or solitary breeding. These have confirmed that Hamilton's rule is fulfilled for helpers in some contexts and not others (e.g. in some years and not others), so accounting for the existence of both helper behaviour and solitary breeding within these populations (Metcalf and Whitt 1977; Noonan 1981; Grafen 1984; Stark 1992; Bourke 1997; Gadagkar 2001). Other tests of inclusive fitness theory have sought to demonstrate the predicted effects of relatedness on social traits by experimental manipulations of relatedness, or by comparative analyses that assume (reasonably) that costs and benefits of social behaviour do not vary systematically with relatedness. These studies confirm that relatedness influences social traits as inclusive fitness theory predicts (Figs 2.3, 2.4; Table 2.5).

It is worth noting that the predictions of inclusive fitness theory are often truly a priori, in that they have foretold the discovery of previously unrecognized social phenomena consistent with the theory (Ratnieks et al. 2001). These include, in the eusocial Hymenoptera, female-biased population sex ratios (Trivers and Hare 1976), worker policing of worker-laid eggs (Ratnieks 1988; Ratnieks and Visscher 1989; Wenseleers and Ratnieks 2006a), and split sex-ratios (i.e. the phenomenon in which different classes within a population produce systematically different sex ratios) based on variation in relatedness asymmetry (Boomsma and Grafen 1991; Sundström 1994; Sundström et al. 1996). The theory has also generated new explanations for previously known but unexplained social phenomena, such as culls of virgin queens in stingless bees (Bourke and Ratnieks 1999; Wenseleers et al. 2003). And it has made a priori predictions that have received preliminary support but that remain to be fully evaluated. Examples include the many predictions regarding the distribution of reproduction, or reproductive skew, within animal societies with multiple breeders (Keller and Reeve 1994; Magrath and Heinsohn 2000; Reeve *et al.* 2000; Reeve and Keller 2001; Langer et al. 2004; Hammond et al. 2006; Nonacs 2006). Another set of examples comes from the theory's predictions regarding patterns of genomic imprinting and resource allocation in plants, mammals, and social insects (Moore and Haig 1991; Haig 1992, 2004; Queller and Strassmann 2002; Queller 2003).

Inclusive fitness theory is also recognized as the basis for successful prediction and interpretation in several major areas in evolutionary biology that overlap with, but are frequently regarded as distinct from, the field of social evolution. These include sex allocation (Trivers and Hare 1976; Charnov 1982; West 2009), host–parasite relationships (Frank 1992, 1996a), and aspects of ageing (Alexander et al. 1991; Travis 2004; Bourke 2007). Lastly, as presented in this book, the entire major transitions view of life's history serves as potential evidence for the utility of inclusive fitness theory. Hence, to the extent that the validity of this view is established herein and by earlier works (e.g. Maynard Smith and Szathmáry 1995; Queller 2000), the theory receives further support.

Table 2.5 Selected experimental and comparative evidence for effects of relatedness on social traits as predicted by inclusive fitness theory

Organism(s)	Evidence	Method and sample size for comparative studies (References)	References for prediction and for related studies, if any
Experimental			
Bacterium, *Pseudomonas aeruginosa*	Frequency of altruistic strain increases at high relatedness and decreases at low relatedness; localized competition dampens this effect (Fig. 2.3).	Experimental evolution in laboratory cultures (Griffin et al. 2004)	Prediction: Frank (1995, 1998) Related study: Diggle et al. (2007)
Cellular slime mould, *Dictyostelium discoideum*	Relative fitness of a selfish mutant decreases as relatedness increases.	Experimental laboratory cultures set up at differing levels of relatedness (Gilbert et al. 2007)	Prediction: Frank (1995), Gilbert et al. (2007)
Halictid bee, *Augochlorella striata*	Females rear less female-biased sex ratio when relatedness asymmetry experimentally reduced.	Field experiment (Mueller 1991, Mueller et al. 1994)	Prediction: Boomsma and Grafen (1991)
Allodapine bee, *Exoneura nigrescens*	Reproductive skew varied inversely with experimentally altered relatedness as 'tug-of-war' skew model predicts.	Field experiment (Langer et al. 2004)	Prediction: Reeve et al. (1998)
Comparative			
Eusocial Hymenoptera	Mean population sex ratios in monogynous ants are significantly female-biased as predicted by relatedness asymmetry.	Comparative analysis (not phylogenetically controlled) of 16 species or populations of monogynous ants (Bourke 2005)	Prediction: Trivers and Hare (1976) Related studies: Trivers and Hare (1976), Nonacs (1986), Pamilo (1990)
Eusocial Hymenoptera	Colony sex ratios covary significantly with relatedness asymmetry and queen number as predicted by split sex ratio theory.	Meta-analysis of 22 studies of ants, bees, and wasps (Meunier et al. 2008)	Prediction: Boomsma and Grafen (1991) Related studies: Queller and Strassmann (1998), Bourke (2005), Kümmerli and Keller (2009)

Table 2.5 *Cont.*

Euocial Hymenoptera	Worker policing of other workers' reproduction is significantly more common in species in which workers are more closely related to queens' than to workers' sons.	Comparative analysis (phylogenetically controlled) of 14 species of ants, bees, and wasps (Wenseleers and Ratnieks 2006a)	Prediction: Ratnieks (1988)
Eusocial Hymenoptera	Percentage of worker-produced males increases significantly as relatedness of workers to workers' sons (relative to queens' sons) increases (Fig. 2.4).	Comparative analysis (phylogenetically controlled) of 90 species of ants, bees, and wasps (Wenseleers and Ratnieks 2006a)[1]	Prediction: Trivers and Hare (1976), Ratnieks (1988) Related studies: Hammond and Keller (2004), Bourke (2005)
Eusocial Hymenoptera	Percentage of reproductive workers decreases significantly as relatedness among workers increases in queenless groups lacking worker policing.	Comparative analysis (phylogenetically controlled) of 10 species of bees and wasps (Wenseleers and Ratnieks 2006b)	Prediction: Frank (1995), Wenseleers et al. (2004a)
Stingless bees (*Melipona* spp.)	Percentage of females developing as queens decreases significantly as females' relatedness to males increases as predicted by inclusive fitness model of conflict over caste fate.	Comparative analysis (not phylogenetically controlled) of 12 studies of 4 *Melipona* species (Wenseleers and Ratnieks 2004)	Prediction: Wenseleers et al. (2003)
Cooperatively-breeding birds and mammals	On average, helpers preferentially aid relatives and probability of such kin discrimination occurring increases significantly as benefit of aiding relatives increases.	Meta-analysis of studies of 18 species (Griffin and West 2003)	Prediction: Hamilton (1964) Related study: Griffin et al. (2005), Cornwallis et al. (2009)

[1] Hammond and Keller (2004) found no significant association of percentage of worker-produced males with relatedness of workers to workers' sons (relative to queens' sons). The study of Wenseleers and Ratnieks (2006a) had a larger sample size and greater statistical power.

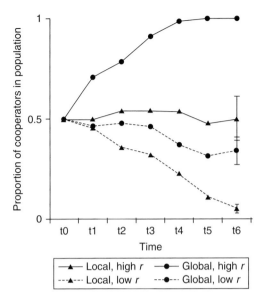

Fig. 2.3 Change in frequency with time of an altruistic strain of the bacterium, *Pseudomonas aeruginosa*, in experimental cultures of the altruistic strain and a non-altruistic strain. Cells of the altruistic strain produce group-beneficial iron-binding compounds (siderophores) at a cost to their growth rate. Cells of the non-altruistic strain do not produce siderophores. Time steps are 24 h each. Under global competition, the frequency of the altruistic strain increases at high relatedness levels (black circles, solid line) and decreases at low relatedness levels (black circles, dotted line). Under localized competition, high relatedness (black triangles, solid line) and low relatedness (black triangles, dotted line) have the same comparative effects, but their magnitude is dampened. All results are as predicted by inclusive fitness theory (Boxes 2.2, 3.1). Reproduced from Griffin et al. (2004) with kind permission of the authors.

Recent critiques of inclusive fitness theory

Despite widespread acceptance of inclusive fitness theory, approval of the theory is not universal. Some recent authors have joined earlier ones in criticizing the theory, especially, but not exclusively, as it applies to the social insects (Alonso and Schuck-Paim 2002; Wilson 2005, 2008; Wilson and Hölldobler 2005; Hunt 2007; Wilson and Wilson 2007, 2008). The voice of the eminent biologist E.O. Wilson has been particularly influential in this context, both on account of his scientific distinction and because, among experts on social evolution, he was an early supporter of the theory (Wilson 1966, 1971). In turn, several authors have responded to the criticisms of E.O. Wilson and others (Foster et al. 2006a; Thompson 2006; Helanterä and Bargum 2007; Crozier 2008; Hughes et al. 2008; Monnin and Liebig 2008). In my view, the criticisms of the theory do not survive analysis, largely through being based upon misconceptions about it (Box 2.3). Moreover, both E.O. Wilson and his distinguished co-author B. Hölldobler appear to have altered their views to some, though not the

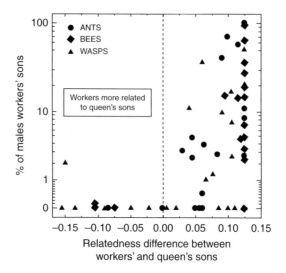

Fig. 2.4 The percentage of worker-produced males (logarithmic scale) reared is significantly greater in eusocial Hymenoptera ($N = 90$ species) in which workers are more closely related to workers' sons than to queens' sons (right-hand side of figure) than it is in species in which workers are more closely related to queens' sons than to workers' sons (left-hand side of figure), which is as inclusive fitness theory predicts. Reproduced from Wenseleers and Ratnieks (2006a) with kind permission of the authors and The University of Chicago Press. © 2006 by The University of Chicago.

same, extent (Reeve and Hölldobler 2007; Hölldobler and Wilson 2009; Pennisi 2009). As a scientific orthodoxy, albeit an underappreciated one (Sections 1.4, 2.2), inclusive fitness theory requires continued scrutiny from new concepts and new data. To guard against its models hardening into accepted fact without passing through the essential filter of empirical test, it doubtless benefits from a vigorous shaking-up. However, taken together, past and recent criticisms have not met the necessary standards for any successful challenge to existing theory. They have neither invalidated the basis of the theory, nor provided a cogent replacement theory, nor explained otherwise inexplicable data, nor made new and unique predictions that have then been confirmed.

There is in fact considerable common ground between the critics and advocates of inclusive fitness theory, given that the group selection for altruism espoused by the critics is not fundamentally distinct from kin selection (Box 2.3). To try to extend the common ground, it is worth examining some general reasons underlying the dissatisfaction of inclusive fitness theory's critics. These appear to reduce to three points. The first is that inclusive fitness theory seems too abstract to be especially useful (Wilson and Hölldobler 2005; Hunt 2007; Wilson 2008; Hölldobler and Wilson 2009), i.e., in the case of the social insects, too far removed from the real-life natural

Box 2.3   A defence of inclusive fitness theory

In this box, I present a response to recent criticisms of inclusive fitness theory (kin selection theory). Dawkins (1979) addressed early criticisms of the theory, some of which more or less resurface among the recent criticisms. The latter have also been ably addressed by several previous authors (Foster et al. 2006a; Thompson 2006; Helanterä and Bargum 2007; Crozier 2008; Hughes et al. 2008; Monnin and Liebig 2008). The numbering of the points is for convenience.

Criticism 1: The important selective force in social evolution is group selection (Wilson and Hölldobler 2005; Hunt 2007; Wilson 2008; Hölldobler and Wilson 2009).

Response: Conditions for the evolution of altruism can be modelled using either inclusive fitness theory or an equivalent form of group selection theory, namely intrademic group selection (in which members of the population re-assort into groups each generation). Both sets of conditions boil down mathematically to Hamilton's rule (Wade 1980; Grafen 1984; Queller 1992; Reeve 2000a; West et al. 2007b). They differ only in how fitness is partitioned. In inclusive fitness theory, fitness is the sum of direct and indirect fitness. In intrademic group selection, fitness is the sum of a within-group and a between-group component of selection. Both formulations yield the same solutions to given problems (Wenseleers et al. 2003, 2004a). Hence there is no fundamental theoretical distinction between the two sets of models (Bourke and Franks 1995; Foster et al. 2006a). It follows that no empirical finding can support intrademic group selection but not inclusive fitness theory, or inclusive fitness theory but not intrademic group selection (Dugatkin and Reeve 1994). For the same reason, conceptual criticisms of inclusive fitness theory must fundamentally apply to intrademic group selection. For example, it has been argued that inclusive fitness theory is 'largely empty of content' because its 'abstract parameters can be jury-rigged to fit any set of empirical data' (Wilson 2008). If this were correct, which it is not, it would also be true of intrademic group selection theory. The present book uses the methodology and language of inclusive fitness theory because they have proved useful for the issues it addresses, especially in focussing attention on the effects of relatedness and in facilitating simultaneous analyses of cooperation and conflict.

Criticism 2: In contrast to group selection, which is a 'binding force' in social evolution, kin selection is 'weakly binding' or 'dissolutive' because kin selection is 'the preferential favoring of collateral relatives (i.e., not including offspring) within groups according to their degree of relationship' (Wilson and Hölldobler 2005; Wilson 2008).

Response: As in the previous response, there is no fundamental distinction between kin selection and group selection (of the appropriate kind), so one cannot be more or less 'binding' than the other. In fact each is 'binding' in proportion to the level of relatedness within social groups, since increasing relatedness ensures an ever-greater coincidence of interest among group members (Section 2.3). Moreover, in inclusive fitness theory, kin selection (for altruism) does not require preferential treatment of relatives within groups according to their degree of relatedness. If social groups of relatives form and group members exclude non-groupmates, then kin selection occurs even if within-group kin discrimination is absent, because social benefits are still preferentially directed at related members of the population (Hamilton 1964; Bourke and Franks 1995; West et al. 2007b).

Criticism 3: Eusociality can evolve in the absence of kinship if the bearers of a 'eusociality allele' associate (Wilson and Hölldobler 2005).

Response: A gene that caused its bearers to associate and interact socially regardless of kinship would effectively be a green-beard gene. In principle, a social group could originate via a green-beard effect, but because green-beard genes for altruism have unstable

Box 2.3 *(Cont.)*

evolutionary dynamics in the absence of kinship (Section 2.4), a eusocial society formed on the basis of such a process would not be stable and none are known (Foster et al. 2006a; Thompson 2006; Helanterä and Bargum 2007; Crozier 2008; Gardner and West 2009).

Criticism 4: Eusociality does not require relatedness to evolve (Wilson 2005; Wilson and Hölldobler 2005).

Response: Hamilton's rule finds that any form of social group based on altruism, including a eusocial society, can only evolve if relatedness at the locus for altruism is greater than zero (Sections 2.1, 2.2). Positive relatedness could come about through a green-beard effect or through kinship (Section 2.4). As the previous point states, a green-beard route to eusociality is not impossible, but it is implausible. Hence, the evolution of eusociality does not, in theory, require relatedness stemming from kinship, but no other cause of positive relatedness is as likely (Section 2.4). Furthermore, as detailed in the next point, a role for relatedness stemming from kinship in the origin and maintenance of euso-cial societies is what the evidence supports.

Criticism 5: Relatedness (brought about by kinship) is a consequence of eusocial evolution, not its cause (Wilson 2005; Wilson and Hölldobler 2005). In addition, relatedness cannot cause eusocial evolution because many species occur in conditions of high related-ness but eusocial evolution is relatively rare (Wilson 2008).

Response: There is ample evidence that positive relatedness is not secondary in eusocial societies and other social groups of altruists, and that, on the contrary, they only evolve in clones or families and hence under conditions of above-zero and usually high relatedness (Alexander et al. 1991; Boomsma 2007, 2009; Helanterä and Bargum 2007; Hughes et al. 2008) (see also Section 4.2). Furthermore, inclusive fitness theory does not assert that positive relatedness causes altruism to evolve, but that it is a necessary precondition. There must be an opportunity for altruism to occur, and the other terms in Hamilton's rule must take appropriate values as well (Sections 2.1, 2.2). Hence the absence of eusociality in every case where relatedness is high is not evidence against the theory.

Criticism 6: Ecological benefits of group life are likely to have been important at the origin of eusociality, so the emphasis of inclusive fitness theory on exceptional degrees of relatedness is misplaced (Wilson and Wilson 2007).

Response: Hamilton's rule contains terms for relatedness and for the expected changes in the offspring numbers of social partners (benefit and cost in the case of altruism). Conditions increasing the benefit term, including ecological ones, are therefore recognized by Hamilton's rule as integral to eusocial evolution (West-Eberhard 1975; Bourke and Franks 1995; Foster et al. 2006a) (such conditions are discussed in Sections 4.3 and 4.4). Relatedness need not be exceptionally high for altruism to evolve, only greater than zero. Relatedness remains crucial because no level of ecological benefit can bring about altruism if relatedness is not above zero.

Criticism 7: The failure of Hamilton's (1964) 'haplodiploidy hypothesis' represents evi-dence against inclusive fitness theory (Wilson 2005; Wilson and Hölldobler 2005; Hunt 2007; Wilson and Wilson 2007).

Response: Hamilton (1964) proposed that the unusual relatedness values brought about by haplodiploidy accounted for the prevalence of eusocial evolution in the haplodiploid Hymenoptera (the 'haplodiploidy hypothesis' or 'three-quarters relatedness hypothesis'). Whether the hypothesis is correct remains an open question to some (Bourke and Franks 1995; Bourke 1997; Crozier 2008), while others think it has been proven wrong (Wilson and Hölldobler 2005; Hunt 2007; Wilson and Wilson 2007). Either way, the hypothesis is a subset of inclusive fitness theory, not the theory itself, which does not require the

Box 2.3 (*Cont.*)

conditions assumed by the hypothesis to be present to account for the evolution of eusociality (West-Eberhard 1975; Bourke and Franks 1995; Bourke 1997; Foster et al. 2006a). Inclusive fitness theory can readily account for eusocial evolution in diploids, provided the conditions specified in Hamilton's rule are met.

Criticism 8: Lack of evidence for within-colony kin discrimination in the eusocial Hymenoptera is evidence against inclusive fitness theory (Wilson 2008).

Response: It is true that inclusive fitness theory predicts, other things equal, within-group kin discrimination. In social insects, where such nepotism is rare, the conclusion is that other things are not equal. Nepotism might be costly, or cues permitting nepotism might be absent, or genetic variation for nepotism might erode with time (Ratnieks 1991; Bourke and Franks 1995; Keller 1997; Boomsma et al. 2003). In principle, these arguments are all testable. Furthermore, in social birds and mammals, there is evidence for within-group kin discrimination (Emlen 1995; Wahaj et al. 2004; Silk 2009). There is also evidence from these taxa that, across groups, helpers are more likely to help or increase their level of helping as relatedness to the group increases (Russell and Hatchwell 2001; Richardson et al. 2003b). Why social birds and mammals differ from social insects in these respects is unclear.

Criticism 9: A single-locus change can bring about the change from a non-social to a eusocial population, especially in the presence of pre-adaptations such as parental care, and this is (by implication) an alternative explanation to that offered by inclusive fitness theory (Wilson 2008).

Response: Given pre-existing parental care, a single-locus change can also bring about the change from a non-social to a eusocial population under inclusive fitness theory, and indeed scenarios based on such a process have been presented (Dawkins 1979; Bourke and Franks 1995). For example, starting with an ancestral population of solitary Hymenoptera with maternal care, a change at a locus affecting timing of daughters' dispersal from the nest could lead to non-dispersing daughters expressing care towards larval siblings in the natal nest and so acting effectively as workers. If care were costly, the gene for non-dispersal would then be a gene for altruism and would spread or not according to whether Hamilton's rule was satisfied (Crozier 1992; Bourke and Franks 1995).

Criticism 10: The distinction between helper and reproductive group members is phenotypically flexible but (by implication) inclusive fitness theory does not account for this (Hunt 2007; Wilson 2008).

Response: Inclusive fitness theory assumes facultative expression of genes for social actions, such that bearers of the same gene can adopt the phenotype of either actors or recipients (Section 2.4). Hence, inclusive fitness theory requires there to be phenotypic flexibility in the helper–reproductive distinction (Bourke and Franks 1995; Queller and Strassmann 1998). Scenarios for the origin of eusociality in the Hymenoptera that invoke workers originating through the expression of a flexible phenotype in females, but assert the absence of kin-selected genes for altruism (Hunt 2007; Wilson 2008), are effectively revivals of West-Eberhard's (1987, 1988, 1992) 'epigenetic theory' of eusocial evolution. Arguments against this were presented by Crozier (1992) and Bourke and Franks (1995). Bourke and Franks (1995) argued that there is no reason to assume absence of genetic variation for a female's decision whether to disperse or not. As in the previous response, a gene for non-dispersal that leads to the expression of costly worker behaviour would be a gene for altruism.

Criticism 11: Inclusive fitness theory 'assumes that behavioral similarity is proportional to genetic similarity; the only way for a group to be behaviorally uniform, for example, is to be genetically uniform' (Wilson and Wilson 2008).

Box 2.3 (*Cont.*)

Response: All biological theories of social evolution assume a genetic basis to social behaviour (Section 2.4). Hence they all assume that behavioural phenotype depends on genotype in some manner. But, according to inclusive fitness theory, a gene for altruism at fixation in a population of eusocial colonies would lead, because of the point made in the previous response, to colonies each containing conditionally expressed helper and reproductive phenotypes. Therefore, the theory does not assume that behavioural similarity is proportional to genetic similarity or that genetic monomorphism leads to behavioural monomorphism.

Criticism 12: Evidence that worker social insects manipulate sex-ratios and police eggs as predicted by inclusive fitness theory is not evidence for the theory's explanation of the evolution of eusociality (Alonso and Schuck-Paim 2002).

Response: Evidence for inclusive fitness theory from the study of within-group conflicts is evidence that social behaviour evolves according to the theory's precepts and hence, since the theory is a general one, represents indirect evidence for the theory's applicability in other contexts (Bourke 2005; Foster et al. 2006a). The origin of eusociality is hard to study directly because in fully eusocial lineages it occurred in the past. It could, in principle, be studied in insect populations with facultative or weakly-developed sociality (Bourke 1997). Such studies remain relatively rare, but several exist and they find support for inclusive fitness theory (Stark 1992; Hogendoorn and Leys 1993; Packer and Owen 1994). In addition, some models of the origin of eusociality in the Hymenoptera invoke sex-ratio biasing (Grafen 1986; Pamilo 1991), so the connection between within-group manipulations and the origin of Hymenopteran eusociality could in fact be close.

history of the bustling nest to represent a helpful aid in understanding it. For example, Hunt (2007) argued that Hamilton's rule, while correct, represents as useful a guide to social evolution as the fundamental equations of flight (e.g. thrust must exceed drag) are to understanding the evolution of flight. Everything must obey the fundamental conditions, but, Hunt (2007) maintained, establishing this does not get one very far in dissecting the details of the process. However, this analysis does not reflect actual practice in research within the inclusive fitness framework. Researchers have extended the basic expressions of inclusive fitness theory to derive hypotheses that incorporate the detailed biology of specific social phenomena. This way, in the eusocial Hymenoptera, the theory has been extended to predict the sex ratio at both the population and colony levels (Trivers and Hare 1976; Boomsma and Grafen 1991), the length of dominance hierarchies (Monnin and Ratnieks 1999), the frequency of female larvae eclosing as queens (Wenseleers et al. 2003), and the frequency of reproductive workers (Wenseleers et al. 2004a). As outlined above, many tests of these predictions have then met with success (Ratnieks et al. 2001, 2006; Bourke 2005). Furthermore, even the most basic permutations of Hamilton's rule yield fruitful insights into the conditions required for the different types of social behaviour (Section 2.2) and, by extension, the different types of major transition (Section 3.1). So considering inclusive fitness theory as too abstract to be useful severely underestimates what the theory has achieved.

Hunt (2007) advocated an approach to researching social evolution based on identifying the likely behavioural, physiological, and life-historical antecedents of social traits in the solitary relatives of social species, along with the genetic pathways underlying them. He and colleagues have produced a series of valuable studies of sociality in the Hymenoptera based on this approach (Hunt and Amdam 2005; Amdam et al. 2006; Hunt 2007; Hunt et al. 2007; Toth et al. 2007). But all evolutionary processes must work by modification of existing systems, so in this sense it is expected that social traits would be foreshadowed in traits of solitary ancestors. Moreover, identifying these antecedents, while highly important in establishing the proximate basis of social evolution, does not of itself address its ultimate causes (e.g. Monnin and Liebig 2008). Hunt (2007) argued against the value of distinguishing between proximate and ultimate explanations in biology. However, if one holds otherwise, then there can be no clash between a commitment to uncovering the proximate details of social evolution and a commitment to inclusive fitness theory.

The second point underlying the criticisms of inclusive fitness theory is that the theory is preoccupied with relatedness, yet many social traits do not covary with relatedness (Wilson and Hölldobler 2005; Hunt 2007; Hölldobler and Wilson 2009). Relatedness of course plays a key role in social evolution according to the theory. Indeed, that does so has been one of the theory's central insights (Sections 2.1, 2.2). It is also true that, while relatedness is a successful predictor of social variation in many contexts (Table 2.5), it does not explain all social variation (Keller 1997; Gadagkar 2001; Korb and Heinze 2004, 2008a). But this does not mean that inclusive fitness theory is inadequate. Instead, from the theory, we would conclude that, in these cases, variation in the benefits and costs of social actions experienced by interacting partners is more important to the outcome of social evolution (Keller 1997; Gadagkar 2001; Helanterä 2007; Korb and Heinze 2008a). In addition, in the eusocial Hymenoptera, there is evidence that lineages in which workers have lost the ability to adopt a queen-like role as adults undergo secondary decreases in within-colony relatedness (Hughes et al. 2008), perhaps through productivity benefits of increased genetic variation within the social group (Section 5.2). This might decouple some linkages of social organization with relatedness in complex eusocial societies in comparison with those obtaining in simple societies (Korb and Heinze 2004). In principle, these explanations could be tested, and should be. In practice, it has usually proved difficult because measuring changes in the expected productivity of group members and groups is harder than measuring relatedness. But it is not impossible, as shown both by the direct tests of Hamilton's rule in bees and wasps described above, and by estimates from long-term field data of the direct and indirect components of inclusive fitness in cooperatively-breeding birds (e.g. Richardson et al. 2002, 2003a, 2007; Maccoll and Hatchwell 2004).

The third point motivating its critics involves the notion that inclusive fitness theory has concentrated research on within-group conflict while neglecting the most salient feature of animal societies and, particularly, of insect societies. This is held to be group-level cooperation, the intricately coordinated behaviour of numerous group members in pursuit of common goals such as foraging, nest-building, or defence.

And, the point continues, the theory neglects the competition between groups that seems to underpin the evolution of such behaviour (Hunt 2007; Hölldobler and Wilson 2009).

Again, however, inclusive fitness theory does not ignore group-level traits or between-group competition. If members of a social group, having an appropriate coincidence of fitness optima, cooperate in such a way that group output is increased relative to that of competing groups, this would be reflected in the benefit term of Hamilton's rule (Sections 2.1, 2.2). The critics of inclusive fitness theory prefer to express this statement in the language of intrademic group selection, which effectively involves attributing group success to an elevated between-group component of selection (Wade 1980; Bourke and Franks 1995). But, given the complementarity of inclusive fitness theory and intrademic group selection, this is not to propose an alternative explanation to the one offered by inclusive fitness theory (Box 2.3). The reason there have been relatively few empirical studies of the fitness consequences of between-colony competition in social insects in the natural environment is not through inclusive fitness theory discouraging such studies. It is again because, in social insects (more so than in social vertebrates), the long-term field studies of individually-identified social groups that are required to measure differential group productivity are hard to conduct, though not absent (Matsuura and Yamane 1990; Itô 1993; Gordon 1999; Gadagkar 2001; Hunt 2007). More widely, to address the evolution of the major transitions with inclusive fitness theory is, of course, to address the cooperative aspects of social evolution on the grandest scale. And as for within-group conflict, inclusive fitness theory stresses this topic because any theory of social evolution needs to explain the balance between conflict and cooperation that allows social groups to evolve and remain stable.

Stepping back, one may read the arguments of inclusive fitness theory's critics as a plea for more research to be carried out on topics they hold to be important but neglected. These include the proximate basis of social evolution and the ecology of cooperative behaviours in the natural environment. More research in these areas should certainly prove valuable from any standpoint. Provided misconceptions over inclusive fitness theory can be cleared up (Box 2.3), the debate over the theory is, in this light, largely a debate about emphasis, not absolute right and wrong. Overall, while the research conducted explicitly within the framework of inclusive fitness theory might not match its critics' judgement as to what is most important, it can perhaps be agreed, from the foregoing analysis, that no fundamental matters of evolutionary logic are at stake.

Concluding that inclusive fitness theory has emerged conceptually unscathed by recent criticisms risks leaving the impression that the theory is unfalsifiable. This is emphatically not the case. Like any scientific theory, inclusive fitness theory is essentially provisional. The best there is at present, it remains in principle capable of being toppled at any time by logical undercutting or by weight of evidence, should that evidence be forthcoming. Specifically, what would falsify the theory is the demonstration that any social action evolved under conditions not specified by the appropriate version of Hamilton's rule (Table 2.3). As already mentioned, such demonstrations may often prove difficult to set up because it is hard to measure all the terms in Hamilton's rule. But it is not impossible, either in principle or in practice. Moreover, as in analogous situations throughout biology, experimental and comparative approaches can be used to side-step this problem to provide robust tests of the theory (Table 2.5). Lastly, for some potential falsifications of the theory, only one term in Hamilton's rule need be measured. For example, Hamilton's rule forbids the evolution of altruism when relatedness is zero, regardless of the levels of benefit and cost. As the theory predicts, the evolution of altruism among non-relatives is conspicuously absent (Section 4.2). In science, if not beyond, the evidence of things not seen can be genuinely revealing.

Inclusive fitness theory has become a scientific orthodoxy precisely because it has not been falsified, but has instead seen raised upon its firm conceptual foundations a large and growing set of empirical successes. It is deeper and more accommodating than often appreciated. Yet it is not an immutable monolith. On the contrary, it is dynamic and expansive in several ways. As we have seen, its central theorem (Hamilton's rule) makes an extensive set of predictions based on relatively few parameters. The implications of the theory continue to be worked out and tested in the context both of the major transitions and, through elaborations of the basic models, of the evolution of cooperation and conflict in general. Applying the theory to new contexts, measuring inclusive fitness in field settings, testing the theory experimentally, determining the phylogenetic history of social groups, and characterizing the effect on social evolution of the genetic architecture of social traits, all still represent challenging research frontiers. Furthermore, as expected when a lively body of ideas meets a world of complex phenomena, unsolved issues in the theory, inconsistencies in the data, and gaps in our empirical knowledge remain (Section 7.2). Although falsification of inclusive fitness theory remains an ever-present possibility, such is the strength of the theory's empirical support that falsification now seems unlikely. Hence, as with other successful scientific ideas, the fate of inclusive fitness theory may be not falsification but assimilation in some larger theory. Until such a larger theory is successfully formulated, inclusive fitness theory remains an orthodoxy that the open-minded can feel at home in.

2.6 Summary

1. Inclusive fitness theory is encapsulated in Hamilton's rule, which specifies the conditions for the spread of a gene for any of four types of social action, namely cooperation, altruism, selfishness, and spite. Hamilton's rule (in one common form) has three parameters: the change in expected offspring number of a social actor; the change in expected offspring number of a social recipient; and the relatedness of actor and recipient. The rule says that a gene for a social action spreads if the change in an actor's expected offspring number, plus the change (weighted by relatedness) in the recipient's expected offspring number, is positive. Informally, the rule states that social evolution proceeds as if social

actors valued the reproduction of others in proportion to their relatedness with them.

2. Hamilton's rule predicts that the four types of social actions differ in the conditions required for their evolution. For example, cooperation can evolve if relatedness is zero but altruism evolves only if relatedness is positive. Selfishness can evolve whatever the value of relatedness, but spite can evolve only when relatedness is negative.

3. In social groups that are not clonal, potential conflict arises because not all members of the group will have identical inclusive fitness optima. Inclusive fitness theory can be used to analyse conflicts both within societies and within multicellular organisms. Examples of conflict within societies include parent–offspring conflict over resource allocation within vertebrate families and queen–worker conflict over male parentage and the sex ratio in eusocial Hymenoptera. An example of conflict within multicellular organisms is conflict between genes in the nuclear genome (intragenomic conflict, as in meiotic drive) and between these genes and those in the genomes of organelles (intergenomic conflict, as in cytoplasmic male sterility). A general form of conflict within social groups involves the tragedy of the commons, whereby group members are selected to exploit public goods for selfish gain, with the result that there is overexploitation and the group as a whole suffers. Mechanisms preventing a tragedy of the commons in biological systems must evolve rather than be imposed in a top-down manner.

4. Inclusive fitness theory assumes (in its usual form) complete dispersal by the products of altruism followed by their competing with the population at large. If dispersal is incomplete, the products of altruism compete with relatives in the natal patch. The theory needs modifying for this case and the modified theory shows that limited dispersal can inhibit the evolution of altruism. Inclusive fitness theory also assumes (correctly) that, when relatedness arises via kinship, genes in the genome other than the genes for social actions 'agree' with the social action being performed should Hamilton's rule be fulfilled, because relatedness between the actor and recipient is then the same across all corresponding loci. For this reason, relatedness arising via kinship, unlike relatedness arising from other causes (e.g. direct recognition of co-bearers by bearers of green-beard genes), does not provoke intragenomic conflict. Lastly, the theory assumes that the difference between actors and recipients involves facultatively expressed phenotypes, and that genes for social actions actually exist in nature, which they do.

5. Inclusive fitness theory is a gene-centred theory of natural selection and so recognizes that the levels of the biological hierarchy are not all alike, the lowest level (genes) being the level of replicators, the other levels being levels of 'vehicles' for replicators (cells, bodies, societies). Adaptations always serve the interests of genes. They serve the interests of vehicles incidentally, when the interests of genes coincide. Each major transition involves the evolution of a new level of vehicle.

Inclusive fitness theory has a robust logical basis in natural selection theory, is well-supported by data from many studies of different taxa, and makes novel a priori predictions about social phenomena, many of which have been confirmed. Past and recent criticisms of the theory are flawed and are mainly based on a lack of appreciation of its scope and flexibility. Nonetheless, the theory remains falsifiable.

3

The major transitions in light of inclusive fitness theory

3.1 Egalitarian versus fraternal major transitions

The aim of this chapter is to apply some of the conclusions of inclusive fitness theory from the previous chapter to make a set of general points about the major transitions in evolution. We start with Queller's (1997, 2000) division of the major transitions into egalitarian and fraternal transitions. As outlined earlier (Section 1.3), this distinction is essentially that between associations of non-relatives and associations of relatives. In fact, it might have been better for Queller just to have used these labels for the two types of transition, and this chapter will do so, when appropriate. This is for clarity and so as not to prejudge what effect the lack or presence of relatedness has on any one transition. (Queller chose 'egalitarian' and 'fraternal' in a witty reference to the French Revolutionary slogan, '*Liberté, Egalité, Fraternité*', and we should certainly treasure the wit of the author who gave us the 'Spaniels of St. Marx' (Queller 1995).) The key point of the section is that, as inclusive fitness theory demonstrates, and Queller (1997, 2000) recognized, associations of non-relatives and associations of relatives taking part in major transitions should differ in the suites of social behaviours that they are able to exhibit.

Interactions within species

To consider the major transitions in light of the preceding point, take first the case of interactions within species. In a social group of conspecific non-relatives, narrow-sense cooperation can evolve, but not altruism, since inclusive fitness theory rules out altruism between non-relatives (Section 2.2). Furthermore, treating both parties as actors, one can see that each must gain a direct fitness benefit for cooperation to remain stable (Table 2.3). These conditions appear to be met in associations of conspecific non-relatives, although it is not always straightforward to decide in a given case the exact nature of the social behaviour being performed (Dugatkin 2002; Clutton-Brock 2009a). Correspondingly, when conspecific non-relatives group together in a major transition, one expects each partner to retain its ability to reproduce. This is the point made by Queller (1997, 2000) in characterizing egalitarian transitions as those observing 'fairness in reproduction' (Table 1.2). An example is the evolution of sexual reproduction, in which genes from both females and males retain their ability to replicate (Keller and Reeve 1999; Bourke 2009).

In a social group of conspecific relatives, both cooperation and altruism can evolve (Section 2.2). Altruism between non-relatives indeed appears absent (Section 2.5) or, as in the unicolonial ants, present only as a condition derived from kin-altruist ancestors (Section 7.3). Hence, in any major transition involving altruism, one expects the partners to be relatives. Moreover, in these cases one need not expect both partners to have retained the ability to reproduce. This is because altruism permits the evolution of facultatively expressed sterility, provided sterile helpers aid relatives (and Hamilton's rule for altruism is quantitatively satisfied; Sections 2.2, 2.4). Examples of such transitions are the transitions to multicellularity and eusociality (Queller 1997, 2000; Grosberg and Strathmann 2007). As will be fully demonstrated later (Section 4.2), these only occur when relatives associate. In addition, each involves the evolution of facultative sterility in helper members, namely somatic cells or non-reproductive workers. Queller (1997, 2000) expressed this point by highlighting the involvement of 'epigenesis' in the fraternal transitions (Table 1.2), by which he meant that the division of labour in the social group, including the reproductive division of labour, is generated through differential (facultative) expression of genes.

Interactions between species

The introduction to inclusive fitness theory in the previous chapter dealt only with social interactions within species. But some of its conclusions can also be extended to between-species interactions. When a social group is formed from members of different species, these members are, by definition, not relatives of one another. In a way, relatedness is undefined in this case, because a focal gene may be completely absent from a non-conspecific rather than present at the population-average frequency. In fact a 'population' of non-conspecifics, in the sense of organisms sharing a common gene-pool, does not exist. Social groups formed of different species consist of interspecific mutualisms. As we will see later in this section, and beyond (Sections 3.2, 4.2, 5.5), they generally form because each partner has some resource or ability that benefits the other, they may achieve a degree of coincidence of fitness interest through shared reproductive fate, and they are stabilized by coercion of one sort or another. The main point for now is that, although non-conspecifics cannot be related in the same way as conspecifics can be, the logic of inclusive fitness theory still applies. Precisely because of the lack of gene-sharing, one can conclude that only narrow-sense cooperation can evolve between non-conspecifics. This predicts that major transitions involving non-conspecifics will always be cooperative and that both parties will retain their ability to reproduce.

Examples of such transitions are the transition that created the eukaryotic cell and the evolution of interspecific mutualisms involving multicellular organisms (Queller 1997, 2000). In eukaryotic cells, as predicted, both nuclear and organellar genomes reproduce. It is true (e.g. Archibald 2009) that many genes have moved from organelles to the nucleus (Section 5.5). This shows that, on occasion, the barrier between the gene pools of separate species can be leaky (defining organelles and host cells as separate species, since they originated from separate ancestors). But the transferred genes have not lost their ability to replicate. Instead, they do so according to the rules of the nuclear genome, rather than those of the genome of the organelle, with the result that (because of their altered relatedness asymmetry) they have a new set of evolutionary interests. In addition, having transferred completely, they do not form a common gene-pool with the remainder of the organellar gene-pool. They are like floor-crossing members of a parliament, who retain voting rights but change their party allegiance. In the case of interspecific mutualisms, of which there are many (Bergstrom et al. 2003; Sachs et al. 2004), both partners likewise retain their ability to reproduce. An example comes from lichens, in which the fungus and its algal or cyanobacterial partner each reproduces, often via specialized structures housing both symbionts (Lutzoni and Miadlikowska 2009).

The corollary of the preceding argument is that, in social groups formed of non-conspecifics, altruism cannot evolve. This requires some examination. Several authors have derived models of interspecific mutualisms and have shown that they partly rely on positive associations of cooperative genotypes or phenotypes across species, which are in some senses analogous to relatedness within species (Frank 1994b; Fletcher and Zwick 2006; Foster and Wenseleers 2006; Gardner et al. 2007a). Interspecific mutualisms can also be promoted by conspecific groups of each partic-ipating species having elevated within-species relatedness (Foster and Wenseleers 2006), but this is different from what is being discussed here, which is the nature of non-random associations between species. If 'relatedness' can be positive across species, it seems to follow that altruism can evolve between species. Fletcher and Doebeli (2009) presented a general model of social evolution in which within-species relatedness was found to be a special case of positive assortment operating either across or within species between altruistic genotypes and phenotypes. From this they concluded that altruism can evolve between species just as it can within species.

In my view, such an implication is incorrect and altruism cannot evolve between species. As pointed out by Foster et al. (2006b), confusion has arisen over this issue because some authors have previously used 'altruism' in this case in the sense of 'reciprocal altruism' (e.g. Fletcher and Zwick 2006). 'Reciprocal altruism' was the name given by Trivers (1971) to a form of what, in the present book, is termed narrow-sense cooperation. A reciprocal altruist may experience a short-term loss of direct fitness, but this is in return for a net gain in direct fitness (via reciprocated aid) over its lifetime. The question being addressed now is whether between-species altruism can evolve in which altruism, following the (standard) definition employed in this book, incurs a net cost in direct fitness to the actor over the actor's lifetime. My argument is that it cannot. Inclusive fitness theory shows altruism evolves within species when a gene in an actor denies itself reproduction in order to channel aid to copies of itself in reproductive bearers of the gene (Sections 2.2, 2.4). Copies of the focal gene will not occur in other gene pools, so altruist and beneficiary must share a gene pool in order for the focal gene to increase in frequency. Hence altruism can evolve within species, but not between them.

To dramatize this point, consider whether members of one species (Species A) could ever evolve to become altruistic to benefit recipients that were all of another species (Species B). This may seem an extreme case, given that, as previously stressed (Section 2.4), some bearers of a gene for altruism (indeed a gene for any trait) must reproduce for the trait to be naturally selected. But a strength of inclusive fitness theory (within species) is that it explains universal altruism in some classes of individuals (e.g. workers in the social insects) on the grounds that they are related to other classes, who make up the beneficiaries (e.g. the reproductive forms in social insects). Therefore, if between-species altruism arising via between-species relatedness is to mean anything distinctive, it should cover the case where one species acts altruistically entirely on behalf of another. Such between-species altruism cannot evolve, however, because the focal gene would decrease in frequency within Species A and the genes aided in Species B would increase in frequency only within the gene pool of Species B. Species B would effectively be a parasite of Species A. It is possible that some members of Species A could altruistically aid members of Species B, who in return aided reproductive members of Species A. But, for this to work, there would have to be positive relatedness between altruists (in Species A) and the ultimate beneficiaries (also in Species A), otherwise the gene for altruism would again decrease in frequency. So the altruistic element would still rely on positive within-species relatedness. Species B would in this case form a sort of living tool for the delivery of within-species altruism.

Such a scenario is similar to what happens in fungus-growing ants (Mueller 2002). In these, the worker ants rear a symbiotic fungus and the fungus, in a sense, rears (by providing food for) the ants, including the reproductives. The worker and reproductive ants are of course nestmates and hence relatives of one another. Since the fungus also benefits by being propagated by the ants, this also resembles the case in which within-species relatedness contributes to interspecific mutualism as envisaged by Foster and Wenseleers (2006). But the conclusion remains that altruism in one species in which members of another species are the sole beneficiaries cannot evolve. A similar point was made by Foster et al. (2006a, 2006b). It was likewise expressed by Queller (1997, 2000) when pointing out that the units within fraternal major transitions (but not within egalitarian ones) are 'fungible', meaning interchangeable (Table 1.2), as are genes within a common gene-pool.

The conclusion that altruism between species cannot evolve is borne out by the data, in that no such instances appear to exist. Interestingly, Darwin himself made similar assertions, writing in *The origin of species* (1859) that: 'Again as in the case of corporeal structure, and conformably with my theory, the instinct of each species is good for itself, but has never, as far as we can judge, been produced for the exclusive good of others.' It is for these reasons that there are no fraternal major transitions involving unions of separate species. Fletcher and Doebeli's (2009) conclusion that between-species altruism can evolve arose because their model included the possibility of direct benefits to the actor (Bijma and Aanen 2010). This makes their model one for the evolution of narrow-sense cooperation, which can of course evolve interspecifically. It is useful to keep altruism and narrow-sense cooperation distinct because they entail very different consequences for the nature of the phenotypes expected in the actor (e.g. Lehmann and Keller 2006a). As we have seen, in extreme cases, these phenotypes could be, respectively, non-reproductive and reproductive. For these reasons, between-species 'relatedness' and within-species relatedness, though both representing varieties of non-random association, cannot be regarded as wholly equivalent.

Shared genes versus shared reproductive fate

The conclusion of this section so far is that inclusive fitness theory illuminates the nature of the major transitions by providing a rationale for the kinds of social action that they embody and for the distribution of reproduction across the relevant social partners (Table 3.1). Essentially, related social partners can undergo fraternal transitions and exhibit altruism, forming groups in which some members become totally non-reproductive. But non-related social partners can only undergo egalitarian transitions and exhibit narrow-sense cooperation, forming groups in which each social partner remains reproductive.

Table 3.1 Major transitions classified by whether they occur between- or within-species and between relatives or non-relatives. Major transitions are defined as in Table 1.3. The transition from separate replicators (genes) to a cell enclosing a genome is omitted because it almost certainly involved both egalitarian and fraternal elements (Section 1.3). Note that narrow-sense cooperation can, in principle, also evolve between relatives within species (Section 2.2), but is omitted in this case for clarity. See Section 3.1 for further explanation

	Evolvable social action, distribution of reproduction, and example of relevant major transition(s)	
	Between species	Within species
Between unrelated partners (egalitarian)	Cooperation (narrow sense), so both partners remain reproductive: • Separate unicells ⇒ symbiotic unicell • Separate species ⇒ interspecific mutualism	Cooperation (narrow sense), so both partners remain reproductive: • Asexual unicells ⇒ sexual unicell
Between related partners (fraternal)	Cannot evolve	Altruism, so one partner may be non-reproductive: • Unicells ⇒ multicellular organism • Multicellular organisms ⇒ eusocial society

In anticipation of the full consideration of social group formation, maintenance, and transformation to come in later chapters, we can now add to this conclusion by asking what, in general, contributes to the evolutionary stability of these two kinds of social group. The answer is a basic principle underlying all social evolution. The principle is that social groups are stable in proportion to the extent to which social partners experience, through sociality, a coincidence of fitness interests. Such a coincidence may come about through two basic methods, namely shared genes or shared reproductive fate (e.g. Dawkins 1982, 1990; Sachs et al. 2004). The shared genes method refers to relatedness and hence applies to fraternal transitions. It emerges from Hamilton's rule in an obvious way (Section 2.2). Shared genes contribute to social stability because increasing relatedness reduces potential within-group conflict (Sections 2.3, 3.2). Shared reproductive fate occurs when the genes of social partners share their route to future generations (Dawkins 1982, 1990), with the result that the reproductive success of one social partner becomes very closely tied to that of the other. As an alternative to shared genes, shared reproductive fate is most relevant to the egalitarian transitions. It likewise emerges from Hamilton's rule but in a more subtle way than the concept of shared genes. Asserting that two social partners have a shared reproductive fate is to assert that, writing down Hamilton's rule for each partner for a given social behaviour, one would find the inclusive fitnesses of each partner to be maximized under the same conditions. How this might arise will be clarified in later chapters. For example, Section 4.2 examines the respective roles of shared genes and shared reproductive fate in social group formation among non-relatives and relatives.

3.2 Conflict resolution

The previous chapter set out the grounds for thinking that social groups will often be subject to within-group conflicts of interest, with the nature of such conflict being interpretable in terms of inclusive fitness theory (Section 2.3). This section examines how such conflicts, including those embodied in the idea of a tragedy of the commons (Hardin 1968; Rankin et al. 2007a), might be resolved. A good operational definition of conflict resolution is that it occurs when the proportion of group resources expended on actual conflict is reduced to a low level (Ratnieks et al. 2006). What classes of factor might allow this to happen, and, in general, what prevents social groups from collapsing under the pressure of cheating or free-loading? This has direct bearing on the occurrence of the major transitions, because achieving social groups that are comparatively free of internal conflict is necessary for any transition to complete all three component stages.

Social group maintenance is also promoted by constraints that prevent potential conflicts becoming actual conflicts, as outlined in Section 2.3. Conflict resolution is more concerned with the situation in which actual conflict already occurs, but does not run out of control and bring down the social group and may indeed evolve to be less costly. Researchers have identified two broad factors contributing to conflict resolution in this sense, namely self-limitation and coercion (Frank 2003; Travisano and Velicer 2004; Ratnieks et al. 2006; Rankin et al. 2007a; Gardner and Grafen 2009). Note that this classification resembles, but is slightly different from, that proposed by Ratnieks et al. (2006), who identified kinship (relatedness), coercion, and constraint as factors in conflict resolution. I include relatedness in the wider category of self-limitation and I omit constraint because constraint is more readily viewed as a factor preventing potential conflicts being realized, rather than as a factor resolving actual conflict once it has arisen (Section 2.3). Models of reproductive skew are essentially extensions of Hamilton's rule combining both kinship and coercion to specify the conditions required for the stable sharing of reproduction within social groups (Vehrencamp 1979; Emlen 1982a, 1982b; Reeve and Ratnieks 1993; Reeve 1998a, 2000b; Johnstone 2000; Reeve and Shen 2006; Nonacs 2007; Buston and Zink 2009; Cant and Johnstone 2009), which may or may not be associated with costly conflict.

Self-limitation

Perhaps the most fundamental form of self-limitation occurs when members of a social group are related. In this case, inclusive fitness theory predicts that selfishness will be held in check because too great a level of cost imposed on relatives harms the fitness of the gene for selfishness (Section 2.2). This is evident from Hamilton's rule for the evolution of selfishness (Table 2.3). It was also formally modelled in the context of the tragedy of the commons by Frank (1994a, 1995, 1996b). He derived the evolutionarily stable strategy (ESS) for a selfish trait in a social group, showing that the ESS frequency of selfishness (z^*) is given by $1 - r$, where r is within-group relatedness (Box 3.1). The model assumed that group productivity fell linearly with rising selfishness, such that, at the ESS, group productivity was proportional to $1 - z^*$. Hence the model showed that, as intuitively expected, when relatedness rises, the ESS level of selfishness falls and group productivity rises. Conversely, when relatedness falls, the ESS level of selfishness rises and group productivity falls. At the extremes, a member of a social group should be maximally self-restrained when in a clone ($r = 1$) and maximally selfish when relatedness is zero. The result for clones does not mean that no group member should reproduce. It means that, genetically, each member is indifferent as to whether it reproduces or a groupmate does. Therefore, Frank's (1994a, 1995, 1996b) result for clonal groups represents a formal way of making the point that in a clone all group members have identical evolutionary optima (Section 2.3). Foster (2004) extended Frank's basic model to cases such as that occurring when selfishness does not reduce group productivity in a linear fashion but has a larger impact as the level of selfishness rises (i.e. there are diminishing returns on investment in cooperation). Frank (2010a) explored the effect of the timing of cooperation and the demographic context on the tragedy of the commons. Both authors identified conditions that reduce the severity of the tragedy, without it being abolished.

Box 3.1 Frank's model for the evolutionarily stable level of selfishness within groups

Frank (1994a, 1995, 1996b) modelled the effect of relatedness on the evolutionarily stable level of selfish behaviour within a social group, i.e. the level representing an evolutionarily stable strategy (ESS). The ESS is defined as a strategy that cannot be invaded by a new, mutant strategy, if all the population are already following it (Maynard Smith 1982). One of Frank's (1994a, 1995, 1996b) central results was that the ESS level of selfishness is given by $1 - r$, where r is within-group relatedness. Though this result is simple and intuitive, it is important because it leads to a key conclusion about how a tragedy of the commons can be resolved (Section 3.2). This box presents a step-by-step guide to deriving Frank's result, based on the treatment of Foster (2004). The modelling approach combines inclusive fitness theory and ESS theory, and is technically known as the 'neighbour-modulated' or 'direct fitness' approach (Taylor and Frank 1996; Frank 1998; McElreath and Boyd 2007; Gardner and Foster 2008; Wenseleers et al. 2010). It represents a powerful tool for the theoretical study of social behaviour, and has generated an important family of models tackling such issues as the evolution of policing (Frank 1995, 1996b) and, in eusocial Hymenoptera, the stable frequencies of reproductive workers and of females attempting development as queens (Wenseleers et al. 2003, 2004a). So the current example also serves as an introduction to this method. Note that direct fitness and alternative inclusive-fitness formulations of the same problems yield the same solutions, and both are also essentially equivalent to allele frequency models (Wenseleers et al. 2004a, 2010; Taylor et al. 2007).

Say there is a social group whose members can behave selfishly with a certain probability, which is set by the frequency with which each carries a gene for selfishness (i.e. selfishness is treated as a continuous trait). An example of selfish behaviour in this context would be taking group resources for personal reproduction. Let the following terms be defined:

g_i = frequency of the gene for selfishness in group member i,
r = within-group relatedness,
w_i = fitness of a focal member i of a social group,
z = probability of an average member of the group acting selfishly,
z_i = probability of focal group member i acting selfishly,
z^* = ESS level of selfishness as a proportion of group behaviour.

Assume that there is a negative linear relationship between the level of selfishness and group productivity, such that a group with no selfishness has maximal productivity and one in which all members are totally selfish has zero productivity. Therefore, group productivity is proportional to $(1 - z)$.

A focal group member's fitness is then modelled as the product of its relative share of group reproduction (z_i/z) and total group productivity, i.e., if total group resources represent a 'pie', its relative share of the pie × the total size of the pie. Hence:

$$w_i = \left(\frac{z_i}{z}\right)(1 - z).$$

(Equation 3.1)

One can imagine that a focal group member is 'tempted' to act a bit more selfishly (rising z_i/z) to increase its fitness. But in so doing it would also decrease its fitness, since, as part of the group, it would cause average group selfishness to rise and hence group productivity to fall, i.e. $(1 - z)$ to fall. What level of selfishness should a group member then choose to maximize w_i? This is the essential dilemma of the tragedy of the commons, neatly captured by Equation 3.1.

Box 3.1 (*Cont.*)

By the definition of an ESS, the ESS level of selfishness (z^*) is the level yielding maximum fitness to the focal group member (if it did not yield maximum fitness, then a mutant strategy yielding higher fitness would invade). The level of selfishness that a group member will be selected to choose is, therefore, the ESS level. So our aim is to find an expression for z^*. To do this, we need to examine how fitness (w_i) varies with changing levels of z_i. Mathematically, this can be done by differentiating w_i with respect to z_i, since differentiating a function yields the rate at which one variable changes with another. Let us proceed in the following steps:

1. Differentiating w_i with respect to z_i is not wholly straightforward because w_i is a function of two variables, z_i and z. Recall from textbook calculus that, if a variable is a function of two variables, then the function can still be differentiated, but one needs to employ partial differentiation. Specifically, if a variable u is a function of two variables x and y, i.e. $u = f(x,y)$, then the partial differential of u with respect to x is written $\partial u/\partial x$ and is calculated by differentiating u with respect to x and treating y as a constant. Likewise, the partial differential of u with respect to y is written $\partial u/\partial y$ and is calculated by differentiating u with respect to y and treating x as a constant.

Equation 3.1 is of the form:

$$u = f(x, y),$$

so we need to calculate the partial differentials of w_i with respect to z_i and z. To do so, it is convenient to rewrite Equation 3.1 as follows:

$$w_i = \left(\frac{z_i}{z}\right)(1 - z) = (z^{-1})(z_i) - (z_i). \qquad \text{(Equation 3.2)}$$

Hence (treating z^{-1} as a constant):

$$\frac{\partial w_i}{\partial z_i} = z^{-1} - 1, \qquad \text{(Equation 3.3)}$$

and (treating z_i as a constant):

$$\frac{\partial w_i}{\partial z} = -z_i(z)^{-2}. \qquad \text{(Equation 3.4)}$$

2. Standard calculus also states that, if a variable u is a function of two variables x and y, and x and y are both functions of a fourth variable v, then the following (the chain rule) is true:

$$\frac{du}{dv} = \left(\frac{\partial u}{\partial x}\frac{dx}{dv}\right) + \left(\frac{\partial u}{\partial y}\frac{dy}{dv}\right).$$

Let us now suppose that z_i and z in Equation 3.1 are both functions of g_i. In the case of z_i, this follows from the fact that the amount of selfishness shown by a focal group member (its phenotype) was assumed above to depend on its personal frequency of the gene for selfishness. It is less clear why z should be associated with g_i, but the reason for assuming this becomes apparent in the next step.

In Equation 3.1, w_i is a function of z_i and z, and both these are now assumed to be functions of g_i. So, from the chain rule above (with u represented by w_i, v by g_i, x by z_i, and y by z):

$$\frac{dw_i}{dg_i} = \left(\frac{\partial w_i}{\partial z_i}\frac{dz_i}{dg_i}\right) + \left(\frac{\partial w_i}{\partial z}\frac{dz}{dg_i}\right). \qquad \text{(Equation 3.5)}$$

Box 3.1 *(Cont.)*

3. Next note that in Equation 3.5 the differential function of z with respect to g_i (i.e. dz/dg_i) measures the slope of the relationship between the phenotype (z) of the members of the social group as a whole and the gene frequency (g_i) of the focal group member. Because phenotype was assumed proportional to genotype, dz/dg_i therefore measures the slope of the relationship between gene frequency in the group as a whole and the focal group member's gene frequency, which is the regression relatedness (Section 2.1) between the focal group member and the group as a whole. (Strictly speaking, because z is the selfishness level of an average group member and hence the focal group member is included in the calculation of recipients' gene frequency, dz/dg_i is regression relatedness only under certain assumptions, such as large group size or vanishingly small genetic variation at the locus for social behaviour (Frank 1996b; Taylor and Frank 1996; Foster 2004; Wenseleers et al. 2010). This is a technical point that makes no important difference to the model outcome.) Hence $dz/dg_i = r$, within-group relatedness (it was to achieve this equivalence that z was assumed a function of g_i in the previous step). Similarly, dz_i/dg_i is the slope of the relationship between the focal group member's own phenotype (proportional to its genotype) and its own gene frequency, and is therefore relatedness to self, which is 1. So Equation 3.5 becomes:

$$\frac{dw_i}{dg_i} = \left(\frac{\partial w_i}{\partial z_i} \times 1\right) + \left(\frac{\partial w_i}{\partial z} \times r\right). \qquad \text{(Equation 3.6)}$$

The right-hand side of this equation is the rate of change in the inclusive fitness of the focal group member with a varying level of selfishness, being (rate with which fitness is gained by the focal group member as its personal level of selfishness varies) + (rate with which fitness is lost by the focal group member as the average amount of selfishness in the group varies, weighted by its relatedness to the group). The second of these terms is negative (hence represents a fitness loss) because, as can be seen from Equation 3.4, $\partial w_i/\partial z$ is negative (an increase in group selfishness decreases the focal group member's fitness). Note the similarity of Equation 3.6 to Hamilton's rule for selfish behaviour affecting relatives in Table 2.3. In fact, technically it represents a 'marginal' Hamilton's rule, being Hamilton's rule for the case when one wishes to evaluate the effect of small changes in an actor's behaviour on inclusive fitness (Frank 1998).

Next, Equations 3.3 and 3.4 are substituted into Equation 3.6 to yield the following:

$$\frac{dw_i}{dg_i} = \frac{(1-z)}{z} - \left(\frac{z_i}{z^2}\right) r. \qquad \text{(Equation 3.7)}$$

4. Finally, recall that, at the ESS level of selfishness, fitness must be at a maximum (by the definition of an ESS), so the rate of change of fitness with the level of selfishness must be zero, i.e. $dw_i/dg_i = 0$ (by the assumption of a one-to-one relationship of phenotype with gene frequency). In addition, at the ESS, every group member must adopt the same strategy (again by the definition of an ESS). So, at ESS, $z^* = z_i = z$. Substituting these values into Equation 3.7 and simplifying yields:

$$0 = \frac{(1-z^*)}{z^*} - \left(\frac{1}{z^*}\right) r;$$

$$0 = 1 - z^* - r;$$

$$z^* = 1 - r. \qquad \text{(Equation 3.8)}$$

This is the result of Frank (1994a, 1995, 1996b) that we set out to prove. The meaning of this result is discussed in Section 3.2. Frank (1995, 1996b) went on to explore the effect of within-group policing on social evolution, using a more elaborate version of the present model.

The overall conclusion, based on inclusive fitness theory, is that the tragedy of the commons can be mitigated if relatedness is high, or rising, within a social group. More broadly, selfishness may similarly be held in check by self-limitation when a mutant cheat genotype within a social group becomes so frequent that it damages its host group and hence itself (Section 5.3).

Coercion

As Hardin (1968) argued, the forcible prevention of selfish behaviour is an obvious way of restraining it and so allowing conflict to be held at less costly levels. However, recall from Section 2.3 that, in natural systems, the only plausible forms of coercive control are those capable of evolving by natural selection. Planned, top-down control is not permissible. Researchers have identified several evolvable forms of coercion; for example, dominance, punishment, and policing (Section 5.5). The key point for now is again that these phenomena themselves arise according to the tenets of inclusive fitness theory.

This point can be illustrated using policing as an example. Policing is coercive behaviour directed at social cheats that reduces the benefits of cheating and so holds it in check, while not necessarily behaviourally preventing further attempts at cheating (Ratnieks 1988; Ratnieks and Wenseleers 2008). A classic example is worker eating of other workers' male eggs in the Honey bee (Section 5.5). Frank (1995, 1996b) extended his basic tragedy of the commons model (Box 3.1) to examine the evolution of policing within social groups in general. The conclusion was that the evolution of policing depends on the degree of within-group relatedness (Frank 1995, 1996b). In a commentary on Frank (1995), Hammerstein (1995) extended this conclusion with the following analysis. Policing was assumed by Frank's (1995, 1996b) model to have a cost to the policer. This means that policing qualifies as an altruistic act, given the benefit of policing falls on the group as a whole and not just on the policer. Therefore, when relatedness is zero, policing does not evolve, for the same reason that altruism of any sort does not evolve at zero relatedness (Section 2.2). In Hammerstein's (1995) formulation, group members are always under selection to refrain from policing and leave it to others in the group, so creating a second-order tragedy of the commons. At zero relatedness there would be no counterbalancing effect of relatedness to stop this. Similarly, when relatedness is low, policing needs to be relatively cheap to evolve. When relatedness is very high or unitary, policing again does not evolve, but in this case because there is no need for it. Here, relatedness alone ensures a lack of cheating, as in Frank's (1995, 1996b) original model. Hence policing evolves most readily at intermediate relatedness levels, when there is both a need for it and its personal cost can be counterbalanced by the beneficiaries of policing being related.

Frank (1995, 2003) went on to suggest that relatedness and coercion are effectively factors of equal status in social evolution, with each able to act independently to keep within-group selfishness pegged down to relatively low levels. However, analyses like the one above imply that it is more accurate to regard coercion as a secondary factor, which itself is selected for (or not) on the basis of its inclusive fitness payoffs (Queller 2006; Gardner et al. 2007a; Lehmann et al. 2007c; Ratnieks and Wenseleers 2008). It is hard to see how this could be otherwise, given that coercion is itself a form of social behaviour. The point can be brought into sharp focus by asking whether altruism within a social group can be sustained by coercion alone when coercion has a cost to the actor and relatedness is zero. The logic of Hammerstein's (1995) analysis above is that it cannot. But this was a verbal argument, not a formal, mathematical one. Although they have not posed the question quite this way, formal models of costly coercion, or 'strong reciprocity' as it is sometimes termed (Fehr and Fischbacher 2003), also suggest that the answer is no. Either coercion requires a direct benefit to invade (Lehmann et al. 2007c; West et al. 2007b), in which case the situation does not involve altruism, or it invades provided Hamilton's rule for altruism is satisfied (Gardner et al. 2007a), in which case relatedness cannot be zero. Costly coercion reinforcing altruistic behaviour therefore evolves only on a substrate provided by relatedness.

Despite not operating independently of relatedness within groups of altruists, coercion remains a highly important factor in such groups. Take again the example of policing. Wenseleers et al. (2004a) used an extension of Frank's (1995) basic tragedy of the commons model to derive the ESS frequency of reproductive workers in colonies of eusocial Hymenoptera. They showed that, assuming a linear relationship between the frequency of reproductive workers and colony productivity, this frequency was given by:

$$(1 - r_F)/(1 + r_F),$$

where r_F = relatedness of workers to sister females (new queens). In many species, this predicts a frequency of reproductive workers much higher than the one actually observed. For example, in the Honey bee, r_F is approximately 0.3 (assuming 10 matings by the single, multiply-mating queen), so the predicted frequency of reproductive workers is 54% (Wenseleers et al. 2004a). The observed frequency is 0.1% or below (Wenseleers and Ratnieks 2006b; Ratnieks and Wenseleers 2008). As mentioned earlier, worker Honey bees are well known to police one another's reproduction by eating worker-laid eggs (Ratnieks and Visscher 1989), suggesting that policing accounts for the frequency of reproductive workers being far less than expected. The effect of policing in driving down the frequency of reproductive workers may be indirect, as well as direct. When policing is efficient, the benefits of attempting reproduction are reduced, so, in theory, policing can then select for reproductive self-restraint or acquiescence in workers (Wenseleers et al. 2004a, 2004b).

The quantitative size of the gap between the frequencies of reproductive Hymenopteran workers predicted by tragedy of the commons models and the observed frequencies is uncertain. This is because the shape of the relationship between the frequency of reproductive workers and colony productivity is not precisely known. However, the qualitative point is well supported, especially in social bees and wasps. In ants, lack of relevant data means that it is not clear how generally the rarity of worker reproduction within colonies with a queen (e.g. Bourke 1988; Choe 1988) stems from active policing of workers' reproduction, although policing clearly occurs in some species (Wenseleers and Ratnieks 2006a). In social bees and wasps, widespread policing behaviour, coupled with a smaller frequency of reproductive workers being found than expected purely on the basis of relatedness levels, suggests that relatedness combined with policing often generates a greater level of reproductive altruism than relatedness alone (Wenseleers and Ratnieks 2006b; Ratnieks and Wenseleers 2008; Ratnieks and Helanterä 2009).

Conflict resolution, as we have just seen, can occur in social groups of related members via the effects of relatedness with or without coercion. Clearly, in social groups of unrelated members, relatedness cannot help resolve conflicts, although this does not preclude other forms of self-limitation from having an influence (Sections 5.3, 5.4). As discussed, however, what coercion can achieve among non-relatives is itself dictated by the absence of relatedness. Nonetheless, one therefore expects coercion to be a prime factor in resolving conflicts arising in major transitions involving either non-conspecifics or unrelated conspecifics. Queller (1997, 2000) made this point indirectly by referring to 'control of conflicts' as the greatest hurdle that egalitarian transitions needed to overcome (Table 1.2). Consistent with it, there is clear evidence of coercive control of social partners in the egalitarian transitions (Frank 2003; Ratnieks and Wenseleers 2008). For example, a fair meiosis enforced by suppressors of meiotic drive alleles helps stabilize the evolution of sexual reproduction. Physical destruction of mitochondria by nuclear genes controls conflict between nuclear and organellar genomes, once sex has evolved. Sanctions against defecting partners help stabilize interspecific mutualisms (Section 5.5). Overall, therefore, coercion of various kinds occurs in social groups representing both fraternal and egalitarian transitions, so contributing substantially to the maintenance of cooperation in its broad sense at several hierarchical levels.

3.3 Life cycles and the major transitions

Living things have an astonishing variety and complexity of life cycles, which implies that life cycles can be classified in many ways. As regards the evolution of individuality, a critical feature of any life cycle is whether or not it features a 'bottleneck' (Dawkins 1982; Buss 1987; Maynard Smith 1988; Maynard Smith and Szathmáry 1995; Grosberg and Strathmann 1998, 2007; Queller 2000; Roze and Michod 2001; Wolpert and Szathmáry 2002; Frank 2003; Lachmann et al. 2003; Szathmáry and Wolpert 2003; Boomsma 2009). This refers to whether the social group develops from a unitary propagule (a bottlenecked life-cycle) or a group propagule (a non-bottlenecked life-cycle). Here, a 'propagule' is the stage in the life cycle from which a new 'adult' (e.g. organism or society) develops and 'group propagule' refers to a propagule that is itself a group. For example, a zygote is a unitary propagule because it is a single cell. A group of foundress ant queens represents a group propagule because it contains several members. Group propagules may themselves arise in one of two ways. The first is through the aggregation of previously separate units and the

Table 3.2 Classification of life cycles, with selected examples

Life cycle	Example of propagule among multicellular organisms	Example of propagule among eusocial insects
(a) With unitary propagule (bottleneck)	Haploid spore or diploid zygote (Maynard Smith 1988)	Colony-founding mated pair of reproductives[1] (Maynard Smith 1988)
(b) With group propagule (no bottleneck)		
(i) Group propagule formed by aggregation	Aggregation of cells forming 'slug' and then fruiting body in cellular slime moulds (Bonner 2003a)	Foundress association of unrelated queen ants (Bourke and Franks 1995)
(ii) Group propagule formed by budding-off from parent	Budded-off polyp in *Hydra* (Buss 1987)	Swarm of swarm-founding wasps (Jeanne 1991)

[1] Queen carrying in her sperm receptacle the sperm of her mate in the eusocial Hymenoptera, a queen plus accompanying king in the termites.

second is through the budding-off of a set of units from their 'parent'. These considerations yield the simple, three-way classification of life cycles shown, with examples, in Table 3.2.

The canonical multicellular organism with a group propagule that forms by aggregation is the cellular slime mould *Dictyostelium discoideum* (Bonner 2003a). This and related species exist for part of the life cycle as a collection of separate unicellular amoebae in the soil, which then aggregate to form first a mobile multicellular 'slug' and then a multicellular fruiting body, some cells of which are non-reproductive (Fig. 3.1). Some species of bacteria known as the Myxobacteria (e.g. *Myxococcus xanthus*) have convergently evolved the same habit of forming, by cell aggregation, multicellular fruiting bodies composed of reproductive and non-reproductive cells (Shimkets 1990, 1999; Velicer and Vos 2009).

The presence or otherwise of a bottleneck stage in the life cycle, along with the manner of formation of group propagules, has several consequences for the nature of a social group. In essence, unitary propagules carry few genomes in them and group propagules carry several. Therefore, unitary propagules give rise to social groups with relatively high within-group relatedness and reduced potential conflict, whereas group propagules give rise to groups with relatively low within-group relatedness and greater potential conflict (Dawkins 1982; Maynard Smith 1988; Maynard Smith and Szathmáry 1995; Grosberg and Strathmann 1998, 2007; Queller 2000; Wolpert and Szathmáry 2002). Achieving greater within-group relatedness is likely to have been at least part of the reason why a bottlenecked life-cycle evolved, but there could have been additional reasons too. Sexual reproduction seems almost impossible without a unicellular gamete stage (implying a unicellular zygote), since multicellular gametes would somehow have to fuse genomes across all their cells

Fig. 3.1 A multicellular organism with a life cycle with a group propagule that forms by aggregation: the cellular slime mould, *Dictyostelium discoideum*. At the far left, separate cells (amoebae) exist in the soil. These aggregate to form first the star shapes then the mobile 'slugs' (centre and centre-right). Each slug then develops into a fruiting body with a stalk of non-reproductive, 'somatic' cells and a fruiting head of reproductive spores.   Drawing by Patricia Collins reproduced from Bonner (1969) with kind permission of the author.

(Grosberg and Strathmann 1998; Ridley 2000; Wolpert and Szathmáry 2002). In multicellular organisms, development into a complex adult through coordinated differentiation of cells and tissues might also only be achievable starting with a uni-cellular stage (Dawkins 1982; Wolpert and Szathmáry 2002; Szathmáry and Wolpert 2003). Finally, propagule size presumably depends partly on the trade-off between selection for dispersal ability in propagules, which would drive their size down, and selection for independence, defensive ability, and so on, which would drive their size up (Grosberg and Strathmann 1998). The remainder of this section is concerned less with why a bottleneck may or may not have evolved in the first place, and more with the inclusive-fitness consequences of its presence or absence.

Unitary propagule (bottleneck present)

Take the case of a diploid, multicellular organism developing from a zygote, as in humans, or a haploid, multicellular organism developing from a haploid spore, as in the multicellular alga *Volvox* (Kirk 2004; Michod 2007). The point is that, since development proceeds from a single cell by asexual cell division (mitosis), the result-ing adult is a clone of cells and hence harbours no potential between-cell conflicts of interest. (This is true as regards the interests of both the nuclear genomes as a class and the organellar genomes as a class, provided organelles are clones of one another within the zygote or spore.) This is because, as discussed in the previous section, clonality guarantees social harmony. That the clonal nature of multicellular organ-isms underpins their stability, as expected from inclusive fitness theory, was pointed out by Hamilton (1964) himself. One can now see that it is the presence of the bot-tleneck that leads to clonality and hence stability. As development proceeds, there is a chance of somatic mutation generating a selfishly over-replicating cell lineage (a cancer), threatening the internal harmony of the organism. Somatic mutation refers to any mutation occurring in the soma after the start of development of the organism from the propagule stage. But, given mutation rates are generally low, the level of genetic disruption from somatic mutation is likely to be negligible when the organism is relatively small in terms of its total cell number, since there are then fewer cell divisions in which mutations can occur (Queller 2000). 'Relatively small' in this context means having of the order of 10^6 to 10^9 cells or fewer (Queller 2000). This encompasses many multicellular organisms (Bell and Mooers 1997). For example, the adult nematode *Caenorhabditis elegans* has 959 cells (Kenyon 1988). By contrast, vertebrates have vastly more cells, an estimated 10^{11} to 10^{14} cells in fish, dog, and mouse (Bell and Mooers 1997), with the estimate for an adult human being 10^{13} to 10^{14} (Buss 1983; Ridley 2000; Burt and Trivers 2006; Frank 2010b). At such enormous cell numbers, somatic mutation becomes a significant generator of potentially damaging cancerous cell lineages (Section 6.4). The earliest multicellular organisms were almost certainly relatively small (Pál and Szathmáry 2000; Queller 2000). Therefore, it is likely that unitary development from a single zygote or spore would by itself have guaranteed a very high level of internal stability during the early phases of multicellular evolution.

The high relatedness within multicellular organisms developing from a single zygote or spore is also important because, in permitting altruism, it is associated with facultative gene expression leading to helper and reproductive phenotypes (Section 2.4). Cells in many multicellular organisms are morphologically differentiated into either germline or somatic cells. This raises two questions. The first concerns the reason for having such a reproductive division of labour based on morphology to begin with. The second concerns the reason why differentiation to somatic cells is often irreversible (cells lose totipotency, i.e. the ability to express the full repertoire of possible states), with the germline being concentrated in one or a few cell lineages physically confined to one part of the body (a segregated germline), which diverge from the somatic lineages early in development.

Both these questions can be elucidated using inclusive fitness theory. Regarding the first, in a clonal group of cells there is no conflict between cell lineages as to whether they are germline or somatic (barring mutations), for the same reason that all clones lack internal potential conflict. Hence, if efficiency in reproduction were best served by having a reproductive division of labour among cells based on morphology, all cells would 'agree' to this (Section 2.3). So, in a clone, germline and somatic cell lineages may readily evolve, without enforcement, simply to bring about an efficient reproductive division of labour based on conditionally-expressed altruism (Queller 2000). Regarding the second issue, as described above, the expected number of somatic mutations in an organism rises as its total size in terms of cell number increases. Selfish cell lineages would disrupt the organism even more if they could differentiate into germline cells *in situ*, or if cells from them could physically migrate to the germline lineage. Hence, as multicellular organisms grew larger in evolutionary time in terms of total cell number, the need to prevent this happening might have led to selection for loss of totipotency in somatic cells and the early segregation of the germline (Buss 1987; Grosberg and Strathmann 1998; Michod 2000; Frank

2003). This idea is discussed further in Chapter 6 on social group transformation (Section 6.4).

Note that these interpretations differ to a degree from some in the literature. Buss (1987) argued that, to control takeovers by selfish cell lineages, germline segregation would have been enforced in multicellular organisms from the beginning of their evolution. But this overlooked the role in generating the germline–soma division of the factors discussed above, namely (a) high relatedness arising from a bottleneck promoting kin-selected altruism between conditionally-determined germline and somatic cells; and (b) low cell number rendering somatic mutations a negligible factor (Maynard Smith 1988; Raff 1988; Wolpert 1990; Koufopanou 1994; Maynard Smith and Szathmáry 1995; Pál and Szathmáry 2000; Queller 2000; West-Eberhard 2003). Michod and colleagues have extensively modelled the conditions required for selfish cell lineages to threaten the emergence of multicellularity (Michod 1996, 1997a, 1997b, 1999, 2000, 2003, 2007; Michod and Roze 1997, 2001; Roze and Michod 2001; Michod et al. 2003). Although these studies emphasized the potential importance of control of selfish cell lineages when there are many cell divisions and mutation rate is high, they also effectively confirmed that the threat from such lineages is minimized when cell number is small and mutation rate is low (Queller 2000). Finally, Rainey (2007) hypothesized that germline cells originated as within-group selfish cheats capable of dispersal, but were favoured because they allowed somatic cells lineages a route to onward transmission. This idea was supported by Hochberg et al. (2008) but is problematic from an inclusive fitness theory perspective. Cheating cells necessarily differ from somatic cells at the locus or loci controlling cheating behaviour. Moreover, if they are cheats at all, they must be costly to somatic cells. Somatic cells would, therefore, be selected to suppress cheating cells, not promote them. Instead, as already discussed, the original germline–soma distinction is straightforward to evolve, if somatic cells favour clonally identical germline cells with a conditionally-expressed dispersing phenotype.

Similar reasoning to that applied to multicellular organisms can be applied to eusocial societies. From the analogy with a zygote, it is evident that a relatively high level of social harmony within eusocial societies is facilitated by the bottleneck in the colony cycle represented by colony foundation by a single mated pair (Maynard Smith 1988; Queller 2000; Boomsma 2009). Similarly, there is a strong analogy between an irreversible morphological distinction between germline and somatic cells inside multicellular organisms and an irreversible morphological distinction between reproductive and worker phenotypes inside some eusocial societies, with both being based on kin-selected altruism and differential gene expression (Bourke and Ratnieks 1999; Queller 2000; Boomsma 2009).

Analogies between the life cycle of multicellular organisms and the colony cycle of eusocial societies need qualifying in two important respects. First, although some eusocial societies are clonal, most are not (Section 4.2). This is because offspring of a founding pair are usually generated by sexual reproduction (true only for females in social haplodiploids). For example, in the Hymenoptera, clonal societies are infrequent and almost certainly derived (e.g. Himler et al. 2009; Mikheyev et al. 2009;

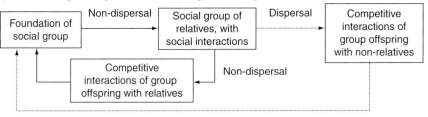

Fig. 3.2 (a) A life cycle in which dispersal and non-dispersal phases alternate. The dispersal phase often involves a unitary propagule (bottleneck) and dispersal is associated with population-wide competition, i.e. competition with non-relatives. Examples are a seed developing into an adult plant (the social group) from which new seeds then disperse by wind, or a mated queen eusocial Hymenopteran founding a colony (the social group) from which winged reproductives disperse. (b) A life cycle in which non-dispersal is followed by a mixture of dispersal and non-dispersal. The non-dispersal phase often involves a group propagule (lack of bottleneck) budding off from the parent and non-dispersal is associated with competition with relatives in the natal patch. Examples are a seed developing into a plant producing both vegetative offshoots and wind-dispersed seeds (Cousens et al. 2008), or a mated queen eusocial Hymenopteran founding a colony that both buds off daughter colonies in the neighbourhood and produces dispersing winged reproductives (Bourke and Franks 1995).

Rabeling et al. 2009). Hence some potential conflict would have been present in emerging eusocial societies even when a bottleneck was present. Second, the number of expected 'somatic' mutations in eusocial societies does not scale with worker number in the same way as with total cell number in multicellular organisms, because workers are generally the product of one generation of reproduction rather than being produced in branching lineages in which workers successively give rise to other workers (Queller 2000). In addition, even the largest insect societies only reach of the order of 10^7 workers in size (Bourke 1999), which is ten million times less than the number of cells found in some multicellular organisms (from above, as many as 10^{14} in vertebrates). Therefore, in eusocial societies, one does not expect an increasing problem from lineages of selfish workers arising through mutation as colony size rises. However, because of non-clonal kin structures within eusocial societies, there can still be selection to prevent colony members designated for development as workers veering off towards development as reproductives instead, which represents an analogue of germline segregation (Bourke and Ratnieks 1999). Again, further discussion of this point appears in the context of social group transformation in Sections 6.4 and 6.5.

A final point about life cycles with a bottleneck is that they tend to be associated with alternations of non-dispersal and dispersal (Fig. 3.2a), with the unitary propagule (spore, seed, egg, or alate social insect) acting as the dispersive stage. Hence they involve population-wide competition for nest-sites and other resources among the extra offspring that altruism generates, so avoiding the kin competition that can obstruct the evolution of altruism (Section 2.4; Box 2.2). This represents another way in which a bottlenecked life-cycle contributes to social group formation and maintenance.

Group propagule (bottleneck absent)

As already mentioned, life cycles without a bottleneck stage are less conducive to social group formation and maintenance than bottlenecked life-cycles because social groups start off with lower within-group relatedness, so increasing their internal level of potential conflict (e.g. Maynard Smith 1988; Queller 2000). In addition, they are at higher risk of having selfish lineages present within them from the outset. These may either introduce themselves during the formation of an aggregation, or, in the case of group propagules that form from a parental bud, already be present as the descendants of selfish lineages that have arisen in the parent. By contrast, within-group selfishness in a bottlenecked life-cycle must arise by recurrent mutation. At least, this follows if one assumes that, through lacking cooperative groupmates to exploit, a unitary propagule with a selfish genotype could not develop (Queller 2000; Grosberg and Strathmann 2007).

Evidence suggests that multicellular organisms forming by aggregation can indeed pick up disruptive selfish mutants during aggregation, examples being the cellular slime moulds and Myxobacteria (Crespi 2001; Grosberg and Strathmann 2007; Santorelli et al. 2008; Velicer and Vos 2009; Vos and Velicer 2009). It is almost certainly for these reasons that, in multicellular organisms, development by aggregation is relatively rare (Grosberg and Strathmann 1998, 2007), with Bonner (1998) estimating that it has evolved only four times independently. Furthermore, it has never led to complex multicellularity (Szathmáry and Wolpert 2003; Grosberg and Strathmann 2007). The most it has achieved in the way of complexity is something on a level with the cellular slime moulds (Fig. 3.1), which have two to four cell types (Bonner 2003b; Schaap et al. 2006). Why could not, say, a wolf arise in the way a cellular slime mould does, from an aggregation of numerous, unicellular 'wolf amoebae' scattered in the woodland soil? The question seems absurd, but only because it suggests something we know not to happen (it might add a fresh terror to our own team-building sessions in the woods if it did). The current arguments tell us, in large part, why it does not happen. The reason is that a multicellular organism requiring the degree of internal coordination and interdependence needed for the development of a large eukaryote could not survive the level of cheating among cell lineages that results from aggregative development.

In eusocial societies, group propagules are more frequent than in multicellular organisms, being represented by foundress associations of unrelated queens in several ant species or by groups of queens and workers departing from the parent nest in ants, bees, and wasps (Table 3.2). But, through lacking (usually) a clonal structure, eusocial societies already tolerate a higher level of potential within-group conflict than multicellular organisms. Nonetheless, in this context it is significant that swarm-founding wasps, whose colonies are founded by large groups of relatively undifferentiated females, undergo periodic episodes in their colony cycle that bring the number of reproducing queens down to one or a few (Strassmann et al. 1992; Queller et al. 1993; Henshaw et al. 2000; Kudô et al. 2005). This suggests that there has been a limit on the number of reproductive lineages represented in group propagules even in eusocial societies.

Lastly, unlike bottlenecked life-cycles, life cycles without a bottleneck are more liable to be associated with a degree of population viscosity (Fig. 3.2b). This could arise through group propagules formed by aggregation being composed of relatives, which therefore (to be available to aggregate) must remain relatively near their natal patch. A possible example in multicellular organisms is again the cellular slime mould *Dictyostelium discoideum*, in which it was recently discovered that aggregations are normally close kin (Gilbert et al. 2007). In eusocial societies, an example comes from foundress groups of the paper wasp *Polistes dominulus*, which tend to be relatives but are not necessarily from the same nest (Zanette and Field 2008). Alternatively, population viscosity could arise through group propagules formed by budding not travelling as far as unitary propagules from the parent, as in the case of buds of ant colonies that nest near their parent colony, so creating a pattern of fine-scale spatial genetic differentiation (Sundström et al. 2005). Population viscosity tends to retard the evolution of altruism (Section 2.4; Box 2.2), although from cases like the ones just mentioned it clearly does not totally preclude the evolution of social groups. This is almost certainly because population viscosity is rarely absolute, i.e. propagules in many life cycles exhibit some mixture of dispersal and non-dispersal (Fig. 3.2). But, to the extent that the absence of a bottleneck is associated with a high degree of population viscosity, it will again tend to be comparatively unfavourable to social group formation and maintenance.

3.4 Summary

1. Inclusive fitness theory predicts that several aspects of the major transitions depend on the relatedness of the social partners involved. Within species, transitions involving unrelated social partners (egalitarian transitions) will be cooperative (in the narrow sense) and so both partners will retain the ability to reproduce. An example is the evolution of sexual reproduction, in which unrelated genomes unite in a zygote, but genes from both parents replicate. By contrast, transitions

involving related social partners (fraternal transitions) can be altruistic and one partner (the altruistic partner) may lose the ability to reproduce. Examples are the evolution of multicellularity, in which somatic cells lose the ability to reproduce to aid related germline cells, and of eusociality, in which helper phenotypes lose the ability to reproduce to aid related reproductive phenotypes. Social interactions between members of different species always involve non-relatives and hence can only be cooperative (in the narrow sense), never altruistic. Examples are found in interspecific mutualisms. A basic principle underlying all social evolution is that social groups grow more stable the more their members achieve a coincidence of fitness interests. In groups of relatives this occurs via shared genes, and in groups of non-relatives via shared reproductive fate, i.e. a common route to future generations.

2. Conflicts within social groups can be resolved (rendered less costly) by two broad mechanisms. The first is self-limitation, including the effect of being related to other group members. Increasing relatedness leads to decreasing within-group selfishness, since a gene for selfishness damages itself if it harms relatives. The second is coercion, including dominance, punishment, and policing. Coercion evolves according to the rules of inclusive fitness theory, so represents a secondary factor in social evolution. In social groups of relatives, coercion combined with relatedness can bring about greater levels of altruism within groups than related-ness alone. In social groups of non-relatives, relatedness is absent and so cannot resolve conflicts but, as expected, there is good evidence that coercion is a major factor in stabilizing the egalitarian transitions.

3. A life cycle with a bottleneck (e.g. development of a multicellular organism from a single cell, or of a eusocial society from a pair of founders) gives rise to social groups with high within-group relatedness and hence reduced potential conflict. To the extent such life cycles involve alternations of non-dispersal and dispersal, they also prevent competition between kin from impeding altruism. At the outset of multicellular evolution, altruism based on clonality of the social group alone can account for differentiation into germline and somatic cells. By contrast, a life cycle without a bottleneck (e.g. development of a multicellular organism from a group of cells, or of a eusocial society from a group of founders) gives rise to social groups with reduced within-group relatedness and hence increased poten-tial conflict. To the extent such life cycles involve population viscosity, they lead to a degree of kin competition that could hinder the evolution of altruism.

4

Social group formation

4.1 Pathways of social group formation

Social group formation concerns the origin of social groups (Section 1.3). Specifically, it concerns the factors that lead to genes for social life spreading through populations of initially non-social organisms. As a first step in considering these factors, it is helpful to examine the pathways that organisms undergoing social group formation may have followed. Pathway in this context refers to the stages through which social groups passed as they first arose. In this section, and throughout the chapter, my treatment will be divided according to Queller's (1997, 2000) classification of the major transitions into those involving unrelated social partners (egalitarian transitions) and those involving related social partners (fraternal transitions). This is because, although inclusive-fitness considerations underpin both types of transition, the manner in which the inclusive-fitness interests of the social partners can be met during social group formation differs between them (Sections 2.2, 3.1).

Pathways of social group formation among non-relatives

Relatively little is known about the pathways followed in the origin of social groups of unrelated partners. A basic distinction within such groups concerns whether the relationship is, in Leigh's (1995) term, 'open' or not. By this is meant whether each side has to find a new partner from the environment in each generation or whether partners remain together across generations (Maynard Smith and Szathmáry 1995; Sachs et al. 2004; Queller and Strassmann 2009). For example, the mutualism represented by the eukaryotic cell is not open because eukaryotic cells do not acquire fresh mitochondria or chloroplasts from the environment each generation. Instead, their organellar partners are descended from those first acquired at the origin of the eukaryotic cell (Box 1.1). But the mutualism of male and female genomes in sexual reproduction in eukaryotes is open, because the sexes need to seek one another out for zygote formation each generation (although courting humans prefer not to put it this way). Similarly, many interspecific mutualisms are open, an example being the mutualism between the Bobtail squid *Euprymna scolopes* and the light-emitting bacterium *Vibrio fischeri*, in which each newborn squid must acquire its bacteria from the environment for itself (Sachs et al. 2004). An interesting contrast between open and 'closed' mutualisms is found in the associations between social insects and fungi. In fungus-growing termites the mutualism is usually open, with new nests of most species of termite acquiring symbiotic fungus from the environment (Aanen et al. 2009). In fungus-growing ants, by contrast, the mutualism is usually closed, with each dispersing ant queen taking a fungal inoculum from the parental nest (Mueller 2002).

The distinction between open and closed mutualisms overlaps closely with that between horizontal and vertical transmission of symbionts, be they mutualists or parasites, where horizontal transmission is potentially to any member of the population and vertical transmission is from parent to offspring. (The overlap is not total, because symbionts acquired from the environment, i.e. horizontally, need not be actively infectious.) For reasons to be discussed in the following section, vertical transmission facilitates a greater coincidence of fitness interests between symbiotic partners. In general, symbiotic relationships between species have originated in many taxa and at many times, and range from parasitic to mutualistic and from facultative to obligate (Sachs et al. 2004; Thompson 2005). Hence, as regards the origin of closed mutualisms, it is likely that the pathway followed has often involved an initially more open, parasitic relationship changing to a more closed, mutualistic one through a switch from horizontal to vertical transmission (Thompson 2005). The origin of mutualism from parasitism could also occur in open mutualisms, and indeed this has been experimentally demonstrated (Marchetti et al. 2010), but in this case it cannot be associated simply with a change in the mode of transmission. The pathway followed in the origin of the open mutualism represented by sexual reproduction remains particularly obscure, its obscurity being compounded by eukaryotic sex having probably originated just once, in the very distant past (Box 1.1).

Pathways of social group formation among relatives

The two leading examples of major transitions involving related social partners are the origin of multicellular organisms by the grouping of unicellular organisms and the origin of eusocial societies by the grouping of multicellular organisms. In these cases, two pathways are known: the subsocial and the semisocial. These terms were originally developed to classify insect sociality (e.g. Michener 1969), but can be usefully generalized. The subsocial pathway occurs when parents and offspring remain in association through the non-dispersal of offspring. For example, cells might remain near the parental cell after cell division, or multicellular organisms might remain near their parents after their birth. Subsociality, therefore, guarantees relatedness among social partners. The semisocial pathway occurs when same-generation organisms associate. In principle, a semisocial pathway could, therefore, lead to a grouping of either relatives (if same-generation relatives formed a group) or non-relatives (if same-generation non-relatives formed a group). However, as the following section shows, in practice, semisociality leading to either multicellularity or eusociality involves relatives. Let us now examine the evidence regarding the pathways followed in the origins of multicellularity and eusociality.

Evidence regarding the pathway by which multicellularity arose comes from two sources, namely comparative analyses and laboratory observations. Take first the comparative evidence. In general, the predominance of a single-cell bottleneck in the life cycle of multicellular organisms (Grosberg and Strathmann 2007) points to a subsocial pathway in the origin of multicellularity (Section 3.3), since adults in such a life cycle develop through daughter cells remaining together. This reasoning assumes that the current mode of development reflects the ancestral mode of grouping. Subsociality is apparent in specific examples as well. For instance, the freshwater green alga *Scenedesmus acutus* is facultatively multicellular, existing either as a population of unicells or in small multicellular colonies. The colonies arise by retention of daughter cells within the cell wall of the parental cell following cell division (Lürling and Van Donk 2000), i.e. subsocially. The same point can be made by a comparative analysis across species in another taxon of freshwater algae, the volvocines. Volvocine algae represent a particularly illuminating group with which to investigate the origins of multicellularity (Kirk 1998, 1999, 2003; Herron and Michod 2008). This is because some species are unicellular, whereas others consist of multicellular clusters of varying degrees of complexity (Fig. 4.1). Herron et al. (2009)

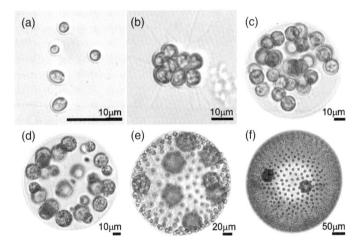

Fig. 4.1 Examples of unicellular and multicellular volvocine algae. (a) *Chlamydomonas reinhardtii*, a unicellular species; (b) *Gonium pectorale*, a sheet of 8–32 undifferentiated cells; (c) *Eudorina elegans*, a globular multicellular species with 16–64 undifferentiated cells held in a sphere by extracellular matrix; (d) *Pleodorina californica*, a multicellular globe with *c.* 32 cells divided into somatic and (larger) reproductive cells; (e) *Volvox carteri*, similar but with 2000–4000 cells; (f) *V. aureus*, similar with c. 2000 cells. *V. barberi* (not illustrated) is the largest *Volvox*, reaching a size of 10,000–50,000 cells. Information on cell number from Michod et al. (2003), Michod (2007), Herron and Michod (2008), and Solari et al. (2008). Image by Christian Solari reproduced from Michod (2007) with kind permission of C. Solari, R. Michod, and the publisher.

estimated that multicellularity in the volvocines arose once (Box 1.1). In extant species, all the multicellular forms develop by asexual cell division of a single haploid cell (Kirk 1998, 2004; Michod 2007), the cells keeping together through production of an extracellular matrix (Fig. 4.1); this again suggests that the origin of multicellularity was via the subsocial route.

Take now the evidence on pathways to multicellularity from laboratory studies. Remarkably, several studies have demonstrated the experimental evolution of multicellularity in the laboratory (Table 4.1), re-creating today events that in many other lineages occurred before the Cambrian. In each case, the evolved multicellular form arose through cells remaining in proximity following cell division (Table 4.1). This alone does not prove subsociality, and in several cases the actual within-group relatedness of the multicellular form, knowledge of which would help establish the pathway followed, was not reported (Table 4.1). However, it certainly suggests mechanisms by which multicellularity can arise by the subsocial pathway.

In one particularly striking example of multicellular evolution in the laboratory, which deserves to become better known, a subsocial pathway was definitely followed. This was in the study of Boraas et al. (1998) on the green alga, *Chlorella vulgaris*, a member of a normally unicellular genus of green algae familiar from their use in classic studies of the biochemistry of photosynthesis. The species is obligately asexual and each cell measures 5–6 µm in diameter. Boraas et al. (1998) added cultures of a predatory unicell, *Ochromonas vallescia*, which has larger cells (8–15 µm in diameter), to cultures of *C. vulgaris*. This induced the evolution of multicellular clusters of *C. vulgaris*. Initially, each cluster contained tens to hundreds of cells, but

(a)   (b)

Fig. 4.2 Evolved multicellularity in laboratory cultures of the green alga *Chlorella vulgaris* subjected to predation by a predatory unicell. (a) A large multicellular colony, as formed in the initial stages of multicellular evolution in this system. (b) The stable, eight-celled form of multicellularity achieved after many generations of selection; the membrane (parental cell wall) enclosing the cells in visible. Reproduced from Boraas et al. (1998) with kind permission of Springer Science and Business Media.

after many generations a stable, eight-cell form of *C. vulgaris* evolved (Fig. 4.2). This conformation was heritable, even in the absence of the predator. Furthermore, its evolution was replicable in multiple trials of the experiment. The mechanism by which cells formed multicellular clusters was, as in the earlier example of *Scenedesmus acutus*, via daughter cells failing to break out of the cell wall of the parental cell following cell division (Fig. 4.2). In short, the pathway to multicellularity was subsocial. The reason why multicellularity evolved was that cells in the multicellular form of *C. vulgaris* were protected against being eaten by the predatory unicell. In addition, Boraas et al. (1998) hypothesized that the final size of the multicellular form was an evolutionary compromise between defence against predation and the need for single cells to maintain sufficient access to nutrients in the medium. So this study not only beautifully illustrates a particular pathway to multicellularity, but also reveals the operation of non-genetic factors that influence social group formation. We return to such factors later in the chapter (Sections 4.3, 4.4).

Not all multicellular organisms develop subsocially. Development by the aggregation of cells, as found in the cellular slime moulds and Myxobacteria (Bonner 2003a; Velicer and Vos 2009), is essentially semisocial development, but, as discussed in Section 3.3, development by aggregation is relatively rare (Grosberg and Strathmann 2007). Bonner (1998) pointed out that, in lineages in which multicellularity originated in aquatic environments, it probably did so via retention of daughter cells close to the parent cell, i.e. subsocially. But in lineages such as the cellular slime moulds and Myxobacteria in which multicellularity originated in terrestrial environments, it did so by aggregation, i.e. semisocially. This difference presumably arises because aggregation is physically more difficult in an aqueous medium, since both cells and the chemical attractants that are the proximate cause of aggregation (Bonner 2009) would be much more easily dispersed than on land. Of course, many lineages (e.g. animals and plants) have become terrestrial secondarily, so these are likely first to have become multicellular subsocially in the aquatic environment. The overall conclusion, given the rarity of aggregative multicellularity, is that the most frequent pathway taken in the origin of multicellularity was subsocial.

The subsocial retention of daughter cells close to parent cells can mechanistically occur either by cells remaining within the parental cell wall or by cells sticking together via surface adhesion molecules or a binding extracellular matrix (Table 4.1). Cell-adhesion molecules and the production of extracellular matrix are also important in the formation of cell aggregations by the semisocial pathway (Bonner 2003a; Queller et al. 2003; Velicer and Vos 2009). It follows that a key step in the evolution of multicellularity was the invention of such prosaic items as bags or glue. It is not difficult to imagine that this could have come about through one or a few mutations (Bonner 1998), such as the mutation delaying the breaking open of the parental cell wall following cell division that must have arisen in the experimental evolution of multicellularity in *Chlorella vulgaris* (Boraas et al. 1998). A major class of cell-adhesion molecules in animals (cadherins) occurs in unicellular species of the choanoflagellates, the sister group of animals. This suggests that co-option of existing

Table 4.1 Examples of evolution of multicellularity from unicells in the laboratory

Organism	Multicellular phenotype	Mechanism of non-dispersal of cells	Factor(s) selecting for multicellularity	Genetic basis (if known)	References
Myxobacterium, *Myxococcus xanthus*	Branching swarm of cells on agar medium	Extracellular fibril matrix binding cells together	Possibly foraging advantages	Unknown, but likely to have involved multiple mutations in an asocial laboratory strain	Velicer and Yu (2003); Velicer and Vos (2009)
Bacterium, *Pseudomonas fluorescens*	Mat of cells floating at surface of liquid medium	Adhesion of daughter cells owing to overproduction by 'wrinkly spreader' mutant of adhesive polymer	Access to gaseous oxygen	Recurring mutations in *wspf* locus leading to emergence of 'wrinkly spreader' strain from ancestral 'smooth' strain	Rainey and Rainey (2003); Bantinaki et al. (2007)
Green alga, *Chlorella vulgaris*	Clusters of eight to hundreds of cells in liquid medium	Retention of daughter cells within membrane following division of parent cell	Addition of unicellular predator (*Ochromonas vallescia*) selects for multicellular clusters that avoid predation	Unknown, but involved heritable mutation	Boraas et al. (1998)
Budding yeast, *Saccharomyces cerevisiae*	Clusters of thousands of cells in liquid medium	Cell adhesion via cell-surface protein coded for by *FLO1*	Protection against toxins in the medium	*FLO1* green-beard locus	Smukalla et al. (2008); Table 2.4
Fission yeast, *Schizosaccharomyces pombe*	Branching multicellular, septate mycelia penetrating agar medium	Cell adhesion leading to failure of dividing cells to separate	On some media, possibly access to nitrogen	At least 12 genes deletion of which disrupts formation of mycelia	Amoah-Buahin et al. (2005); Dodgson et al. (2009)

molecules for a novel role in intraspecific cell adhesion in multicellular development has also occurred (King 2004; Abedin and King 2008). Once a mechanism for multicellularity has arisen by mutation, selection can then act to promote or suppress the new multicellular form. Which of these alternatives is followed depends, of course, on whether or not Hamilton's rule is fulfilled.

Turning to eusociality, one finds from the comparative evidence that the pathway to eusociality, in both insects and other taxa, has been overwhelmingly subsocial (Table 4.2). This conclusion has also been reached by several previous authors (Alexander *et al.* 1991; Thorne 1997; Jeon and Choe 2003; Boomsma 2007, 2009; Helanterä and Bargum 2007; Hughes et al. 2008). Possible exceptions occur in the allodapine bees and in social wasps of the genus *Microstigmus* (Table 4.2). Members of these two lineages have relatively simple societies. In allodapine bees, both subsociality and semisociality occur, but the changeability of the social structure of colonies over time makes inferring their ancestral social structures particularly difficult (Schwarz et al. 2007). Among *Microstigmus* wasps, Ross and Matthews (1989) demonstrated subsociality coupled with monandry and monogyny in the species *M. comes*. This was the only species whose colony genetic structure was known until recently, when Lucas (2009) described a looser colony genetic structure in *M. nigrophthalmus* (Table 4.2). With data on only two species, reconstructing the ancestral state is not possible.

In sum, the evidence shows that, across the independent instances of the origin of both multicellularity and eusociality, the subsocial pathway has usually been followed. Known and possible exceptions exist, but they are infrequent.

4.2 Genetic factors in social group formation

The fundamental evolutionary condition for social group formation is that Hamilton's rule must be fulfilled (Sections 2.1, 2.2). Put another way, the party or parties whose behaviour leads to social group formation must receive a net gain in inclusive fitness. Hamilton's rule contains both genetic terms (relatedness) and non-genetic terms, the latter being terms for the changes in the expected offspring number of the actor and recipient (i.e. m and n from Box 2.1). It follows that Hamilton's rule is a theorem about how both genetic and non-genetic factors interact to determine the course of social evolution (Box 2.3). This section focuses on genetic factors underpinning social group formation, while the next two sections focus on non-genetic ones. As far as genetic factors are concerned, the key one is, of course, relatedness. The lack or otherwise of relatedness determines whether social group formation proceeds through social partners approaching a coincidence of fitness interests via, respectively, either shared reproductive fate or shared genes (Section 3.1).

Genetic factors in social group formation among non-relatives

Social group formation occurs in non-relatives (including partners in interspecific mutualisms) when Hamilton's rule for narrow-sense cooperation is fulfilled (Sections

Table 4.2 Social and genetic structures inferred to have been present at the origin of eusociality in eusocial lineages. The eusocial lineages are taken from Table 1.4. Inference of ancestral state is from ancestral state reconstruction (Hughes et al. 2008) and in other lineages is made by analogy with the known biology of present-day taxa. Relatedness is that between helpers and aided nestmates, unless otherwise stated. In haplodiploid Hymenoptera, relatedness is the average of females' relatedness to sisters and brothers unless otherwise stated. See also Alexander et al. (1991), Thorne (1997), Helanterä and Bargum (2007), Hughes et al. (2008), and Boomsma (2007, 2009)

Taxon	Inferred ancestral state			Inferred relatedness level	References
	Subsociality	Monandry	Functional monogyny		
Blattodea Termites	Yes	Yes	Yes	0.5	Nalepa and Jones (1991); Thorne (1997)
Coleoptera Ambrosia beetle (*Austroplatypus incompertus*)	Yes	Probable	Yes	0.25–0.5[1]	Kent and Simpson (1992); Smith et al. (2009)
Hemiptera Aphids (Pemphigidae, Hormaphididae)	Yes	n/a (foundress female produces offspring asexually)	Yes	0.71–0.90[2]	Johnson et al. (2002); Abbot (2009)
Hymenoptera Ants (Formicidae)	Yes	Yes	Yes	0.5	Bourke and Franks (1995); Hughes et al. (2008)
Bees (Allodapines)	Subsociality and semisociality both present	Unclear	Unclear	0.38–0.60[3]	Langer et al. (2004); Schwarz et al. (2007)
Bees (Corbiculate social bees: Apini, Bombini, Meliponini)	Yes	Yes	Yes	0.5	Hughes et al. (2008)

Taxon	Inferred ancestral state			Inferred relatedness level	References
	Subsociality	Monandry	Functional monogyny		
Bees (Halictidae)	Yes	Yes (or weak polyandry)	Yes (in two of three lineages)	c. 0.5	Hughes et al. (2008); Boomsma (2009)
Wasps (Social Vespidae: Polistinae, Stenogastrinae, Vespinae)	Yes (approximately: alpha foundress reproductively dominates same-generation cofoundresses, with workers being produced later)	Yes	Yes (approximately)	c. 0.5	Hughes et al. (2008)
Wasps (Pemphredoninae: *Microstigmus* spp)	Yes in *M. comes*, not always in *M. nigrophthalmus*	Yes in *M. comes*, weak polyandry in *M. nigrophthalmus*	Yes in *M. comes*, several females breed but high skew in *M. nigrophthalmus*	0.45–0.67[4]	Ross and Matthews (1989); Lucas (2009)
Wasps (Polyembryonic parasitoid wasps, Encyrtidae)	Yes	Yes	Yes	0.25–1.0[5]	Grbić et al. (1992); Gardner et al. (2007b); Giron et al. (2007)
Thysanoptera					
Thrips (Phlaeothripinae)	Yes	Yes	No[6]	0.60–1.0[7]	Chapman (2003); Chapman et al. (2008)
Other groups					
Shrimps (*Synalpheus* spp)	Yes	Yes	Yes (approximately)	c. 0.5	Duffy (2007); Duffy and Macdonald (2010)

Table 4.2 *Cont.*

Taxon	Inferred ancestral state			Inferred relatedness level	References
	Subsociality	Monandry	Functional monogyny		
Mole-rats (Bathyergidae)	Yes	Yes (or weak polyandry)	Yes	0.46–0.81[8]	Reeve et al. (1990); Burland et al. (2002); O'Riain and Faulkes (2008)

[1] Minimum relatedness in the diploid *Austroplatypus incompertus* would be 0.25 under a highly polyandrous foundress and 0.5 under a monandrous one.

[2] Calculated from levels of clonal mixing (10–29% non-natal aphids per gall) in three species of social aphid (*Pemphigus bursarius, P. obesinymphae,* and *P. spyrothecae*).

[3] Range of mean relatedness among co-founding females in two species of allodapine bees, *Exoneura nigrescens* (0.38) and *E. robusta* (0.60).

[4] Based on estimates of between-female relatedness of 0.62–0.67 in *Microstigmus comes* and 0.45–0.49 in *M. nigrophthalmus*.

[5] In polyembryonic wasps, embryos subdivide within hosts to produce clonal groups of larvae some of which develop into a sterile fighter form; a female wasp either lays a single female or male egg into a host, with the embryo within the egg then subdividing to produce a single-sex clonal sibship (relatedness = 1.0), or a female and male egg into the same host, with both embryos subdividing clonally, in which case relatedness ranges from 0.25 (focal female's relatedness to brothers) to 1.0 (focal larva's relatedness to same-sex larvae). It is rare for more than one female to lay eggs in a single host.

[6] Data suggest proto-soldiers in thrips showed 'considerable' reproduction, but the model in Chapman (2003) found that this would have facilitated soldier evolution in thrips.

[7] Range of mean relatedness among female soldiers in six species of social thrips; relatedness may be elevated by inbreeding.

[8] Based on relatedness (0.46) in the Damaraland mole-rat, *Fukomys damarensis*, formerly *Cryptomys damarensis* (Kock et al. 2006), and relatedness (0.81) in the Naked mole-rat, *Heterocephalus glaber*; in the latter, relatedness may by elevated by inbreeding.

2.2, 3.1). But because unrelated social partners cannot (by definition) achieve a coincidence of fitness interests via shared genes, the genetic basis of their sociality must involve at least a degree of shared reproductive fate. So we now need to consider how shared reproductive fate arises when there is narrow-sense cooperation between non-relatives. A well-known way in which this can arise is when two social partners each transmit their genes vertically from parent to offspring in the same propagule (Dawkins 1990). As predicted by the theory of virulence developed for contagious diseases (e.g. Dawkins 1990; Read and Harvey 1993), vertical transmission tends to lead—through shared reproductive fate—to symbionts being less virulent or not virulent at all, whereas horizontal transmission tends to lead—through the absence of shared reproductive fate—to symbionts being more virulent.

As an example of the effect of vertical transmission in the present context, consider—in an idealized form—the origin of the eukaryotic cell (Box 1.1). Let us suppose that, at the origin of this mutualism, there was no sexual reproduction. Then a proto-eukaryotic unicell that engulfed a bacterial cell would, on cell division, have produced daughter cells containing copies of its own, nuclear genome and copies of the genome of the bacterial partner, given asexual reproduction of the bacterial partner within the host cell's cytoplasm. This means that the inclusive fitnesses of host and bacterial genes would each have been maximized simply by maximizing the number of descendant host cells. In turn, this would have led, via shared reproductive fate, to a perfect coincidence of fitness interests. Sexual reproduction would complicate this situation, because a zygote would contain nuclear genes and organellar genes from two parents, leading to potential conflict. The following chapter will explore how this situation has selected for the vertical transmission of organelles to occur via one sex only (Section 5.5). The main point for now is that a social group of unrelated social partners may form that possesses a high degree of shared reproductive fate. The initial association between the ancestors of the eukaryotic cell and its organelles would almost certainly have exhibited elements of conflict (Blackstone 1995). However, the present-day form of the nuclear–organellar mutualism provides a good example of the stabilizing effects of shared reproductive fate (e.g. Sachs et al. 2004). Much research on narrow-sense cooperation within- and between-species is aimed at uncovering the mechanisms, short of being physically bound together as in the case of nuclear and organellar genomes, that allow reproductive fate to be shared. A prominent means by which this could occur is to remain and interact with the same social partner over extended periods, i.e. partner fidelity (Trivers 1971; Axelrod and Hamilton 1981; Bull and Rice 1991; Bergstrom et al. 2003; Sachs et al. 2004).

Shared reproductive fate is also very likely to have had a role in the transition to the first (prokaryotic) cell, the transition to sexual reproduction, and the transition to multicellularity. Nuclear genes at separate loci within a cell and, in sexual organisms under outbreeding, at different allelic positions within a locus, are unrelated to one another. Why then should genes within a cell cooperate? The answer is: if they have a shared reproductive fate. In the case of the first cells, a daughter cell created by asexual cell division (mitosis in eukaryotes) represents the common propagule for vertical gene transmission for all the genes in the nuclear genome.

Asexual cell division therefore brings about a shared reproductive fate. It is less clear how the orderly process of asexual cell division arose (Maynard Smith and Szathmáry 1995). Nonetheless, an experiment in which two independent bacteriophage viruses evolved to share a common protein coat, and hence to reproduce together, hints at how unrelated genes might have evolved to co-replicate within the same cell (Sachs and Bull 2005). Sexual reproduction with a fair meiosis can also create a shared reproductive fate (Dawkins 1982, 1990). In this case, the common propagule is a gamete, in which genes at all loci have the same chance of being represented (Section 2.3). But, once more, how meiosis originated is unclear, especially as it is an even more orderly and complex process than mitosis (Maynard Smith and Szathmáry 1995; Ruvinsky 1997; Leigh 1999; Ridley 2000; Burt and Trivers 2006). Processes that keep meiosis fair after it has arisen are better understood (Section 5.5). The message for now is that social stability is approached in these contexts by the emergence of shared reproductive fate, but the mechanisms by which this came about remain mysterious.

Genetic factors in social group formation among relatives

As stated in the previous section, the leading examples of social groups composed of relatives are multicellular organisms and eusocial societies (a category I will refer to just as eusocial societies but in which I include, as appropriate, societies of cooperative breeders; cf. Box 1.1). Social group formation is predicted to occur in relatives when Hamilton's rule for either narrow-sense cooperation or altruism is fulfilled (Section 2.2). In the origin of multicellularity, altruism in any one cell takes the form of either the costly production of public goods (e.g. polymers for binding cells together) or the adoption of a somatic function associated with loss of the ability to divide, i.e. terminal differentiation (e.g. Herron and Michod 2008). Terminal differentiation can be detected through morphological differences between reproductive and somatic cells, as in the larger forms of multicellular volvocine algae (Fig. 4.1d–f). But in some cases it is possible that all cells within early multicellular clusters had the same expectation of dividing and hence that multicellularity started as a cooperative enterprise. Similarly, some eusocial societies may have formed initially for cooperative reasons, or to provide direct fitness returns to one of the parties involved (Lin and Michener 1972; West-Eberhard 1978). For example, it has been suggested that in termites the earliest subsocial groups formed because they provided opportunities for offspring to inherit the nest (Korb 2009).

In several lineages, however, a degree of altruism among group members is apparent in the relatively early stages of the evolution of both multicellularity and eusociality. The costs to individual cells of producing public goods have been measured in facultatively social bacteria and have been found to be real (Rainey and Rainey 2003; Griffin et al. 2004). In the multicellular volvocine alga *Eudorina*, which has 16–64 undifferentiated cells (Fig. 4.1c), some cells divide more rarely than others (Kirk 1998). This suggests that, in simple multicellular organisms with relatively few, morphologically alike cells, some cells act as an analogue of the behavioural helper caste of eusocial societies, implying an early appearance of altruism. Even in termites, true workers (i.e. sterile workers) arise relatively early within the termite phylogeny (Inward et al. 2007b). In addition, if eusociality arose through 'parental' care being directed at younger siblings by late-dispersing offspring, as has been suggested in the Hymenoptera (Box 2.3), altruism would have been present from the very beginning of social group formation.

Therefore, in considering social group formation among relatives, it is legitimate to focus on altruism as the principal trait to explain. Because altruism requires positive relatedness, the main genetic factor involved in this context is then, of course, relatedness itself. So the question to consider becomes whether it is the case that relatedness is positive in social groups characterized by altruism.

The answer is that, on average, relatedness is indeed positive in these forms of social group. The significance of the frequent involvement of subsociality in the origin of multicellularity and eusociality demonstrated in the previous section is that associations of parents and offspring are automatically associations of relatives. When subsociality is coupled with a single-cell propagule (in multicellularity) or a single, monogamous colony-founding pair (in eusociality), relatedness will be not only positive, but comparatively high. The evidence is that these conditions obtain in nature. Hence, as detailed in the previous chapter (Section 3.3), multicellular organisms arising subsocially from a single-cell propagule are clones. This is the case both when the single cell propagule is a diploid zygote or, as in many cases of simple multicellularity (e.g. Table 4.1), an asexually-produced haploid cell. As stressed by Hughes et al. (2008) and others, eusocial societies usually originate under conditions not just of subsociality, but of monandry and functional monogyny (there is only one female founder or, if more than one, one monopolizes reproduction). The result is that relatedness within the first societies within eusocial lineages can be inferred to have been comparatively high (Table 4.2).

Some members of extant, simple eusocial societies may be non-relatives, but in these, average relatedness still remains positive. An example is the paper wasp *Polistes dominulus*, in which 35% of joint foundresses are non-relatives but average between-foundress relatedness is 0.21–0.43 (Queller et al. 2000). Average relatedness is reduced in some other kinds of extant eusocial societies. In the eusocial Hymenoptera, an example is found in the honey bees (*Apis* species), in which colony queens exhibit high levels of polyandry (multiple mating by females). The record is held by queens in the Giant honey bee *Apis dorsata*, which may each mate with up to 100 males (Wattanachaiyingcharoen et al. 2003). But because every colony is headed by a single queen, within-colony relatedness still cannot fall below 0.25 (Crozier and Pamilo 1996). Less extreme levels of polyandry, but lower relatedness, are found in the unicolonial ants, whose nests contain numerous, weakly-related queens (Section 7.3). However, comparative evidence suggests that high levels of polyandry, along with polygyny (multiple-queening) and unicoloniality, are all derived traits (Bourke and Franks 1995; Hughes et al. 2008).

Therefore, the conclusion remains that within-group relatedness was above zero and indeed generally high at the start of eusocial evolution in insects and other taxa regarded as eusocial. This conclusion is reinforced if one classifies the colonial marine invertebrates as eusocial (Box 1.1), since, as Hamilton (1964) pointed out, the polyps or zooids within these organisms are clones of one another (Dunn 2009). Similarly, but less markedly, positive relatedness is the general rule in cooperatively-breeding birds and mammals (Emlen 1995; Solomon and French 1997; Koenig and Dickinson 2004; Clutton-Brock 2009b; Hatchwell 2009). For example, Hatchwell (2009) estimated that up to 91% of cooperatively-breeding bird species occur in groups of kin.

Tellingly, even when the semisocial pathway has been followed in multicellular or eusocial evolution, within-group relatedness is still positive. For instance, in multicellular organisms forming through aggregation, recent discoveries have detected small-scale genetic structuring in the field environment such that neighbouring cells are generally related. Such genetic structuring occurs in cellular slime moulds below a scale of metres (Gilbert et al. 2009) and in Myxobacteria below a scale of centimetres (Vos and Velicer 2009). In addition, there is evidence that, in both taxa, aggregations of cells actively exclude, via kin discrimination, genetically dissimilar cells from their aggregations (Mehdiabadi et al. 2006; Ostrowski et al. 2008; Vos and Velicer 2009). In natural populations of the cellular slime mould *Dictyostelium. discoideum*, the result is that fruiting bodies consisting of mixtures of clones (chimeras) are rare, while fruiting bodies consisting of single clones are common. In turn, this means that average relatedness within natural fruiting bodies is high, at $r = 0.97$ (Gilbert et al. 2007). *D. discoideum* was long regarded as demonstrating that multicellularity via chimerism is not much harder to achieve than it is within clones. With hindsight, one might suspect that the readiness of *D. discoideum* to form chimeras in laboratory cultures led to an underestimation of the importance of genetic relatedness in multicellular evolution.

Within eusocial lineages where a semisocial route may have been followed, such as the allodapine bees or *Microstigmus* wasps, relatedness inside the colony is also above zero and may be fairly high (Table 4.2). In fact there are no firm cases in insects of semisociality among non-relatives preceding the origin of eusociality (Bourke and Franks 1995; Boomsma 2007, 2009). More specifically, communal nesting in bees and wasps (i.e. in which unrelated females share a nest but care for their own offspring in separate cells) is almost always in clades that lack eusocial species (Wcislo and Tierney 2009). Overall, therefore, under both subsociality and semisociality, social group formation leading to altruism is characterized by positive relatedness. This is as predicted by inclusive fitness theory.

Recall, however, that positive relatedness needs to be accompanied by some degree of dispersal leading to population-wide competition between the products of altruism in order for the origin of altruism to be maximally facilitated. Otherwise the result is population viscosity, which hinders the evolution of altruism (Sections 2.4, 3.3). How might population-wide competition have been achieved in the origin of multicellularity and eusociality?

Consider first the case of multicellularity, where the problem caused by population viscosity might be particularly acute because, other things being equal, single cells or groups of a few cells—through being small and being unable to move far through their own efforts—are poor dispersers. I suggest that the nature of this problem depends on whether multicellularity originated in an aquatic or terrestrial environment. An aquatic environment appears to have allowed only a subsocial pathway to multicellularity (Section 4.1), which is conducive to altruism because it automatically creates positive relatedness. Another feature of the aquatic environment is that the fluid nature of the medium might have led to multicellular groups, once formed, mixing well at the population level and hence experiencing competition with unrelated groups. If so, the retarding effects on altruism of population viscosity would have been avoided. Hence, the aquatic environment is likely to have been doubly conducive to the origin of multicellularity. A terrestrial environment is associated with a semisocial pathway to multicellularity (Section 4.1). Interactions with relatives can nonetheless occur, as evidenced by the small-scale genetic structuring found in cellular slime moulds and Myxobacteria (Gilbert et al. 2009; Vos and Velicer 2009). But, if altruism is to have originated unhindered in the terrestrial environment, it appears some factor other than the mixing effect stemming from the nature of the environment itself needed to be present. In this context, it is significant that multicellularity arising in the terrestrial environment is always associated with the development of fruiting bodies, i.e. multicellular structures specifically adapted for the dispersal of spores, as seen in the cellular slime moulds, the Myxobacteria, and two additional taxa with similar habits (Bonner 1998). This implies that, in these cases, multicellularity persisted only when specific provision occurred for long-distance dispersal, which would have generated at least a degree of population-wide competition between the products of altruism.

Consider now the case of eusociality. In general, the potential problem caused by population viscosity is less acute than in the origin of multicellularity because of the superior dispersive powers of multicellular organisms compared to unicells. In addition, in the majority of cases, a subsocial pathway neutralizes the problem, because it is associated with non-dispersing, subsocial helper phenotypes and dispersing, reproductive phenotypes (Section 3.3). A semisocial pathway, to the extent that it involves related adults re-aggregating after dispersing, is more associated with population viscosity (Bourke and Franks 1995). However, in practice, semisociality (with positive relatedness) still seems capable of leading to eusociality, with allodapine bees and *Microstigmus* wasps being candidate examples, probably because in many systems non-dispersal coexists with a degree of dispersal (Section 3.3).

There is a final way in which genetic factors interact with the pathway by which social groups form. A clone of cells or multicellular organisms is a clone regardless of whether it is formed subsocially or semisocially. In both cases, in genetic terms, each member values its clone-mates as much as it values itself, leading to a complete coincidence of fitness interests. In non-clonal organisms, subsociality—but not semisociality—has the interesting property of producing a similar genetic effect under some conditions. Take the case, which is a likely forerunner of eusociality in many lineages (Table 4.2), of a single, sexually-reproducing diploid or haplodiploid mother associating with her offspring. The offspring have the choice of rearing their siblings or producing their own offspring. As has been discussed by several authors (Charnov 1978; Dawkins 1989; Bourke and Franks 1995; Boomsma 2007, 2009), this makes the offspring genetically indifferent in choosing either to become sibling-altruists or to reproduce themselves, since average relatedness to siblings is 0.5 and relatedness to offspring is also 0.5 (Table 2.2). In other words, altruism can evolve simply if, in the notation of Box 2.1, $n > m$, which is the same condition as under clonality (Table 2.3). Hence, in terms of the effects of relatedness alone, it is easier for altruism to originate under subsociality than under semisociality.

This and the other features associated with subsociality (automatic positive relatedness, association with widespread dispersal) help explain why altruism in social groups of both cells and multicellular organisms has evolved more often under subsociality than not. Boomsma (2007, 2009) extended the argument about genetic indifference to propose that strict lifetime monogamy (i.e. a founding female pairs with a single male for life) under subsociality was an essential precursor of complex eusociality. This idea is further considered in the chapter on social group transformation (Section 6.5).

4.3 Ecological factors in social group formation

This and the following section consider the non-genetic factors that affect social evolution through their influence on the terms in Hamilton's rule other than relatedness. These terms (m and n from Box 2.1) quantify the changes in expected offspring numbers of the social partners. In the case of social groups based on narrow-sense cooperation, non-genetic factors promote social group formation if they increase the offspring output of the actor (m). In the case of social groups based on altruism, they promote social group formation if they either increase the offspring output of the recipient (n) or decrease the cost in lost offspring of the actor (m), for example by decreasing the relative benefit of not joining a social group (Table 2.3). I divide the relevant non-genetic factors into two types: ecological and synergistic. Ecological factors are features of the external environment. Synergistic factors are features inherent in group organization. Ecological and synergistic factors may have overlapping effects, as when the presence of predators in the external environment is met with the evolution of collective defences against predation. But, for simplicity, I treat the two types of factor separately, considering ecological factors in the present section and synergistic factors in the following one.

Ecological factors in social group formation among non-relatives

As will become apparent later in this section, much attention has been paid to establishing the ecological factors leading to social group formation among relatives. By contrast, there have been fewer systematic efforts to establish the ecological factors leading to social group formation among non-relatives. The origin of the eukaryotic cell may have been precipitated by a rise in atmospheric oxygen (Section 1.5), but demonstrating a causal relationship between such a rise, the appearance of aerobic bacteria, and the origin of eukaryotes at such a temporal remove is clearly difficult.

The ecological basis, if any, of the origin of sexual reproduction in eukaryotes is likewise obscure. Several ecological theories of sexual reproduction have been proposed (e.g. Bell 1982). However, in practice, they apply mainly to the maintenance of sexual reproduction, not its origin (West et al. 1999; Agrawal 2006). A logical, empirical approach to investigating the ecological basis of the origin of sexual reproduction would be to determine the ecological correlates of sexual and asexual reproduction in unicellular species that exhibit facultative sexuality, since sex originated in unicells and was presumably facultative when it arose (Dacks and Roger 1999). Such an approach has rarely been taken. A likely reason is the same reason as that suggested earlier for why so little is known about the pathway by which sexual reproduction originated (Section 4.1), which is that sexual reproduction originated once, in the very distant past. This means that the ecological conditions relevant to the origin of sex may not resemble any occurring at present and may in practice be almost unknowable. A second, related reason is that the exact mode of reproduction of many taxa of unicellular eukaryotes is not known and, in many unicells, and perhaps all multicellular organisms, asexuality appears to be secondary (Bell 1982; Dacks and Roger 1999; Schurko et al. 2009). Even taxa famous for being 'ancient' asexuals, such as the bdelloid rotifers, are, as far as is known, secondarily asexual. These taxa seem to have become asexual when they split off from sexual ancestors tens to maybe a few hundreds of millions of years ago (Schurko et al. 2009). Hence they are not anciently asexual relative to the origin of sexual reproduction. This means that investigations of the ecological correlates of sexual and asexual reproduction have not, of necessity, involved comparisons of sexual reproduction with an ancestral form of asexuality, i.e. that found in unicells before the origin of sexual reproduction. Furthermore, even allowing for this, ecological studies of sexual reproduction have tended to concentrate on multicellular, not unicellular, species. In sum, the ecological basis of the origin of sexual reproduction remains little studied and, as a result, is very poorly understood.

What about the ecological basis of interspecific mutualisms? Are there ecological factors that consistently facilitate the origin of such mutualisms? Although ecological correlates of particular types of mutualism have been noted (Boucher et al. 1982), a consensus answer to this question does not seem to exist. This may be partly because the immense diversity of mutualistic interactions (Boucher et al. 1982; Bergstrom et al. 2003; Sachs et al. 2004) means that ecological preconditions are also likely to be very diverse. Nonetheless, several broad generalizations can be made.

First, qualitative if not quantitative data suggest that tropical environments harbour a higher proportion of species engaged in mutualisms than non-tropical ones (Farnworth and Golley 1974). Such a pattern would arise if, in tropical environments, mutualisms either formed at a higher rate, or were maintained at a higher rate, or both. Tropical environments are more species-rich than temperate ones (e.g. Gaston 2007), which could contribute to a greater rate of formation of mutualisms in them. The reason is that greater species richness implies species following a greater diversity of 'trades', which, as Leigh (1999) argued, increases the opportunities for the pooling of resources or capabilities inherent in mutualism. Tropical environments could also favour the maintenance of mutualisms for reasons explored by May (1976). He showed that perturbations to the dynamics of a pair of obligate mutualists would tend to result in extinction of both populations. This is because, below a certain threshold density of one population, the other, being dependent upon it, cannot increase. Addicott (1981) pointed out that this might only be true of a subset of mutualistic relationships between species, depending on the detailed population-dynamic assumptions. Nonetheless, the effect May (1976) highlighted, even if not universal, could help explain the existence of a greater frequency of mutualism in the tropics via the greater stability and persistence of tropical environments, as May (1976) himself suggested.

Second, interspecific mutualisms are promoted by cooperator association, which is defined as the co-occurrence of cooperative genotypes of one species and cooperative genotypes of the partner species (Foster and Wenseleers 2006). This predicts that geographic features of landscapes that lead to spatial structure across the ranges of the partner species, and specifically that promote relatively isolated, non-dispersing sympatric populations of these species, should promote mutualism. This idea comes from the models of interspecific mutualism of Frank (1994b) and Doebeli and Knowlton (1998). More broadly, it is suggested by Thompson's geographic mosaic theory of coevolution (e.g. Thompson 2005), which proposes that populations of coevolving species exist in a geographic mosaic of differing interactions characterized by varying levels of coadaptation. Consistent with these ideas, the same pair of species may exhibit mutualistic and non-mutualistic interactions at different spatial locations in their common geographic range (Bronstein 1994; Thompson and Cunningham 2002; Szilágyi et al. 2009). However, because within-species genetic variation in the propensity to act as a mutualist has been little studied, let alone covariation in this trait across potential interspecific partners, further support for the occurrence of cooperator association is required (Foster and Wenseleers 2006).

Third, an ecology and life-history associated with a sessile habit appears to predispose species to interspecific mutualism, as evidenced by the high frequency of mutualisms entered into by sessile marine invertebrates, such as bryozoans, corals, and hydroids (e.g. Osman and Haugsness 1981), and by plants (Janzen 1985; Leigh 1999). The underlying reasons involve some of the factors invoked in the previous two points. One is that an organism rooted to one spot is likely to benefit, for purposes of dispersal, from an association with any organism whose 'trade' entails mobility. This is suggested by the frequent evolution of animal dispersal of plant gametes (pollen) and seeds (Janzen 1985; Leigh 1999). Since sessile organisms cannot actively forage for nutrients, they would also benefit from mutualists that help them acquire nutrients from their immediate environment, such as the photosynthetic algal partners of corals or the nitrogen-fixing rhizobial bacteria of plants (Sachs et al. 2004). Finally, being sessile potentially renders a mutualistic partner more likely to exhibit cooperator association, and at a finer scale makes the partner fidelity conducive to mutualism easier to achieve (Nowak and May 1992; Sigmund 1992).

Ecological factors in social group formation among relatives

Many authors have discussed the non-genetic factors that might have favoured social group formation in the context of both the origin of multicellularity (Bonner 1974, 1998; Kaiser 2001; Lachmann et al. 2003; Szathmáry and Wolpert 2003; King 2004; Grosberg and Strathmann 2007) and the origin of eusocial societies (Alexander 1974; Alexander et al. 1991; Koenig et al. 1992; Brockmann 1997; Crespi and Choe 1997; Queller and Strassmann 1998; Bennett and Owens 2002; Crespi 2007; Korb and Heinze 2008b; Hatchwell 2009). It is worth pointing out, following Bonner (1998, 2006), that such factors need to have operated immediately social groups arose in order to have been effective. For instance, any advantages of very large body size cannot have influenced the origin of multicellularity, because the earliest multicellular groups almost certainly had low to moderate cell numbers. Therefore, to echo the point made above about investigating the ecological basis of the origin of sexual reproduction, the best case studies demonstrating the involvement of non-genetic factors in the present context are those involving facultatively multicellular or facultatively social species, or comparisons across species within taxa with unicellular and multicellular representatives or social and non-social representatives. These are the cases where species are poised on the border of solitary and group life, as must the ancestors of multicellular or eusocial lineages have been when multicellularity or eusociality first arose. Hence these are the cases in which any non-genetic benefits of group life in current populations most convincingly represent the sorts of benefits that obtained in the past.

A number of ecological factors promoting the origin of multicellular organisms and eusocial societies have been proposed. In Table 4.3, I have tabulated them side by side for the two kinds of social group in order to facilitate comparison. (Again, for present purposes, I include eusocial societies and cooperative breeders together.) I have added supporting case studies of the required type where they exist. Evidence from such studies is still relatively sparse, precluding systematic analysis of the relative frequency with which each proposed factor has been influential. This is because robust evidence for the influence of any one ecological factor ideally involves experimentally manipulating the costs and benefits of group life in a natural setting, or, for comparative analysis, constructing phylogenies on which values for ecological traits measured in the field have been mapped. Each of these is a demanding exercise. Nonetheless, although there is inevitably some subjectivity in comparing across systems that differ in many ways, existing evidence suggests some interesting similarities and differences between the ecological factors likely to have promoted multicellularity and eusociality (Table 4.3).

Taking first the similarities, one can discern shared roles for environmental stresses, variability in the food supply, and the presence of social or brood parasites (intraspecific or interspecific) or predators in favouring social group formation in the origin of both multicellularity and eusociality (Table 4.3). To explore further the role of predation in this context, consider the division of eusocial species into two broad ecological syndromes: 'fortress defenders' and 'life insurers'. This division was proposed

Table 4.3 Ecological factors promoting social groups in the origin of multicellularity and eusociality (including cooperative breeding)

Factor	Multicellularity		Eusociality	
	Mode of operation	Selected examples of supporting evidence	Mode of operation	Selected examples of supporting evidence
Dispersal	Food shortage selects for dispersal from current location and multicellularity aids dispersal (Bonner 1974)	• Cellular slime moulds form multicellular, mobile slug and dispersive fruiting body in response to starvation (Bonner 1974)	n/a	None found
Environmental stress	Environmental stresses select for internally protecting some cells within layers of other cells (Lachmann et al. 2003; Grosberg and Strathmann 2007)	• Budding yeast (*Saccharomyces cerevisiae*) formed multicellular clusters possibly protecting internal cells against toxins in the medium (Smukalla et al. 2008); Table 4.1	Harsh, variable environments select for group life as a buffer against worst periods (Jarvis et al. 1994)	• Across species of African starlings (Sturnidae), phylogenetic analysis shows that cooperative breeding is associated with semi-arid savanna habitats and temporally variable rainfall (Rubenstein and Lovette 2007) • Across species of African mole-rats (Bathyergidae), phylogenetic analysis shows that degree of sociality (group size) is negatively associated with density of tubers (food) and positively associated with variability in rainfall (Faulkes et al. 1997; O'Riain and Faulkes 2008)

Factor	Multicellularity		Eusociality	
	Mode of operation	Selected examples of supporting evidence	Mode of operation	Selected examples of supporting evidence
Food supply	Food shortage selects for larger size for increase in rate of nutrient uptake and for nutrient storage (Bell 1985; Koufopanou and Bell 1993; Kerszberg and Wolpert 1998; Kirk 1998; Szathmáry and Wolpert 2003)	• In the alga *Volvox carteri*, germline cells (gonidia) with somatic cells grew faster than those experimentally deprived of somatic cells (Koufopanou and Bell 1993) • Cannibalism of groupmate cells is likely to occur in response to starvation in some simple multicellular organisms, suggesting self-sacrificial release of stored nutrients (Kerszberg and Wolpert 1998; Szathmáry and Wolpert 2003)	Variability in amount of resources across sites selects for non-dispersal from high-quality sites (Stacey and Ligon 1991)	• In the Carrion crow (*Corvus corone*), experimentally adding food to territories increased level of non-dispersal and helping by offspring (Baglione et al. 2006) • In the Seychelles warbler (*Acrocephalus sechellensis*), offspring remained as helpers on the natal territory if it was high-quality and if vacant high-quality territories were absent (Komdeur 1992)

Table 4.3 *Cont.*

Factor	Multicellularity		Eusociality	
	Mode of operation	Selected examples of supporting evidence	Mode of operation	Selected examples of supporting evidence
Nest-site limitation (habitat saturation)	n/a	None found	Low frequency of unoccupied nest-sites or territories selects for non-dispersal (e.g. Emlen 1982a)	• In an allodapine bee (*Exoneura nigrescens*), females remained in their initial nests more frequently when nest-sites were experimentally removed (Langer et al. 2004) • In a paper wasp (*Mischocyttarus mexicanus*), joint-nesting increased when nest-sites were experimentally removed and decreased when they were experimentally added (Gunnels et al. 2008)[1] • In a cooperatively breeding cichlid fish (*Neolamprologus pulcher*), helpers left groups to breed independently when experimentally offered breeding substrate (Bergmüller et al. 2005) • In the Seychelles warbler (*Acrocephalus sechellensis*), cooperative breeding commenced when population density passed a threshold (Komdeur 1992)

Factor	Multicellularity		Eusociality	
	Mode of operation	Selected examples of supporting evidence	Mode of operation	Selected examples of supporting evidence
Parasitism	n/a	None found	Interspecific or intraspecific brood or social parasites select for group life through groups being better protected (Lin and Michener 1972)	• In a carpenter bee (*Xylocopa sulcatipes*), presence of a guard bee reduced frequency of intraspecific usurpation of nest by conspecific females (Stark 1992) • In the Carrion crow (*Corvus corone*), presence of helpers reduced the reproductive success of a brood parasite, the Great spotted cuckoo, *Clamator glandarius* (Canestrari et al. 2009)

Table 4.3 *Cont.*

Factor	Multicellularity		Eusociality	
	Mode of operation	Selected examples of supporting evidence	Mode of operation	Selected examples of supporting evidence
Predation	Predators promote ecological diversification in prey species including evolution of multicellularity (Stanley 1973), or select for multicellularity through multicellular groups being better protected (Bell 1985; Kessin et al. 1996; Kirk 1998; Kaiser 2001; King 2004)	• In the volvocine algae, multicellular but not unicellular species are above the size threshold for predation by filter feeders (Bell 1985; Kirk 1998) • In the cellular slime mould *Dictyostelium discoideum*, multicellular aggregations but not single amoebae escaped predation by experimentally added nematodes (Kessin et al. 1996) • In the alga *Chlorella vulgaris*, multicellularity evolved when a unicellular predator was experimentally added to unicellular populations (Boraas et al. 1998; Table 4.1	Predators select for group life through groups being better protected (Lin and Michener 1972; Krause and Ruxton 2002)	• In a cooperatively breeding cichlid fish (*Neolamprologus pulcher*), experimental addition of predators reduced dispersal by potential helpers (Heg et al. 2004)
Temporal facilitation	n/a	None found	In annual species, increasing the length of the breeding season facilitates sociality, since the colony cycle requires more time than a solitary life-cycle (e.g. Eickwort et al. 1996)	• In facultatively social halictid bees, northern and high-altitude populations tend to be solitary and southern and low-altitude populations social (Eickwort et al. 1996; Schwarz et al. 2007)

[1] In a stenogastrine wasp (*Liostenogaster flavolineata*), experimentally adding nests to populations failed to result in high uptake of vacant nests, suggesting that in this system nest-site

by Queller and Strassmann (1998), building on ideas of Queller (1989), Gadagkar (1990), Alexander et al. (1991), and Crespi (1994). Fortress defenders include Ambrosia beetle, some species of termites, social aphids, social thrips, social shrimps, and mole-rats (e.g. Duffy 2007; Chapman et al. 2008; Korb 2008; Pike and Foster 2008). Social groups in these species inhabit a nest that provides safe access to their food source (for example, in social aphids, a gall from the inside of which the inhabitants can securely access plant sap). Because this life-style puts a premium on nest defence against both intraspecific competitors and predators, fortress defenders are typified by a soldier caste (Fig. 4.3), i.e. a non-reproductive caste specializing in nest defence (Queller and Strassmann 1998). Life insurers include the eusocial Hymenoptera (e.g. Field 2008). They also usually have a nest but forage outside it. In this case foragers risk suffering from predation and are 'life insurers' in the sense that sociality allows their investments in offspring to be completed by their nestmates should they be killed by predators (Queller and Strassmann 1998). Although a reasonable characterization of syndromes of sociality in the eusocial arthropods (plus the mole-rats), Queller and Strassmann's (1998) classification—as they themselves pointed out—does not easily fit the majority of social vertebrates. Instead, social vertebrates combine features of the two syndromes (Korb and Heinze 2008b). Nonetheless, the broad applicability of Queller and Strassmann's (1998) scheme to eusocial arthropods highlights the fact that predation is likely to have been a key factor influencing the formation of social groups in these species (Queller and Strassmann 1998).

Let us now examine the differences in the ecological factors influencing the origin of multicellularity and eusociality. One is that group formation in response to starvation as a prelude to dispersal, which occurs in the cellular slime moulds (Table 4.3), has no obvious analogue in the formation of eusocial societies. This is probably because multicellular organisms, being larger, are usually capable of dispersing when solitary. A second difference concerns nest-site limitation. Ecological factors favouring social group formation have often been divided into 'benefits of philopatry' and 'ecological constraints' (Emlen 1982a; Stacey and Ligon 1991). These reflect, respectively, the potential benefits of remaining on the natal site and the potential costs, or lack of profitability, of dispersal from it. An example of a benefit of philopatry is the benefit from remaining on a food-rich natal site. There is evidence that such a benefit was implicated in the origin of cooperative breeding (Table 4.3). Queller and Strassmann's (1998) characterization of some eusocial societies as fortress defenders implies that it operated in the origin of these societies as well. An example of an ecological constraint is predation, at least to the extent that it makes dispersal costly. As just discussed, predation is likely to have influenced the origin of both multicellularity and eusociality (Table 4.3). Nest-site limitation or habitat saturation represents another well-known example of an ecological constraint (e.g. Brockmann 1997). When vacant nest-sites are rare, the expected chances of successfully reproducing as a singleton are reduced, which decreases the relative benefit of not joining a social group and hence promotes social grouping. Experimental evidence suggests that nest-site limitation has promoted the origin of eusociality and cooperative breeding (Table 4.3), though its predicted effects depend on its interaction with demography, life history, phylogenetic

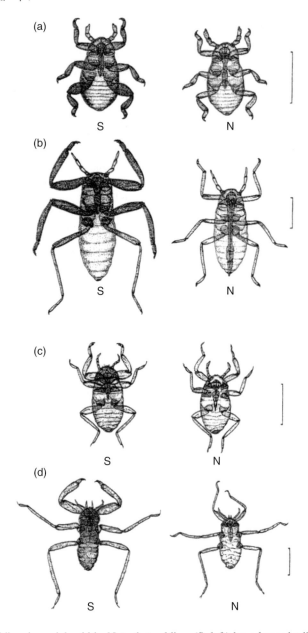

Fig. 4.3 Soldiers in social aphids. Note that soldiers (S, left) have larger bodies, thickened cuticles, larger legs, and, in (d), enlarged horns compared to normal nymphs (N, right). (a) *Pemphigus spyrothecae*; (b) *Colophina monstrifica*; (c) *Pseudoregma bambucicola*, second instars from primary plant host; (d) *P. bambucicola*, first instars from secondary plant host. Scale bar = 0.5mm. Drawing by Thalia Grant reproduced from Stern and Foster (1996) with kind permission of the authors and artist.

history, and the spatial scale of dispersal (Koenig et al. 1992; Pen and Weissing 2000; Bennett and Owens 2002; Kokko and Ekman 2002; Hatchwell 2009). However, there is no indication that anything analogous to nest-site limitation has promoted the origin of multicellularity. The reason is presumably that, because of the very small size of unicellular organisms, finding a vacant space to occupy does not represent a major brake on their dispersal. Likewise, nothing analogous to temporal facilitation of eusociality (Table 4.3) appears to exist for multicellularity, probably because unicellular organisms are not generally time-limited in their reproduction.

To conclude, although there is no overriding reason in theory why the specific ecological factors that promote multicellularity and eusociality should resemble one another, some resemblances do exist. This supports the idea, from the major transitions standpoint, that multicellularity and eusociality have a common basis in processes of social evolution. Where the ecological factors affecting the two transitions differ, it is mainly because the large disparity in size between unicells and multicellular organisms means that they tend to live their lives at very different spatial and temporal scales.

4.4 Synergistic factors in social group formation

When members of a group work together to complete actions that each alone could not complete, or could only complete less efficiently, the group is said to be acting synergistically. As with ecological factors, in seeking synergistic factors that contributed to social group formation, one needs to focus on factors that could have operated straight after social groups arose. Hence, for example, marked morphological differentiation of group members should not be invoked, because the initial distinction between synergistic parts of a social group is likely to have been physiological or behavioural, with marked morphological differentiation arising later.

Synergistic factors in social group formation among non-relatives

When non-relatives unite in a social group, they cannot, by definition, bring shared genes together; synergy and the possibility of achieving shared reproductive fate are all they have. It is entirely to be expected, therefore, that in the major transitions involving non-relatives, social groups are often based on some sort of sharing of functions (Queller 1997, 2000). A good example is found in the origin of eukaryotic cells (Knoll 2003). Here, the host cells gained the biochemical ability to respire aerobically (through endosymbiosis with the proteobacterial ancestors of mitochondria) or to perform photosynthesis (through endosymbiosis with the cyanobacterial ancestors of chloroplasts). Many interspecific mutualisms are based on the exploitation, direct or indirect, of the diverse biochemical abilities of prokaryotes. Examples of such abilities include producing light by chemical means, as in the mutualism between Bobtail squid and the luminescent bacterium *Vibrio fischeri*, or fixing nitrogen, as in the mutualism between leguminous plants and rhizobial bacteria (Sachs et al. 2004). Similarly, mutualisms between marine invertebrates and algae (Sachs et al. 2004) and, in lichens, between fungi and algae (Lutzoni and Miadlikowska 2009), are indirectly based on the photosynthetic abilities of cyanobacteria, as represented by algal chloroplasts.

Unlike the case with interspecific mutualisms, it is not clear whether synergistic factors have been important in the origin of sexual reproduction. Theories of the advantages of sexual reproduction based on diploidy permitting the repair of damaged genes or providing back-up gene copies in the case of irreparable damage, or based on heterozygosity increasing overall fitness (e.g. Maynard Smith and Szathmáry 1995; Kleiman and Tannenbaum 2009), implicitly invoke a synergistic benefit of sex. But the relative importance of such factors among the wealth of others that have been suggested to influence the origin and maintenance of sex remains undecided (West *et al.* 1999; Agrawal 2006).

Synergistic factors in social group formation among relatives

Synergistic factors facilitating the origin of social groups of relatives have been suggested in the context of both multicellular and eusocial evolution. Following the presentation in Table 4.3, Table 4.4 sets out some of these factors side by side, along with supporting evidence, so as to allow comparisons between the two contexts. From this treatment, it is evident that division of labour and advantages of group foraging each represent types of synergistic factor shared across the origins of multicellularity and eusociality (Fig. 4.4; Table 4.4). This is doubtless because specialization of roles can occur at the level of both the cell and the multicellular organism (Oster and Wilson 1978; Bonner 1988, 1993). Likewise, both unicells and multicellular organisms ingest food when living solitarily and so can benefit from a pooled effort in foraging. By contrast, only eusocial societies exhibit the synergistic benefit of assured fitness returns (Table 4.4). This difference arguably stems from members of eusocial societies, unlike cells within multicellular organisms, not being physically stuck together (with the exception of the colonial marine invertebrates). As a result, members of eusocial societies, but not cells within multicellular organisms, can act as independent central-place foragers. While foraging, they therefore expose themselves to mortality risks that can be offset by relying on groupmates to complete investments in offspring, so achieving assured fitness returns (Queller 1989; Gadagkar 1990).

Synergistic benefits of enhanced dispersal, locomotion, or sedentariness appear to be experienced by cells within multicellular organisms and by the zooids or polyps within colonial marine invertebrates but not by members of the more conventional eusocial societies (Table 4.4). As discussed in Section 4.2, adaptations for enhanced dispersal might have been favoured in the origin of multicellularity via semisociality in terrestrial environments because they would have allowed the retarding effects of population viscosity on the origin of altruism to be mitigated. The other differences again appear to depend on physical interconnection. It seems that only when members of a social group are physically stuck together, as are cells within multicellular organisms and zooids or polyps within colonial marine invertebrates, can there be sufficient coordination to achieve enhanced locomotion or sedentariness through grouping. At least, evidence for such benefits exists for multicellular organisms in the case of locomotion and sedentariness,

Fig. 4.4 Naked mole-rat workers in their burrow, in which they carry out group excavation, foraging, and care of the young, so exemplifying synergistic benefits of social group formation through division of labour and teamwork. Image reproduced with kind permission of Chris Faulkes.

and for pelagic colonial marine invertebrates in the case of locomotion (Table 4.4). Given this, and the biology of the sedentary (sessile) forms, one can reasonably conclude that sedentariness benefits for colonial marine invertebrates are also likely to have occurred.

4.5 Hamilton's rule and social group formation

Genetic, ecological, and synergistic factors promoting social group formation are all integrated within Hamilton's rule or, in models of reproductive skew (Section 3.2), its variants. In the case of social interactions between non-relatives, it has not been traditional to invoke Hamilton's rule. This has perhaps been because the direct fitness benefits to the social partners, while assumed present, are hard to quantify in obligate mutualisms, since estimating the fitness of any one partner in the absence of the other is obviously difficult. In addition, research has tended to concentrate on how the evolutionary dynamics of cooperative interactions can generate stability, and game theory has been the natural tool for this purpose. Game theory is not separate from inclusive fitness theory (Section 1.4; Box 3.1), but the connections between the two bodies of theory are not always made explicit. In the case of social interactions between relatives, the role of Hamilton's rule has long been recognized. However, only a few studies have simultaneously quantified all the terms in Hamilton's rule because estimating parameter values for the terms affected by non-genetic factors is empirically challenging. Where all terms have been estimated, the usual result is that Hamilton's rule is found to be fulfilled under some conditions but not others (Section 2.5). Nonetheless, in general, and as expected from Hamilton's rule, relatedness is positive in social groups characterized by altruism and evidence exists for

Table 4.4 Synergistic factors promoting social groups in the origin of multicellularity and eusociality (including cooperative breeding)

Factor	Multicellularity		Eusociality	
	Mode of operation	Selected examples of supporting evidence	Mode of operation	Selected examples of supporting evidence
Assured fitness returns	n/a	None found	If a single organism dies, all investment in its young is lost; but if a group member dies, other group members may rear its young to independence, so groups provide assured fitness returns (Gadagkar 1990)	• In a stenogastrine wasp (*Liostenogaster flavolineata*), experimental removal of wasps demonstrated fitness loss for lone females and assured fitness returns for group-nesting females (Field et al. 2000)
Dispersal	In terrestrial forms, multicellular fruiting body achieves better spore dispersal (Bonner 1974; Kaiser 2001)	• Association exists between origins of multicellularity in terrestrial environments with evolution of fruiting body (Bonner 1998)	n/a	None found

Factor	Multicellularity		Eusociality	
	Mode of operation	Selected examples of supporting evidence	Mode of operation	Selected examples of supporting evidence
Division of labour	Separation of incompatible biochemical functions, e.g. photosynthesis and nitrogen-fixation in filamentous cyanobacteria (Bonner 1998; Kaiser 2001), or mechanical functions, e.g. cell division and locomotion in flagellated cells (Buss 1987; Koufopanou 1994; Grosberg and Strathmann 2007)	• Unicellular cyanobacteria photosynthesize by day and fix nitrogen by night; filamentous cyanobacteria (e.g. *Anabaena*) spatially separate these functions, with a second cell type (heterocyst) specializing on nitrogen fixation (Bonner 1998; Kaiser 2001; Golden and Yoon 2003) • In volvocine algae, only multicellular species with a cell number greater than 32 ($= 2^5$) develop unflagellated germline cells, consistent with flagella being lost in cell division of undifferentiated single cells after five rounds of cell division (Koufopanou 1994)	A single organism can perform several tasks only in series, but a group can perform them in parallel or series-parallel, so improving efficiency and robustness of overall performance; furthermore, in a group, group members can become task-specialists, again improving efficiency (Oster and Wilson 1978)	• In a carpenter bee (*Xylocopa sulcatipes*), non-usurped social nests were more productive than non-usurped solitary nests, because one bee specialized on foraging and reproducing and the other on guarding (Stark 1992; Bourke 1997) • In African lions (*Panthera leo*), lionesses adopted consistently different roles in group hunting (Stander 1992)

Table 4.4 *Cont.*

Factor	Multicellularity		Eusociality	
	Mode of operation	Selected examples of supporting evidence	Mode of operation	Selected examples of supporting evidence
Foraging	Foraging advantage, e.g. reach or digest otherwise inaccessible food (Bonner 1974)	• In the cellular slime mould *Dictyostelium discoideum*, amoebae shed from migrating slugs could reach food sources inaccessible to single amoebae (Kuzdzal-Fick et al. 2007); • Aggregations of Myxobacteria collectively produce digestive compounds (Velicer and Vos 2009)	Grouping increases the predictability of food acquisition through decreasing the variance in the amount of food gathered (Wenzel and Pickering 1991; Stevens et al. 2007); grouping allows more efficient foraging, e.g. more frequent capture of prey or capture of larger prey by group-hunting predators (Alexander 1974; Creel and Macdonald 1995)	• In an allodapine bee (*Exoneura nigrescens*), variance in cumulative food gathered (indexed by brood weight) fell as group size increased (Stevens et al. 2007); • In African wild dogs (*Lycaon pictus*), prey size, capture probability, and per capita net rate of energy intake increased with group size (Rasmussen et al. 2008b)
Locomotion	Faster, more efficient, or more coordinated movement, i.e. swimming in aquatic forms (Bonner 1998; Lachmann et al. 2003) or terrestrial locomotion in terrestrial forms (Foster et al. 2002)	• In the cellular slime mould *Dictyostelium discoideum*, larger slugs moved faster than smaller ones (Foster et al. 2002)	Faster, more efficient, or more coordinated movement (Mackie 1986)	• In a pelagic colonial marine invertebrate, the tunicate *Salpa fusiformis*, zooids expended less energy per zooid swimming in groups than swimming singly, because drag per zooid was reduced (Bone and Trueman 1983; Mackie 1986)
Sedentariness	Multicellularity facilitates remaining in one place, e.g. in sessile, aquatic forms (Bonner 1974, 1998) or in mat-building forms (Rainey and	• In a bacterium (*Pseudomonas fluorescens*), multicellularity allowed floating mat with access to gaseous oxygen to form (Rainey and Rainey	n/a	None found

a wide range of non-genetic factors promoting such groups, some being shared across the two main cases, namely multicellular organisms and eusocial societies.

It is striking that, although Hamilton's rule is routinely invoked in analyses of the origin of eusociality, it is rarely explicitly invoked in analyses of the origin of multicellularity. There are some exceptions, such as Grosberg and Strathmann's (2007) review. In a clone of cells, the form of Hamilton's rule relevant to the origin of multicellularity with altruism is simply $n > m$, i.e. that the benefits of grouping should exceed its costs (Table 2.3). Given unitary development, subsociality, mitotic cell division, and the right combination of ecological and synergistic factors, this seems a relatively easy condition to achieve, perhaps thereby accounting for the multiple independent origins of multicellularity (Box 1.1) and its repeated emergence under laboratory conditions (Table 4.1). It was in this sense that Grosberg and Strathmann (2007) suggested that the origin of multicellularity was a 'minor' major transition. Despite this, formal models of multicellular evolution have tended not to stress the simplicity of the conditions required for the origin of multicellularity (e.g. Michod 1996, 2007; Pfeiffer and Bonhoeffer 2003; Hochberg et al. 2008; Willensdorfer 2009). Instead, by examining the control of within-organism conflict or the evolution of morphological differentiation among cells, several models have concentrated on stages in the transition downstream from the origin of multicellularity itself. In the terminology of the present book, they have effectively dealt less with social group formation and more with social group maintenance and transformation. Hence, with the notable exception of work investigating social group formation in bacteria (e.g. Rainey and Rainey 2003; Griffin *et al.* 2004), research estimating the parameters of Hamilton's rule in laboratory systems that permit the experimental analysis of the origin of multicellularity (e.g. *Chlorella vulgaris* and other eukaryotic examples in Table 4.1) appears lacking. It would be highly valuable.

More broadly, the field of social evolution needs to reach the stage at which it can explain why a given lineage has become social (i.e. initiated a major transition) whereas other similar lineages have not. At present, such an explanation has been approached in a few taxa only, all involving the transition to eusociality or cooperative breeding (e.g. Faulkes et al. 1997; Hunt 1999; Rubenstein and Lovette 2007; Duffy and Macdonald 2010). Carefully estimating the terms of Hamilton's rule in tractable cases, and combining such studies with a broader assessment of genetic and non-genetic parameters in further phylogenetically-controlled comparative analyses, would go a long way to achieving this important goal.

4.6 Summary

1. The pathways by which social groups formed are poorly known in the cases of the origin of the eukaryotic cell, of sexual reproduction, and of many interspecific mutualisms. Pathways of social group formation are much better known in the cases of the origin of multicellularity and eusociality. Here, social group formation has usually occurred via a subsocial pathway (i.e. via the association of parents and

offspring), with a semisocial pathway (i.e. via the association of same-generation organisms) occurring in a minority of cases.

2. Social group formation occurs in non-relatives when Hamilton's rule for narrow-sense cooperation is fulfilled. Such social groups are favoured when the partners achieve a shared reproductive fate, which can be brought about by each transmitting their genes via the same propagule. Social group formation occurs in relatives when Hamilton's rule for narrow-sense cooperation or for altruism is fulfilled. Most social groups of relatives exhibit some form of altruism, which requires positive relatedness. Given a subsocial pathway, relatedness can be inferred to have been positive in the origin of multicellularity. Ancestral state reconstruction, and comparisons with extant species, show that relatedness was also positive at the independent origins of eusociality via the subsocial pathway. Even when the semisocial pathway has been followed in the origin of multicellularity or eusociality, the evidence suggests that relatedness was also positive. Population-wide competition between the products of altruism, which is required for the unhindered evolution of altruism, could have been achieved in the origin of multicellularity either, for aquatic origins, by mixing caused by the fluid medium itself, or, for terrestrial origins, by the development of fruiting bodies adapted for spore dispersal.

3. Non-genetic factors can promote social group formation by affecting the terms in Hamilton's rule for changes in the offspring numbers of the actor and recipient. These factors may be divided into ecological and synergistic ones. It is unknown what ecological factors, if any, contributed to the origin of sexual reproduction. Ecological conditions that might contribute to the origin of interspecific mutualisms include species-rich and stable habitats, and conditions promoting spatial structure in sympatric populations of potential mutualists. Ecological factors leading to the origin of both multicellularity and eusociality include environmental stresses, variability in the food supply, and pressure from parasites or predators. These two transitions also exhibit idiosyncratic differences in the ecological factors affecting them, stemming largely from the very different spatial and temporal scales at which unicellular and multicellular organisms live.

4. Mutualisms are built on synergy and so many mutualisms involve an obvious pooling of complementary abilities. The main synergistic factors affecting the origin of multicellularity and eusociality are division of labour and advantages of group foraging.

5. Hamilton's rule applies in principle to social group formation by both non-relatives and relatives but in practice has been mainly applied to the origin of eusociality. Overall, in both this case and that of the origin of multicellularity, the evidence shows that relatedness is positive and that a wealth of non-genetic factors potentially favour social group formation. This suggests that Hamilton's rule for altruism is fulfilled even when, as is the usual situation, its terms have not been fully quantified. The terms of Hamilton's rule need to be quantified in cases representing all types of social group in order for the field to approach a full understanding of the distribution of sociality in its broadest sense.

5

Social group maintenance

5.1 Limitation of exploitation: principles and processes

Social group maintenance refers to the processes that maintain group stability once social group formation has occurred (Section 1.3). A social group, like a bank, represents a repository of lifetimes of personal investments to its members, but to some elements, both on the outside and the inside, it is just a resource to be exploited if at all possible. In the biological world, exploitative elements are those that would use the resources of the social group for their own reproduction. Those from the outside include parasites and pathogens. Those from the inside include selfish genetic elements that subvert the machinery of their host cells for their own propagation. A major part of social group maintenance is, therefore, the limitation of exploitation from both without and within (e.g. Nunney 1999a). Examining the principles that underpin how this occurs will be the chief aim of this chapter.

Within the processes of social group maintenance, it is possible to define a set of subprocesses (Table 5.1). For example, as regards limitation of exploitation from within, I identify two important subprocesses (Table 5.1), namely self-limitation (i.e. selfish group members limit their own reproduction) and coercion (i.e. the reproduction of selfish group members is limited by coercion by other group members). Self-limitation and coercion represent the two principal means of conflict resolution (Section 3.2). Their appearance in the current context is to be expected, since conflict resolution is essentially what limits exploitation from within. Put another way, the general problem addressed in considering the limitation of exploitation from within is the problem of what keeps cheating in check in social evolution (Section 1.4). Any solution must involve something that can evolve by natural selection, rather than

Table 5.1   A classification of processes of social group maintenance

Processes and subprocesses

- Limitation of exploitation from outside
    - ○ Recognition of self versus non-self
- Limitation of exploitation from inside
    - ○ Self-limitation (i.e. effects of self)
        - ▪ Self-limitation through negative frequency-dependence
        - ▪ Self-limitation through excessive costs to the group
    - ○ Coercion (i.e. effects of rivals)

something imposed from the top down as in human society (Sections 2.3, 3.2). The overall approach of this chapter is closely based on the set of principles governing the suppression of within-group conflict outlined by Bourke and Franks (1995), which was itself derived from a number of sources (e.g. Leigh 1977; Alexander and Borgia 1978; Dawkins 1982; Werren et al. 1988; Ratnieks and Reeve 1992). The chapter explores the operation of these principles in social groups in the wide sense of 'social group' adopted throughout the present book.

5.2 Limitation of exploitation from outside

The essential defence against exploitation of a social group by external parties is the recognition of self versus non-self, followed by the exclusion of non-self from group benefits (Buss 1987; Hamilton 1987; Queller 2000; Queller and Strassmann 2002; Grosberg and Strathmann 2007). Systems involving defence of this kind include immunity in multicellular organisms and nestmate or groupmate recognition in eusocial societies. Immune systems are typically highly effective, but some eusocial societies are more 'open' than others. Indeed, operationally, the level of integration of a social group can be defined both by the degree of security of the border around it, where the border is one within which social benefits are shared and outside of which they are not, and by the degree to which group members cooperate to defend breaches of the border. For example, people are very selective over whom they let through the front door of their family home and will collectively exclude strangers, but few concern themselves with who enters their city or cooperate to maintain its borders. Hence, by these criteria, a family of people is like a well-defined social group but a city is not. The evolution of recognition systems in nature represents a very large topic in its own right. This section therefore focuses on a few key points.

Recognition of self versus non-self in social groups of non-relatives

As we will see, some form of recognition of self versus non-self, followed by rejection of non-self, is almost universal in social groups composed of relatives (products of fraternal transitions). In interspecific mutualisms (products of egalitarian transitions), the self is harder to define, precisely because in egalitarian transitions the partners, being non-relatives, are selected to reproduce separately and hence tend to retain their identities (Section 3.1). But if the self is assumed in this case to be embodied in the multi-species collective formed by the mutualistic partnership, one can discern that processes of recognition and rejection of non-self occur in some interspecific mutualisms as well, although they are not always viewed this way. Interspecific mutualisms are certainly prone to parasitism by other parties (Yu 2001). Evidence exists for cooperative defence in some cases. Fungus-growing ants (Attini) and their mutualistic fungal partners (Lepiotaceae) need to defend themselves against a parasitic fungus (*Escovopsis*) that attacks the fungus garden. The ants weed the fungus garden to try to exclude the parasite, while a third mutualist, an actinomycetous bacterium (*Pseudonocardia*), produces antibiotics active against *Escovopsis*. The bacteria are housed in specially modified crypts on the ants' cuticle (Currie et al. 1999, 2006; Currie and Stuart 2001). Hence, the ants and bacteria cooperate to defend the mutualistic association against the parasite. In a sense, the ant–fungus mutualism self-medicates using drugs supplied by the bacteria. Similarly, some vertically transmitted endosymbiotic bacteria confer resistance against third-party pathogens on their invertebrate hosts (Haine 2008). Again, the hosts harbour the symbionts, and presumably add their own immune defences, so host and endosymbiont cooperate in defending their collective against other parties.

In both these examples, the partner providing the primary defence (*Pseudonocardia* bacteria or endosymbiotic bacteria) arguably does so because it benefits directly from the survival of the association as a whole. Hence, other things being equal, one might expect mutualists to be more cooperative in defence of their partnership under vertical transmission than under horizontal transmission, since vertical transmission leads to greater coincidence of fitness interests (Sections 3.1, 4.2). Consistent with this view, in both of the present examples the bacteria are indeed vertically transmitted (Currie et al. 1999; Haine 2008).

In general, whether it is usual or unusual for mutualistic partners to cooperate in recognizing and excluding parasitic intruders appears poorly known. For example, some lichen-forming fungi specialize in invading other lichens and taking over the algal partners of the resident fungus (Richardson 1999). In this case, cooperative defence could hypothetically involve both the resident fungus and the alga secreting toxins effective against the intruder. But whether this happens seems not to have been studied. A lack of joint defences in interspecific mutualisms could arise because the absence of relatedness between the partners precludes behaviours as fully social as joint defence, always allowing for the possibility that shared reproductive fate through vertical transmission may instead bind the interests of the parties together. Alternatively, cooperative defence may be counterselected if only one party bears the cost of exploitation. In the lichen case, the fitness of the alga may be unaffected by which of the two fungi is its mutualistic partner. From the alga's viewpoint, takeover by a new fungus could just represent a horizontal alternative to its usually vertical mode of transmission, with the predictable consequence that it fails to invest in defending its current partner.

Recognition of self versus non-self in social groups of relatives

In social groups of relatives (multicellular organisms and eusocial societies), recognition systems have been extensively studied and several broad types occur (Table 5.2). They typically represent a cooperative effort of the social group and are directed against different sorts of would-be external exploiter. Because they have been studied by separate communities of researchers, different recognition systems have attracted different terminologies. However, immunity, nestmate or groupmate recognition, and allorecognition all refer to systems that resemble one another in each being directed at discriminating against non-self (Table 5.2).

Table 5.2 Systems for the recognition and exclusion of potential external exploiters of social groups of relatives

Level of organization	External exploiters	Recognition system	Examples and notes
Organism	Interspecific: parasites, pathogens	Immune system	Universal defence in vertebrates (Janeway and Medzhitov 2002; Pancer and Cooper 2006) and invertebrates (Rolff and Reynolds 2009)
	Intraspecific: transmissible cancers (Burt and Trivers 2006)	Immune system	Immune response via major histo-compatibility complex (Murgia et al. 2006)
	Intraspecific: selfish cell-lineages (Mehdiabadi et al. 2006; Ostrowski et al. 2008)	Kin recognition	Cellular slime moulds exclude genetically dissimilar conspecific cells from aggreg-ations (Mehdiabadi et al. 2006; Ostrowski et al. 2008)
Society	Interspecific: parasites and pathogens of social insects (Schmid-Hempel 1998)	Social immunity	Communal defences such as communal 'fevers' that kill invasive bacteria and the exclusion of infected nestmates from the colony in the Honey bee (Cremer et al. 2007; Cremer and Sixt 2009; Wilson-Rich et al. 2009)
	Interspecific: interspecific social parasites of social insects (Hölldobler and Wilson 1990)	Nestmate and groupmate recognition	Nestmate recognition in social insects (Breed and Bennett 1987; D'Ettorre and Lenoir 2010)[1]
	Intraspecific: intraspecific social parasites of social insects (Field 1992; Beekman and Oldroyd 2008); conspecific competitors	Nestmate and groupmate recognition	Nestmate recognition in social insects (Breed and Bennett 1987; D'Ettorre and Lenoir 2010); recognition of, and aggression towards, non-nestmates in the Naked mole-rat (O'Riain and Jarvis 1997)[1]

Table 5.2 *Cont.*

Intraspecific: conspecific competitors	Allorecognition	Sessile colonial marine invertebrates reject genetically dissimilar conspecifics and somatic fusion only occurs when the partners are genetically very alike (Buss 1987; Grosberg 1988; Burt and Trivers 2006; Aanen et al. 2008)

[1] Some cooperatively breeding birds are exploited by brood parasites both interspecifically, e.g. Carrion crows, *Corvus corone*, parasitized by Great spotted cuckoos, *Clamator glandarius* (Canestrari et al. 2009), and intraspecifically, e.g. Moorhens, *Gallinula chloropus* (Gibbons 1986; McRae 1997) and White-winged choughs, *Corcorax melanorhamphos* (Heinsohn 1991). But there appears to be little evidence for group defence against the brood parasites (Heinsohn 1991; Canestrari et al. 2009).

Non-self to a social group of relatives includes either unrelated or less related conspecifics, or non-conspecifics, from outside the group. So recognition systems in this context are essentially tools for between-group kin discrimination. Such discrimination may be achieved through recognition of intrinsic genetic cues, as in allorecognition in colonial marine invertebrates (Buss 1987; Grosberg 1988; Aanen et al. 2008). Alternatively, it may be achieved via the 'learning' of cues characteristic of self through early exposure to such cues. This is seen both in immune systems and in nestmate recognition. Vertebrate immune systems can be manipulated to accept foreign cells as self by exposing them to foreign antigens in early development (e.g. Wood et al. 2010). Likewise, workers of social insects can be manipulated to accept foreign workers as nestmates by exposing them to foreign workers as young adults (Cremer and Sixt 2009; D'Ettorre and Lenoir 2010).

The nature of the cues used for recognition varies widely. In social insects, nestmate recognition is generally based on chemicals borne on the cuticle, which may be endogenous or influenced by the local environment (e.g. Akino et al. 2004; Martin et al. 2008; Van Zweden et al. 2009). In social vertebrates, recognition of groupmates may be based on visual and auditory cues, as well as chemical ones (Sherman et al. 1997; Komdeur et al. 2008). In insects, eusocial societies that are exceptions in lacking strong nestmate recognition, such as unicolonial ants, are conspicuous through doing so (Section 7.3). Correspondingly, fusion between entire social insect colonies appears rare. Where it does occur, it often results from a specific set of ecological conditions, namely severe ecological competition for space (Foitzik and Heinze 2001; Johns et al. 2009).

The evidence is, therefore, overwhelming that social groups of relatives act as if concerned to repel genetically alien elements, whether of the same species or not, that seek to enter and exploit them. This contributes to social group maintenance in two ways. The first is, of course, that it allows group members to use group resources for their own reproduction. Safeguarding the inclusive fitness of group members in this way probably represents the primary reason for the evolution of recognition systems. The second way in which exclusion of non-self contributes to social group maintenance occurs through an interaction with the presence of a bottleneck in the life cycle. A bottlenecked life-cycle generates elevated relatedness within the social group, so reducing potential within-group conflict (Section 3.3). Combined with rigorous exclusion of non-self, it also creates predictability in the group's kin struc-ture. In social insects, Boomsma (2007) argued that lifetime monogamy of the colony founders is a pre-requisite of complex eusociality through ensuring a consis-tently high within-group relatedness (Section 4.2). Preventing the entry of alien genes over the lifetime of the colony is essential if this initially high relatedness is to be kept high. If it is, then group members can commit irreversibly to a helper role, i.e. become sterile adult workers, with the firm expectation that the kin structure of the colony will stay as it began. This is likely to help maintain group stability and specifically to facilitate the evolution of an irreversible helper caste (Section 6.5).

Recognition systems are imperfect

Immune systems in multicellular organisms and nestmate-recognition systems in social insects are frequently highly effective in screening out would-be external exploiters, to the extent that many social groups preserve their genetic integrity throughout their lifetimes. Nonetheless, evolution by natural selection does not pro-duce perfection, and recognition systems fall short of being fully effective in all cases. There are two basic reasons for this. One is that recognition systems almost always have to distinguish between welcome and unwanted guests, or, in the termi-nology of Reeve (1989), desirable and undesirable recipients. For example, immune systems in multicellular organisms should tolerate beneficial, symbiotic micro-organisms but exclude pathogenic micro-organisms. Likewise, nestmate-recognition systems in social insects should tolerate foragers returning to their own colony but exclude invaders from other colonies. If the cues used in discriminating between desirable and undesirable recipients overlap, then no recognition system can be per-fect (Getz 1981; Reeve 1989). If all undesirables are rejected, so will some desirables be. In other words, there will be what, in immune systems, is termed autoimmunity. If all desirables are accepted, some undesirables will be accepted too (Fig. 5.1). Both outcomes are potentially costly, since rejection of desirables represents a direct cost to the social group, while acceptance of undesirables leaves the social group open to exploitation. The fine-tuning of the recognition system should then depend on the relative costs of each outcome (Reeve 1989; Sherman et al. 1997).

A second reason why recognition systems are imperfect is that would-be exploit-ers are selected to circumvent them. This sets up selection on hosts to prevent this, leading to coevolutionary arms races of adaptation and counter-adaptation bet-ween exploiters and hosts. Indeed, brood and social parasites have a wide array of stratagems for hoodwinking their hosts, ranging from chemical mimicry or adoption of hosts' recognition compounds in the social parasites of insects (Turillazzi et al. 2000; Lenoir et al. 2001; Sledge et al. 2001), to mimicry of hosts' eggs in the brood parasites of birds (Davies 2000). But hosts are not defenceless, and evidence exists that, as expected, social or brood parasites and their hosts engage in coevolutionary arms races over evolutionary time (Davies 2000; Brandt et al. 2005). However, the strength of selection on exploiters to circumvent the recognition systems of their hosts generally exceeds the strength of selection on hosts to recognize and exclude them. This is particularly true for obligate exploiters specializing on one or a few hosts, since, in this case, all successful exploiters must overcome their hosts' defences, whereas not all hosts will necessarily be exposed to an exploiter (Dawkins and Krebs 1979; Dawkins 1982). Therefore, in the arms race of deception versus detection, one might expect exploiters to be one step ahead of their hosts. For these reasons, some level of exploitation of social groups by outside elements is almost unavoidable.

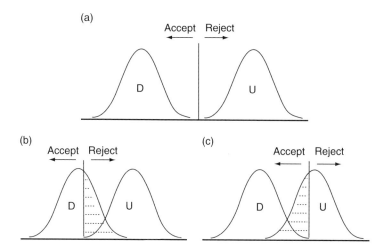

Fig. 5.1 Overlapping recognition cues prevent fully accurate groupmate or nestmate recognition. The normal curves are frequency distributions of recognition cues (horizontal axis) for desirable recipients (D) or undesirable recipients (U). Vertical lines are acceptance thresholds, to the left of which recipients are allowed into the group (accept) and to the right of which recipients are not allowed into the group (reject). (a) When recognition cues do not overlap, 100% acceptance of desirable recipients and 100% rejection of undesirable recipients can be achieved. (b) If recognition cues overlap and all undesirable recipients are rejected, some desirable recipients will be rejected too (stippled area). (c) If recognition cues overlap and all desirable recipients are accepted, some undesirable recipients will be accepted too (stippled area). Hence, if recognition cues overlap, 100% acceptance of desirable recipients and 100% rejection of undesirable recipients cannot be achieved wherever the acceptance threshold is placed. Based on the model in Reeve (1989).

Some forms of social group defence against external exploitation select for genetic variation within groups

Multicellular organisms with unitary development are composed of clonally related cells (Section 3.3). There is, nonetheless, a relationship between complex multicellularity and sexual reproduction, since complex multicellularity does not originate in the absence of sex (Section 6.2). Therefore, complex multicellular organisms have, through typically arising from a zygote composed of two unrelated gametes, an in-built supply of genetic variation. A highly influential theory attributes the maintenance of sex to the need to combat, in a never-ending coevolutionary arms race, the attentions of parasites (Hamilton 1980, 1982; Hamilton et al. 1990). The basic idea is that, if resistance to parasites requires specific alleles, a continually renewed supply of allelic variation is required to maintain it, since means of overcoming resistance will rapidly evolve within parasite lineages. This idea is supported by very high allele numbers at the loci underpinning immunity in vertebrates, i.e. the loci of the major histocompatibility complex or MHC (Edwards and Hedrick 1998; Woelfing et al. 2009).

A parallel theory proposes that eusocial societies may also be under selection for within-group genetic variation to combat parasites, so explaining such phenomena as the high rate of polyandry by queens in species of honey bee (Hamilton 1987; Sherman et al. 1988; Schmid-Hempel 1998). Experimental evidence supports a link between within-colony genetic variation through polyandry or polygyny and enhanced resistance to parasites in ants (Hughes and Boomsma 2004; Reber et al. 2008), the bumble bee *Bombus terrestris* (Baer and Schmid-Hempel 2001), and the Honey bee (Seeley and Tarpy 2007). There is also evidence that, in eusocial societies, within-group genetic variation confers other sorts of group-level benefits, such as a more efficient division of labour among the workforce (Mattila and Seeley 2007; Oldroyd and Fewell 2007; Hölldobler and Wilson 2009). Species of ants and termites in which, unusually, queens are produced by asexual reproduction (thelytoky), while workers are produced in the normal way by sexual reproduction, also suggest an advantage to genetic heterogeneity in the workforce of social insects (Pearcy et al. 2004; Fournier et al. 2005; Ohkawara et al. 2006; Matsuura et al. 2009). Similarly, it has been proposed that, in multicellular organisms, some organism-level benefits arise through within-organism genetic heterogeneity stemming from either somatic mutations or the formation of chimeras (Pineda-Krch and Lehtilä 2004a). However, the case for the benefits of within-group genetic heterogeneity is not as strong as it is for eusocial societies. Chimeras within multicellular organisms appear rare, matching the expectation discussed earlier that selection acts against the fusion of unrelated cell lineages (Strassmann and Queller 2004; Burt and Trivers 2006), and the evidence for organism-level benefits from somatic fusion or chimerism remains scanty (Pineda-Krch and Lehtilä 2004b). Nonetheless, in some eusocial societies, and possibly in some multicellular organisms, genetic variation within the social group is advantageous to the social group as whole.

Increasing genetic uniformity within social groups is associated with lower levels of potential kin-selected conflict (Section 3.2). Selection on within-group genetic variation may, therefore, pull in opposite directions. Selection for decreased within-group conflict would promote greater within-group genetic uniformity. Selection for increased group-level performance through genetic variation *per se* would promote lower within-group genetic uniformity. Social groups would then exhibit levels of within-group genetic variation that reflect the compromise between these opposing forces. Little is currently known about the relative strength of each side within this system of balancing selection. However, clonal ant societies occur sporadically within largely non-clonal lineages across taxa of ants (Himler et al. 2009), suggesting that they are not evolutionarily durable. This implies the existence of a long-term cost to clonal societies, as the idea of between-group selection for within-group genetic variation would suggest. Analogously, with rare exceptions, secondarily asexual lineages of complex multicellular organisms, such as asexual plants, insects, fish, or lizards, are taxonomically patchily distributed and short-lived (Bell 1982; Maynard Smith 1984). In addition, in eusocial Hymenoptera as a whole, Hughes et al. (2008) found that high levels of polyandry and polygyny evolved after the appearance of a morphological worker caste (Section 2.5). This suggests that selection to increase within-colony genetic variation for colony-level benefit gains traction only after levels of actual conflict have been sufficiently reduced by worker specialization (Hughes et al. 2008).

5.3 Limitation of exploitation from inside: self-limitation through negative frequency-dependence

Recognition systems help maintain the genetic integrity of social groups, but, as mentioned at the beginning of the chapter, a serious source of social instability also lies within. Essentially, following a major transition, some of the subunits of which the social group is composed do not always give up their independence totally, but may retain the capacity for rogue reproduction, contrary to the interests of the other group members. Alternatively, within existing social groups, novel selfish mutants might arise that are able to proliferate at a cost to the group. Potential examples include selfish organelles within cells or multicellular organisms, selfish cells within multicellular organisms, and selfish organisms within societies. Such lower-level selfish cheating needs to be kept in check at the group level if the social group is not to collapse. This keeping in check may happen in several ways (Table 5.1). The present section considers how, when the reproductive success of cheats is negatively frequency-dependent (i.e. a cheat's reproductive success falls as the frequency of cheats rises), they limit their own spread.

Ross-Gillespie et al. (2007) highlighted the fact that conditions exist in which, under inclusive fitness theory, the relative success of cheaters and cooperators is frequency-independent. However, these conditions exclude the case in which there is population structure, i.e. relatedness exceeds zero (Ross-Gillespie et al. 2007), because, in this case, a rising within-group frequency of selfish cheats would increasingly damage relatives. They also exclude the case in which groups with more cooperators outcompete groups with a high frequency of cheaters (Ross-Gillespie *et al.* 2007), as would occur if groups rich in cooperators were more productive or survived better, since, in this case, cheats would only gain greater representation in group offspring at the expense of fewer offspring being produced in total. These two cases encapsulate conditions that prevail in many social groups in nature, leading to an expectation of generalized negative frequency-dependence of cheats (Burt and Trivers 2006). In Section 3.2, it was shown that relatedness *per se* acts as a brake on over-exploitative within-group cheating. The cases below involve cheating being self-limited by negative effects either on group productivity or group survivorship (the social bacteria and social insect cases), or on fertilization success (cytoplasmic male sterility case). The examples come from simple multicellular organisms, complex multicellular organisms, and eusocial societies.

Social bacteria

Rainey and Rainey (2003) described the evolution, in the laboratory, of a simple multicellular form of the bacterium *Pseudomonas fluorescens* (Table 4.1). In beakers of unstirred medium, a form called 'wrinkly spreader' invades the ancestral, non-social 'smooth' strain to produce a floating, multicellular mat of cells (Fig. 5.2). Mechanistically, this occurs via a subsocial pathway (Section 4.1) whereby wrinkly spreader cells produce an adhesive polymer that causes daughter cells to stick together. In evolutionary terms, wrinkly spreader is kin-selected because related cells cooperate to form the mat, which gains them access to gaseous oxygen. Producing the polymer is an altruistic trait because it reduces the replication rate of individual wrinkly spreader cells but benefits surrounding cells. In pure cultures of wrinkly spreaders, mutants arise that resemble the ancestral smooth cells in not producing the adhesive polymer. They are cheats because they gain from the presence of the mat but do not pay the costs of contributing to it. Mats in the absence of the cheats eventually collapse because they become too heavy and sink under their own weight. However, in the presence of the cheats, the over-replication of the cheats, combined with the lack of polymer production, causes an earlier, precipitate collapse of the mat (Rainey and Rainey 2003), to the detriment of all the cells (Fig. 5.2). The cheating bacteria literally bring about their own downfall. This occurs through their reducing group survivorship, so creating negative frequency-dependent selection.

Cytoplasmic male sterility

Cytoplasmic male sterility occurs when genes present in the mitochondria of hermaphroditic flowering plants cause a failure of pollen production and an increase in ovule production (Burt and Trivers 2006). Since mitochondria are transmitted in ovules but not in pollen, such genes maximize their own spread (Section 2.3). Although the success of genes for cytoplasmic male sterility relies on mitochondria acting altruistically towards clone-mates (Section 2.3), they are cheats from the viewpoint of genes within the plant's nuclear genome. For one thing, unlike mitochondrial genes, nuclear genes favour equal investment in female and male function. Considering them

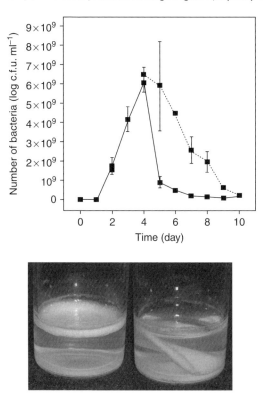

Fig. 5.2 In cultures of the bacterium, *Pseudomonas fluorescens*, numbers of wrinkly spreader cells in pure culture (dotted line) rise to a peak and then gradually decline as they form a floating mat that eventually sinks. In the presence of cheating cells that fail to produce adhesive polymer (solid line), cell numbers plummet sharply after Day 4 through premature mat collapse induced by lack of polymer. Lower Image: Mat of wrinkly spreader form of *P. fluorescens* floating on surface (left-hand figure) and during collapse (right-hand figure). Main figure reproduced from Rainey and Rainey (2003) with kind permission of the authors and lower image reproduced with kind permission of Paul Rainey.

as cheats, we can now ask what, if anything, limits the spread of genes for cytoplasmic male sterility. The answer is several factors, one of which is negative frequency-dependence. This comes about because, as a gene for cytoplasmic male sterility spreads through a population, more and more plants are effectively converted from hermaphrodites into females (male steriles). This reduces the total supply of pollen, making fertilization of ovules less likely. Affected ovules include those containing the gene for cytoplasmic male sterility, particularly because self-fertilization is not possible in male-sterile plants, so the reproductive success of the trait must fall as it becomes more frequent (Burt and Trivers 2006).

Social insects

The Cape honey bee of South Africa, *Apis mellifera capensis*, is among those few, scattered cases in the eusocial Hymenoptera in which the workers are capable of asexually producing daughter workers by thelytoky. This ability is associated with dominance behaviour in workers and has a genetic basis (Table 2.4). Hillesheim et al. (1989) experimentally bred colonies composed of 100% dominant workers, 50% dominant workers and 50% subordinate workers, and 100% subordinate workers. They found that, on three measures of worker productivity, namely brood rearing, comb building, and hoarding of syrup, colonies of 100% dominant workers performed significantly worse than colonies of 100% subordinate workers. In fact, colonies of 100% dominant workers barely registered above zero for each of the performance measures. Colonies with 50% dominant and 50% subordinate workers showed intermediate performance. These findings suggested that reproductive workers in nature would be kept at low frequencies within colonies through the failure of colonies of all-dominants (Hillesheim et al. 1989), that is through negative frequency-dependence. Reproductive *A. m. capensis* workers also have the option of selfishly parasitizing other colonies (Section 5.4). In this case, the workers' reproduction is at the expense of non-relatives, so the evolutionary dynamics differ in that the spread of the trait is not subject to the same degree of self-limitation.

A reproductive system resembling that of *Apis mellifera capensis* occurs in the ant *Pristomyrmex punctatus* (formerly *P. pungens*). Workers reproduce by thelytoky and both morphological queens and males are absent or rare, such that about half of colonies are clonal and the other half are mixtures of clones (Nishide et al. 2007). Tsuji (1994) carried out an experiment similar to that of Hillesheim et al. (1989) in which he created colonies of *P. punctatus* with varying proportions of reproductive workers. In this case he found a more complex outcome than in *A. m. capensis*, in that colonies with an intermediate frequency of reproductive workers were the most productive. Colonies with 100% reproductive workers reared very little brood, but, unexpectedly, so did colonies with 100% non-reproductive workers. Colonies with roughly 80 to 95% reproductive workers reared the most brood. The reason why colonies with 100% non-reproductive workers performed so poorly was unknown (Tsuji 1994). Nonetheless, this example shows that the proportion of selfish reproducers in a colony may still be subject to frequency-dependent selection, although the relationship between colony performance and the frequency of selfish reproducers may not be straightforwardly linear.

5.4 Limitation of exploitation from inside: self-limitation through excessive costs to the group

Self-limitation can occur if selfish members of a social group damage group growth, survival, or productivity to such an extent that they impair their own reproductive success (Table 5.1). This means of self-limitation could, of course, interact with self-limitation through negative frequency-dependence, since, if each selfish group member inflicts a high cost on the group, the overall damage to the group will rise should the frequency of selfishness rise. But it is worth considering separately because its occurrence suggests that selfish group members might evolve so as to cap their per capita group-level costs. However, there are also cases when, if they derive no benefit, selfish group members are not selected to limit their group-level costs. The following examples come from multicellular organisms, eusocial societies, and interspecific mutualisms.

Non-transmissible cancers in multicellular organisms

A cancer is a set of mutant somatic cells inside a multicellular organism that divide in a manner controlled by the mutated loci in the cancer cells themselves, and not by genes elsewhere in the organism. Hence a cancer may be regarded as a lineage of selfishly reproducing cells (Buss 1987). In plants, cancers are more localized within the organism and so are less injurious than in animals. This is thought to be because plant cells, with their rigid cell walls, are less able to proliferate uncontrollably and cannot travel around within the organism to cause descendant cancers at new sites, i.e. to metastasize (Buss 1987; Michod and Roze 2001). In short, fibre cuts cancer even in plants. On top of this, plants may be less prone to damage by cancer because of their modular structure, combined with their lack of a central nervous system and vital organs.

Most cancers cannot be transmitted from one organism to another. The corollary is that the mutations causing them must arise afresh each generation (Queller 2000). A cancerous cell-lineage is unrelated to other cells in the body at any locus at which a cancer-causing mutation has occurred. Non-transmissible cancers potentially proliferate within the body, irrespective of the cost to the body, for two reasons. First, mutant genes causing cancer, precisely because they are no longer related to corresponding loci in the gametes, have no evolutionary interest in gamete propagation. (Genes at non-mutant loci in the cancer cells remain related to their counterparts in the gametes, but, if the cancer grows, they are not the genes in control of the cancer.) Second, when onward transmission by any means is impossible, the mutant genes place no value on the survival of the affected organism. Hence a non-transmissible cancer is an example of a within-group selfish subunit that is not selected to limit its level of group-level cost. Lacking an evolutionary future, it derives no benefit from doing so. These factors explain why, of all diseases, cancers are a byword for tragically relentless destructiveness.

Once arisen, cell lineages of non-transmissible cancers are expected to undergo selection and evolution in competition with other somatic cell lineages within the body, rather as if they were separate asexual species in some closed community, like an island. This evolutionary perspective on non-transmissible cancers has been explored by a number of authors (Nowell 1976; Nunney 1999b; Frank and Nowak 2004; Crespi and Summers 2005; Frank 2007b, 2010b). There is even a possibility that different cancerous cell-lineages cooperate with one another (Axelrod et al. 2006) or that, within a lineage, cancer cells exhibit coordinated, collective behaviours similar to those of other simple multicellular organisms (Deisboeck and Couzin 2009). In short, cancerous cell-lineages may act as social groups, in which case they would be expected to follow the general rules of social evolution (Chapter 2). But such evolution would still be within the host organism alone. Non-transmissible cancers cannot evolve to become better adapted to exploit hosts through exposure to multiple hosts in a manner analogous to that shown by parasites. Whatever self-promoting mutations they have accumulated through within-organism selection must be lost when the affected organism dies (Burt and Trivers 2006). Bodies are selected to combat cancers because cancers, if malignant, reduce the expected reproductive success of all the other genes in the genome. Moreover, bodies can accumulate anti-cancer adaptations because genes within a lineage are exposed to multiple instances of cancer across the generations. The only circumstances under which bodies would not evolve to combat malignant cancers would occur when the cancers struck late in life. In this case, as the evolutionary theory of ageing shows (e.g. Kirkwood and Austad 2000), the harmful effects of cancer would be relatively invisible to between-organism selection.

Transmissible cancers in multicellular organisms

A minority of cancers are unusual in being transmissible, and these cancers will have evolutionary dynamics different to those of non-transmissible cancers. Some transmissible cancers are caused by viruses, as in the case of cervical cancer in humans, which is sexually transmitted and is caused by human papillomavirus (Walboomers et al. 1999). More revealing, from the viewpoint of social evolution, are transmissible cancers in which the infectious agents are the cancerous cells themselves. These are very rare, being known from three or four species of mammal, including humans (Table 5.3). But they are particularly interesting because they represent cases in which selfish cell-lineages have acquired autonomy and become transmissible cellular parasites (Burt and Trivers 2006; Grosberg and Strathmann 2007; Aanen et al. 2008; Queller and Strassmann 2009). In other words, they demonstrate that, on rare occasions, apparently terminally differentiated cells can break free from their somatic prison to lead a new life in the outside world, albeit a parasitic one.

The two best-known cases of this type of transmissible cancer are devil facial tumour disease in the Tasmanian devil and canine transmissible venereal tumour in the Domestic dog (Table 5.3). These involve tumours on the surface of the body (on the face and genitals, respectively) that at first sight resemble any other malignant growth. But, within each species, tumours on all affected animals descend from a single mutant cell that arose in a conspecific animal at some point in the past. This ancestral cell has formed a monophyletic, clonal cell-lineage that proliferates asexually and spreads horizontally from host to host through tumour cells attaching to uninfected hosts during physical contact. The result is that every tumour is genetically unrelated to its host animal, a phenomenon that provided the clue to the unusual

Table 5.3 Transmissible cancers in which the cancer cells are the infectious agent

Species	Cancer	Mode of transmission	Notes, references
Tasmanian devil (*Sarcophilus harrisii*)	Devil facial tumour disease (DFTD)	Horizontally through aggressive physical contact of host animals	A monophyletic clonal cell lineage of 'recent' origin; highly virulent, causing population declines of Tasmanian devils in the wild (Pearse and Swift 2006; Siddle et al. 2007; Jones et al. 2008; McCallum 2008; Murchison et al. 2010)
Domestic dog (*Canis familiaris*)	Canine transmissible venereal tumour (CTVT)	Horizontally through sexual and other physical contact of host animals	A monophyletic clonal cell lineage with a single common ancestor in a wolf or dog a few hundred years ago, but earliest ancestor may have originated several thousand years ago; non-virulent (Murgia et al. 2006; Rebbeck et al. 2009)
Syrian hamster (*Mesocricetus auratus*)	Contagious reticulum cell sarcoma	Horizontally	Observed in laboratory populations only; experimentally transmissible by mosquitoes; but cellular transmission not formally proven and some hamster cancers appear transmitted by viruses or viroids (Banfield et al. 1965; Coggin et al. 1981)
Human (*Homo sapiens*)	ALL/lymphoma	Vertically from pregnant mother to foetus via placenta	One definite occurrence and sixteen other similar occurrences known from humans (Isoda et al. 2009)

nature of these cancers. Burt and Trivers (2006) characterized a cancer like canine transmissible venereal tumour as a 'highly degenerate mammal'. The logic of their point is that, given such tumours reproduce autonomously, are descended from their hosts, and are reproductively isolated, we should classify them as new mammalian species, namely ectoparasitic, asexual sister species of their respective hosts (Frank 2007c). It is only lack of familiarity that instead makes us think of transmissible tumours as intraspecific pathologies (and, in the case of the canine tumour, dog-lovers would probably not welcome the extra breed to their showrooms). Horizontally transmissible, cancerous cell-lineages are fortunately unknown in humans (Dingli and Nowak 2006). However, rare cases of mother-to-offspring transmission of cancerous cells have long been suspected and have now been proven (Isoda et al. 2009). Such vertical transmission demonstrates that, where there is internal fertilization followed by direct maternal provisioning of developing embryos, selfish cell-lineages can potentially overcome the limitations imposed by a single-cell bottleneck in the life cycle (Section 3.3) and achieve direct transmission to the next generation.

A cancerous cell-lineage that transmits from host to host would be expected to accumulate adaptations improving its ability to exploit the host population, although the asexuality of devil facial tumours and canine transmissible venereal tumours may impair their ability to adapt in the long run (e.g. Murgia et al. 2006; McCallum 2008). In any event, hosts are expected to mount an immune response when infected with tumour cells, which represent natural allografts. The transmissible tumours are able to circumvent hosts' immunity through a mixture of selection and chance. First, cells in canine transmissible venereal tumours reduce expression of DLA genes (genes encoding dog leukocyte antigen, part of the MHC), so failing to provoke immune rejection by the host (Murgia et al. 2006). This is presumably an evolved adaptation of the tumour. Second, in the example of human mother-to-foetus transmission studied by Isoda et al. (2009), the infant's tumour had mutated so as to lack part of the HLA (human leukocyte antigen) locus, which again rendered the tumour immunologically invisible to the host. Siddle et al. (2007) found that, by contrast, devil facial tumours express their full complement of MHC antigens. However, these authors also established that extant populations of Tasmanian devils have low diversity at their MHC loci, which prevents the animals from mounting an effective immune response.

In the terminology of the present chapter, transmissible cancers represent exploiters from within ('insiders') that have evolved into exploiters from without ('outsiders'). Hence, they exemplify that the distinction between exploiters from within and without is not necessarily rigid. As in the case of contagious diseases in general, the evolutionary theory of virulence (Section 4.2) helps explain the costs inflicted by transmissible cancers on their hosts. In epidemiological terms, a high level of virulence (high cost to host) is quite possible in diseases that are transmitted horizontally, depending on factors such as the contact rate and the frequency-dependence or independence of transmission (McCallum 2008). The two best-known horizontally-transmitted transmissible cancers differ in their virulence. Canine transmissible venereal tumours do not kill their hosts, regressing after several months (Murgia et al. 2006). This is presumably because each tumour should not allow its host to be destroyed before copulation by the host with a new partner, which is necessary to allow onward transmission of tumour cells (Murgia et al. 2006). Devil facial tumour disease differs in that it is almost always fatal to its Tasmanian devil victims. As a result, its spread is seriously endangering the survival of these unique marsupial carnivores (McCallum 2008), although populations are responding evolutionarily by exhibiting breeding at an earlier age (Jones et al. 2008). Tasmanian devils, unlucky once in lacking substantial genetic variation at their MHC loci (Siddle et al. 2007), are unlucky a second time in exhibiting a high level of aggressive contact between animals, even when uninfected (Jones et al. 2008; McCallum 2008). This double misfortune appears to have tipped the balance in favour of greater virulence of the tumours (McCallum 2008).

Unfortunately, devil facial tumour disease, therefore, illustrates the point that, whether from the inside or the outside, selfish elements imposing excessive costs on their host groups risk driving themselves and their hosts extinct. Extant social groups at all hierarchical levels are the ones in which processes of social group maintenance have proved effective (Rankin et al. 2007b), but there is nothing dictating that these processes prove so in every case.

Selfish reproduction within eusocial societies

In many species of eusocial Hymenoptera, the workers are not totally sterile but can lay viable eggs. Because of the haplodiploid genetics of the Hymenoptera and workers' inability to mate, the eggs of such reproductive workers typically develop into males (Bourke 1988). Reproductive workers represent selfish group members. They are not fully analogous to non-transmissible cancers within multicellular organisms, because selfish workers are able to transmit their genes via the sons they produce. Inclusive fitness theory predicts that the frequency with which workers successfully rear worker-produced adult males should depend both on the kin structure of the colony and on the level of worker policing of other workers' reproduction, which itself depends on colony kin-structure (Trivers and Hare 1976; Starr 1984; Woyciechowski and Lomnicki 1987; Ratnieks 1988; Wenseleers et al. 2004a). Put simply, when the average worker is more closely related to worker-produced males than to queen-produced ones, as occurs in a colony headed by one, singly-mated queen (Fig. 2.2), workers should raise worker-produced males. When the average worker is less closely related to worker-produced males than to queen-produced ones, as in a colony headed by one queen with an effective mating frequency greater than two, workers should police one another's reproduction and hence raise queen-produced males. At a comparative level, this prediction is upheld (Table 2.5). However, there are many species in which a lower level of worker male-production occurs than is predicted (e.g. Bourke 2005), a finding that has been explained as stemming from costs of worker male-production to colony-level productivity (Cole 1986; Ratnieks and Reeve 1992). Such costs might be expected because reproductive workers tend to work less than non-reproductive ones (Wenseleers et al. 2004a) or might mistakenly destroy queen-produced offspring while replacing existing brood (Ratnieks and Reeve 1992; Nonacs 1993).

Although plausible, it has proved difficult to test this explanation, because colony-level costs have proved harder to quantify than levels of relatedness or policing. Methods have included quantifying time and brood care lost as a result of the dominance behaviour of reproductive workers (Cole 1986), measurement of the additional energy used by such workers (Gobin et al. 2003), and experimentally assessing the effect on colony productivity of varying the frequency of reproductive workers (Lopez-Vaamonde et al. 2003; Dijkstra and Boomsma 2007). Results have been mixed, with two studies failing to detect significant costs of worker male-production (Lopez-Vaamonde et al. 2003; Dijkstra and Boomsma 2007) and two detecting moderate costs (Cole 1986; Gobin et al. 2003). Hence the explanation of workers' reproductive restraint through colony-level costs, though possibly correct, remains something of an open issue at present, and needs additional studies.

Worker social parasites of intraspecific origin in eusocial societies

Recently, investigators have found that workers in the eusocial Hymenoptera are able to leave their natal colonies, enter other conspecific nests, and produce their own sons within them as intraspecific social parasites (Beekman and Oldroyd 2008). So-called reproductive 'drifter' workers are now known from several species of bumble bees (Birmingham et al. 2004; Lopez-Vaamonde et al. 2004; Takahashi et al. 2010) and honey bees (Nanork et al. 2005, 2007). As described in the previous section, workers of a South African subspecies of honey bee, the Cape honey bee, are capable of producing worker daughters by asexual reproduction. Genetic analyses reveal that workers of this unique honey bee also enter other nests of the same subspecies and produce their own offspring (Härtel et al. 2006). Similar phenomena occur in social aphids, in which some aphids enter conspecific clonal groups that they then exploit by metamorphosing prematurely into reproductive forms (Abbot et al. 2001; Grogan et al. 2010). These cases present an obvious parallel with the behaviour of transmissible cancers described earlier in this section.

An even stronger analogy with transmissible cancers occurs in a subset of workers of the Cape honey bee (Martin et al. 2002; Oldroyd 2002). These workers occur as social parasites in colonies of a different subspecies of honey bee found in South Africa, *Apis mellifera scutellata*. The parasite workers produce new parasite workers within the host colony, with the result that eventually the host colony dies through lack of a productive workforce and the parasite offspring move on to new hosts (Neumann and Hepburn 2002; Moritz et al. 2008). Genetic studies show that the parasite workers from different host colonies form a clonal lineage that is traceable to a single ancestral worker of very recent origin (Baudry et al. 2004; Beekman and Oldroyd 2008). Furthermore, parasite *A. m. capensis* workers lay eggs that somehow escape the normal egg-policing system of the hosts, whereby worker-laid eggs are nearly always destroyed by other workers (Martin et al. 2002). Overall, therefore, this lineage of *A. m. capensis* workers represents a clonal, intraspecific social parasite of high virulence that evades the immunity of its hosts. At the level of the eusocial society, it mirrors extraordinarily closely the behaviour of transmissible cancers of multicellular organisms. Like them, it is arguably also a newly-evolving descendant species of its hosts (Neumann and Moritz 2002). Intriguingly, a genetically-distinct parasitic lineage of reproductive workers may also exist in the thelytokous ant *Pristomyrmex punctatus* (Dobata et al. 2009).

The existence of worker social parasites of intraspecific origin in eusocial societies reinforces the point made earlier about transmissible cancers, namely that internal exploiters may sometimes break free from the confines of their natal social group to become exploiters from the outside. Because exploiters from the outside are unrelated to their hosts, they do not stand to gain any indirect benefits from their hosts and they are not subject to the same sort of restraint on their levels of exploitation as are internal exploiters. Hence, parasitism of conspecifics in other social groups offers an opportunity to achieve unhindered direct reproduction to group members that, in the natal group, would otherwise form part of the helping caste. Such an alternative avenue for inclusive fitness maximization will, in general, have a retarding effect on social evolution, since it will prevent full commitment of would-be helpers to a helper role.

The extent to which this avenue remains open must depend on the effectiveness of the recognition systems of social groups. For example, if recognition and exclusion of strangers is extremely effective, then attempting intraspecific social parasitism would not yield any benefits. Suggestively, Chapman et al. (2009) found that, in the Honey bee, queenless colonies excluded foreign conspecific workers more effectively than colonies with a queen, consistent with queenless colonies being more prone to attempted infiltration by reproductive foreign workers. This finding represents some of the first evidence that, as the above considerations predict, levels of intraspecific social parasitism by workers and the effectiveness of recognition systems coevolve.

Limitation of exploitation by excessive costs in interspecific mutualisms

Limits to the costs inflicted by a defecting partner within interspecific mutualisms have not been well explored, but a couple of examples show that they exist just as in other social groups. The first example involves an ingenious experiment in which Sachs and Wilcox (2006) manipulated the mutualism between the Upside-down jellyfish, *Cassiopea xamachana* and its algal symbiont, *Symbiodinium microadriaticum*. This is an open mutualism, the chief mode of transmission of the alga being horizontal. The experimenters set up treatments in which transmission of the alga was allowed to be horizontal or vertical, respectively. As predicted by the evolutionary theory of virulence (Section 4.2), the alga evolved to be relatively beneficial in the vertical treatment and relatively harmful in the horizontal treatment. However, the horizontal treatment did not produce unfettered virulence. Sachs and Wilcox

(2006) detected trade-offs in the impact of the alga on its jellyfish hosts, in that faster-dividing algal populations within hosts stunted host growth and so decreased the number of algae produced for onward transmission. Hence, excessive costs to the hosts were likely to have limited the level of selfishness of the symbiont.

The second example comes from another ingenious experiment, this time using the mutualism between fungus-growing ants and their fungal symbionts. Little and Currie (2009) showed that, in this system, the ants and their fungal symbiont experienced greater fitness loss from attack by the parasitic fungus *Escovopsis* (Section 5.2) when either of them was experimentally forced to act as a cheat. These findings suggest that the level of cost that a cheat within a mutualism may inflict depends on the context. Specifically, an external threat to the survival or productivity of the partnership as a whole reduces the tolerable level of cost. This is presumably because, in such a situation, each partner's self-interest lies in investing resources in overcoming the threat. In addition, the findings lead to the counter-intuitive conclusion that parasites of interspecific mutualisms, as well as threatening their destruction (Section 5.2), sometimes act to stabilize them (Little and Currie 2009).

5.5 Limitation of exploitation from inside: limitation by others through coercion

If selfish group members fail to curb their own spread through their effects on the group as a whole, other group members may nonetheless force them to act more cooperatively. This brings us to the final broad means of limitation of exploiters from inside a social group, namely coercion (Table 5.1). Coercion takes many forms (Ratnieks 1988; Leigh 1991; Clutton-Brock and Parker 1995; Frank 2003; Travisano and Velicer 2004; Ratnieks and Wenseleers 2008). Moreover, as before, examples can be found at several hierarchical levels, including the eukaryotic cell, multicellular organisms, eusocial societies, and interspecific mutualisms.

Enforced uniparental inheritance of mitochondria

Organelles derived from formerly free-living bacteria, such as mitochondria and chloroplasts, have travelled the reverse of the journey taken by transmissible cancers, representing outsiders that have become insiders. As insiders, they have evolutionary interests that overlap heavily with those of the nuclear genes within the eukaryotic cell, largely through vertical transmission conferring a shared reproductive fate on the two parties (Sections 3.1, 4.2). The long association and overlapping interests of organellar and nuclear genomes in turn have led to a massive transfer of genes from organelles to nuclei (Rand et al. 2004). For example, the genomes of animal mitochondria code for only 12–13 proteins and it is estimated that 98% of the original complement of genes has transferred to the nucleus (Ridley 2000; Burt and Trivers 2006). Similarly, chloroplast genomes encode 50–200 proteins and most of the genes originally in chloroplasts have transferred to the nucleus (Martin 2003;

Archibald 2009). In *Arabidopsis*, the premier model organism for plant genomic studies, an astonishing 18% of nuclear genes originate from cyanobacteria via chloroplasts (Martin 2003). In some eukaryotic lineages the emigration of organellar genes has been total. Cells in these lineages contain structures called hydrogenosomes and mitosomes that resemble organelles except that they lack DNA. The evidence shows that these structures derive from mitochondria (Embley et al. 2003; Knoll 2003; Embley and Martin 2006), lingering as faint relics of their former inhabitants like crop marks left at the site of an abandoned Roman villa. But despite such comprehensive merging of organellar and nuclear interests, organellar genes that have stayed in place retain distinctive evolutionary interests that, as apparent from the case of cytoplasmic male sterility (Sections 2.3, 5.3), sometimes manifest themselves.

One context in which organellar interests are particularly relevant arises during sexual reproduction in eukaryotes. In outbreeding species, the formation of a zygote brings together not only unrelated nuclear genomes but also unrelated organellar genomes. Potential conflict between the nuclear genomes from each parent is managed through the imposition of a fair meiosis, as described later in this section. What controls potential conflict between the organellar genomes from each parent? Typically, mitochondria and chloroplasts are inherited via one sex of parent only (uniparental inheritance), usually but not exclusively via females (Birky 1995, 2001, 2008; Xu 2005). It has generally been considered that organelles within any single zygote are clones of one another, which, barring mutations, would lead to organelles within any single organism being clones of one another as well (Birky 2001). Some cases of genetic mixtures of organelles within single organisms, or heteroplasmy, are now known (Barr et al. 2005; White et al. 2008; Galtier et al. 2009). Nonetheless, it is reasonable to assume that the usual condition is for genetic diversity of organelles within organisms to be low. In view of these points, it has been proposed that potential conflict between the organellar genes from each parent within zygotes has been actively managed by the imposition of uniparental inheritance (Eberhard 1980; Cosmides and Tooby 1981). The idea in its current form is that the nuclear genomes of both parents conspire, as it were, to enforce uniparental inheritance and that they do this precisely in order to render the organelles within the zygote as genetically uniform as possible, so precluding potentially costly between-organelle conflict (Hurst and Hamilton 1992; Maynard Smith and Szathmáry 1995; Queller 1997; Ridley 2000).

Some of the best evidence for this conflict-reduction interpretation of uniparental inheritance comes from the fate of mitochondria during mammalian reproduction. Remarkably, in mammals, the male's nuclear genes produce a protein, ubiquitin, that tags all mitochondria in the sperm. The tag is masked in some way until the eight-cell stage of the embryo, when it is unmasked and leads to the destruction of all paternally derived mitochondria but none of the maternally derived ones (Sutovsky et al. 2000; Burt and Trivers 2006). The paternal nuclear genome, therefore, sets a time-bomb to destroy its own organellar partners. It is not fully clear why the maternal nuclear genome does not instead destroy its own organelles (Burt and Trivers 2006;

Lessells et al. 2009). Perhaps such a system would lead too often to disastrous consequences if mistakes were made. However, paternal elimination of organelles in mammals appears to have the 'agreement' of the maternal nuclear genome, since the destruction of paternal mitochondria occurs in the embryo. In addition, in mammals, tracing the fate of mitochondrial heteroplasmy down lines of descent within pedigrees suggests that genetic uniformity of mitochondria within individuals is soon restored (Cao et al. 2009). This could happen through a variety of ways. One is via a bottleneck in the population of mitochondria occurring in the female germline cells (White et al. 2008). Recent experimental evidence points instead to selective cloning of mitochondrial variants in those cells (Wai et al. 2008; Cao et al. 2009). Either way, the imposition of genetic uniformity among the mitochondria would again serve the interests of nuclear genes (Ridley 2000).

The conclusion is that nuclear genes take active steps to ensure genetic uniformity of organelles in the zygote or embryo. Uniparental inheritance is an evolved trait, since, across eukaryotes as a whole, several methods of eliminating one parent's organelles have arisen independently (Birky 1995; Xu 2005; Burt and Trivers 2006). One might consider uniparental inheritance to be a by-product of the existence of large gametes or eggs and small gametes or sperm (anisogamy), given sperm cannot physically contain many organelles. For example, mammalian eggs contain 1000 to 100,000 times as much mitochondrial DNA as mammalian sperm, and salmon eggs contain 100,000,000 times as much mitochondrial DNA as salmon sperm (Wolff and Gemmell 2008). That anisogamy alone does not account for uniparental inheritance is shown by the fact that, in sexually reproducing algae with isogamy, i.e. having gametes of equal size, one parent's mitochondria (and chloroplasts) are still eliminated (Maynard Smith and Szathmáry 1995; Ridley 2000; Burt and Trivers 2006). Nonetheless, some authors have argued that anisogamy, and hence gender itself, evolved from isogamy as a means of facilitating uniparental inheritance of organelles (Hurst and Hamilton 1992; Ridley 2000). Others point to disruptive selection on gamete size in the context of competition between gametes for fertilization success as the primary driver of the evolution of anisogamy (Parker et al. 1972; Lessells et al. 2009). Regardless of which of these viewpoints is correct, the evidence for a self-interested, coercive suppression by nuclear genomes of potential conflict among organellar genomes remains strong.

Enforced fairness in meiosis

Conflict between genes within eukaryotic cells can be intragenomic instead of intergenomic as in the previous example (Section 2.3). Intragenomic conflict takes many different forms (Burt and Trivers 2006). A classic example is the conflict between genes for meiotic drive (those achieving greater then 50% representation in gametes) and other genes in the nuclear genome (Section 2.3). To fix ideas, take the case when a drive allele is sex-linked, i.e. it is on one of the sex chromosomes (Jaenike 2001). Assume further that the drive allele arises on the X chromosome in males in a system in which, as in people, XX zygotes develop as females and XY ones develop as males. Therefore, if drive is complete, an adult male bearing the drive allele will, instead of producing 50% X-bearing sperm and 50% Y-bearing sperm, produce 100% X-bearing sperm. Drive genes of this kind are found in several species of *Drosophila* fruitflies (Jaenike 2001). Mechanistically, what happens is that somehow the Y-bearing sperm are destroyed during spermatogenesis. Hence, X-linked drive genes, as well as causing males to produce only X-bearing sperm, pleiotropically reduce the sperm count.

Drive genes of this kind are in conflict with other genes in the nuclear genome for several reasons (Jaenike 2001; Burt and Trivers 2006). As ever, it is instructive to analyse these reasons in inclusive-fitness terms, though they are not always viewed this way. First, if drive is complete, the relatedness of a drive gene in a male to the male's sperm is 1 (since all sperm carry the X chromosome with the drive gene) but the relatedness of the drive gene's allele on the Y chromosome to the male's sperm is 0 (since no sperm carry a Y chromosome). Hence, the drive gene and its allele have differing relatedness asymmetries and are, therefore, in conflict. This point follows directly from the lack of relatedness of alleles within a locus (Section 2.3). Second, because of its X-biased relatedness asymmetry, the drive gene favours an all-female sex ratio. Genes on chromosomes other than the sex chromosomes (autosomes) retain a relatedness asymmetry of 0.5:0.5 and so favour a standard sex ratio of 1:1 females:males. So a potential conflict between the drive gene and the autosomal genes is present that increases in intensity as the frequency of the drive gene rises and the population sex ratio becomes more female-biased. Third, if the drive gene reduces the sperm count to such an extent that male fertility is impaired, this provokes conflict with autosomal genes independently of the drive gene's effect on the sex ratio, since the autosomal genes lose out from the male's lower reproductive success but fail to gain a compensating increase in representation within his sperm.

For all these reasons, the effect of X-linked drive genes is to provoke suppressor (modifier) genes either on the Y chromosome or on the autosomes. Selection for suppression on the Y chromosome will be strong at all frequencies of the drive gene, since the Y chromosome is the immediate victim of drive. Selection for suppression at autosomal loci will be strongest when the drive gene is frequent, because of the sex-ratio effect (Jaenike 2001). As expected, both Y-linked and autosomal suppressors of X-linked meiotic drive occur (Jaenike 2001; Burt and Trivers 2006; Tao et al. 2007). These act to restore normal spermatogenesis in males and hence normal brood and population sex ratios. The molecular mechanisms for the suppression of X-linked drive have not been widely investigated, but a study in *Drosophila simulans* suggest it occurs by RNA interference, with the autosomal suppressor gene's product being a short RNA sequence that inactivates the drive gene (Tao et al. 2007; Jaenike 2008). This and other systems of suppression of X-linked drive therefore represent cases of selfish group members being physically suppressed at the genomic level.

Leigh (1977) suggested that mutations leading to meiotic drive would usually end up being suppressed, since, there being many more non-drive loci than drive ones, suppressor mutations would always be likely to arise. Hence what he famously termed a 'parliament of genes' helps keep meiosis generally fair. The 'parliament' is, however, a group of replicators each in potential conflict but united by a shared reproductive fate. Effectively, each drive suppressor acts for itself, not for its fellow non-drive genes (Dawkins 1982; Bourke and Franks 1995; Maynard Smith and Szathmáry 1995). But every non-drive gene benefits from suppression, because of their shared reproductive fate. Haig and Grafen (1991) further suggested that genetic recombination (crossing-over of chromosomes) itself evolved to keep meiosis fair by disrupting meiotic drive. Drive genes usually work by destroying their alleles indirectly, through causing the destruction of chromosomes labelled by the presence of a target sequence at a locus linked to the drive locus (Ridley 2000). Haig and Grafen's (1991) proposal was that a gene for recombination would be favoured if it broke up the linkage between the drive and target loci, because then a drive gene would find itself recombined with its target sequence and so destroy itself. Hence, both suppressor genes and genes for recombination may act to maintain the stability of organisms produced by sexual reproduction despite the ever-present threat of meiotic drive, which arises through each organism being built from a fused pair of unrelated nuclear genomes.

Enforced suppression of cytoplasmic male sterility and other forms of sex ratio distortion

The evolution of uniparental inheritance of organelles described earlier in this section, while resolving one form of intergenomic conflict, sets up conditions for another (Burt and Trivers 2006). This arises because transmission through only one of the sexes alters the relatedness asymmetry of organelles. If, as is the usual case, organelles are transmitted via females only, they become related to eggs or ovules by 1 and to sperm or pollen by 0 (Section 2.3). Organelles are then selected either to promote the production of female gametes (in hermaphrodites) or to promote the production of daughters (in species with separate sexes). In the latter case, a bias towards daughter production is expected because only female offspring will go on to produce eggs or ovules. Or, as Ridley (2000) put it, 'A male body is one big mitochondrial graveyard'. It is for these reasons that, as we have seen (Section 2.3), mitochondrial genes arise that cause cytoplasmic male sterility in hermaphroditic plants. As we have also seen (Section 5.3), genes for cytoplasmic male sterility limit their own spread through negative frequency-dependence. The new point to be made here is that, in addition, they are countered by suppressor genes in the nuclear genome. Like suppressors of X-linked drive genes, these arise because they favour an equal allocation of investment in female and male function. In the case of cytoplasmic male sterility, several so-called nuclear restorer genes are now known (Schnable and Wise 1998; Bentolila et al. 2002; Burt and Trivers 2006). They act by somehow altering the expression of genes for cytoplasmic male sterility or, in one case, by causing the causative gene to be deleted (Schnable and Wise 1998).

In animals with separate sexes, a large array of non-organellar, maternally-transmitted, cytoplasmic symbionts also act as selfish sex-ratio distorters, creating, as expected, female-biased sex ratios through various means. Some of the best known examples are intracellular *Wolbachia* bacteria (Section 2.3). They are dubbed male-killers, since they cause infected female hosts to abort their male embryos and redirect resources towards the production of daughters (Werren et al. 1986; Hurst 1991; Engelstädter and Hurst 2009). Male-killing *Wolbachia* causing highly female-biased population sex ratios are found in the Blue Moon butterfly, *Hypolimnas bolina*. But the effects of *Wolbachia* are countered by nuclear suppressors, leading to wild swings in the butterfly's population sex ratios (Hornett et al. 2006; Charlat et al. 2007). Hence, coercive control is apparent in many types of intragenomic and intergenomic conflict.

Coercion in eusocial societies

Coercion within eusocial societies can be divided into three types: dominance, punishment, and policing. The distinctions between these are not absolute, but are nonetheless sufficient to be worth making (Monnin and Ratnieks 2001; Ratnieks and Wenseleers 2008).

Dominance is aggressive behaviour aimed at gaining direct fitness for a dominant group member by imposing a within-group reproductive monopoly. It frequently occurs in the context of a dominance hierarchy in which rank correlates with reproductive success (Wilson 1975; Ellis 1995; Beekman et al. 2003). In other words, subordinates within dominance hierarchies are reproductively suppressed. In the classification of Ratnieks and Reeve (1992), which is based on the relative numbers of actors and recipients, dominance is a one-to-one or one-to-many interaction, since, typically, each member of a dominance hierarchy dominates one or more subordinates lower in the hierarchy. Dominance often occurs in small eusocial societies (Alexander et al. 1991), with well-known examples being colonies of polistine paper wasps, in which an alpha female attacks other egg-laying females and eats their eggs (Reeve 1991). It also typifies many kinds of vertebrate society, including those of carnivores and primates (Creel and Macdonald 1995; Ellis 1995).

Dominance is characterized by two kinds of asymmetry. First, there is often some sort of power asymmetry between dominants and subordinates, with dominants being larger, older, or stronger (Wilson 1975; Beekman et al. 2003). Second, there may be an asymmetry in the fitness consequences of escalated fighting. For example, in eusocial Hymenoptera in which mothers (queens) reproductively dominate daughters, a queen killing a subordinate daughter loses one member of the workforce, but a subordinate daughter killing the queen loses all the future siblings the queen would have provided (Trivers and Hare 1976). Both kinds of asymmetry act to reinforce dominants' superior position in the hierarchy. The result is greater broad-sense cooperation, since, if only the dominant reproduces, squabbles over reproduction among the subordinates are minimized.

Punishment is usefully considered as a coercive behaviour that provides a fitness benefit (direct or indirect) to the punisher by deterring its victims from acting selfishly in the future, so again reinforcing broad-sense cooperation (Clutton-Brock and Parker 1995; Monnin and Ratnieks 2001; Ratnieks and Wenseleers 2008). It may be a one-to-one, one-to-many, or many-to-one interaction (Ratnieks and Reeve 1992). An example of punishment occurs in Chimpanzees (*Pan troglodytes*), in which some group members attack former allies that fail to assist them in contests for resources with other group members (Clutton-Brock and Parker 1995). Another occurs in the ant *Dinoponera quadriceps* (Fig. 5.3). In this species, a beta female that challenges the alpha female is first marked for punishment by the alpha with a secretion applied with the sting, then punished by physical immobilization by low-ranking workers. The result is that the beta stops being a challenger (Monnin et al. 2002). In this case, the alpha uses the low-ranking workers as the agents of punishment, but these workers are willing accomplices because, as non-reproductive daughters of the alpha, they are more closely related to the alpha's offspring than to the potential offspring of the beta

(a)

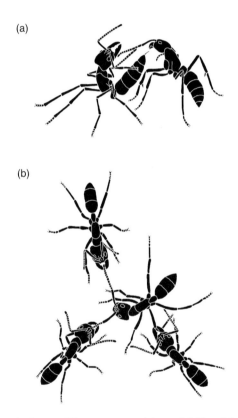

(b)

Fig. 5.3 Punishment in the ant, *Dinoponera quadriceps*. (a) The alpha female (left) smears a challenger, the beta, with a secretion from her Dufour's gland, applied with the sting. (b) Low-ranking workers pin down the smeared beta by her legs or antennae for several days, with the result that the beta loses her high-ranking status. Reproduced from Monnin et al. (2002) with kind permission of the authors.

(Monnin et al. 2002). Punishment has also been intensively studied as a significant means of enforcing cooperation within human societies (Fehr and Gächter 2002; Sigmund 2007).

Finally, policing is a coercive behaviour providing (typically) an indirect fitness benefit to the policer by preventing successful reproduction in social cheats without necessarily deterring cheats from acting selfishly in the future (Ratnieks 1988; Frank 1995; Ratnieks and Wenseleers 2008). Because it usually returns an indirect but not a direct fitness benefit, it need not be performed by socially dominant, reproductive group members. It is generally a many-to-one or many-to-many inter-action (Ratnieks and Reeve 1992). Hence policing enforces broad-sense coopera-tion not only by preventing successful cheating, but also by conferring an indirect benefit on members of the collective from which policers are drawn (the first 'many' in the many-to-one or the many-to-many). Policing has been most widely investigated in the eusocial Hymenoptera (Ratnieks and Wenseleers 2005; Ratnieks et al. 2006). One type of Hymenopteran policing is policing over caste fate (whether a group member develops into a reproductive or a non-reproductive form). This occurs in stingless bees of the genus *Melipona*, in which colonies routinely over-produce virgin queens, nearly all of which are culled by workers (Bourke and Ratnieks 1999). *Melipona* is unusual in that queens and workers develop in cells of the same size, giving developing females nutritional control of their caste fate (whereas, in a species with dimorphic cells, a female larva in a small cell could aspire to become a worker only, excluding unusual cases of dwarf queens). Hence, the bizarre and wasteful massacre of virgin queens makes sense when interpreted as worker policing of the selfish tendency of immature females to develop as queens against the workers' interests (Bourke and Ratnieks 1999; Ratnieks 2001; Wenseleers et al. 2003, 2004b).

The classic type of policing in eusocial Hymenoptera is worker policing of worker egg-laying, first described in the Honey bee. Policing workers are non-laying workers that, in colonies with a queen, eat the male eggs laid by laying workers. Worker policing was predicted in the Honey bee by Ratnieks (1988) on the grounds that, when the single queen is multiply mated, the average non-laying worker is more closely related ($r = 0.25$) to the average queen-produced male than it is ($r \approx 0.125$) to the average worker-produced male (Fig. 2.2). Therefore, a gene for suppressing other workers' reproduction (policing) will spread, because it will result in the average non-laying worker receiving the greater indirect benefit to be gained from rearing queen-produced males. A laying worker remains more closely related to its own eggs ($r = 0.5$) than to any other colony member's eggs, and so is still selected to lay (unless egg destruction becomes very effective). The evidence from the Honey bee suggests that, as expected, non-laying workers selectively destroy worker-laid eggs, with the result that few workers lay eggs and very few adult males are worker-produced (Ratnieks and Visscher 1989; Visscher 1989; Ratnieks 1993).

Subsequently, worker policing of workers' reproduction, both via egg-eating and via attacks on workers with activated ovaries, has been found in many species of eusocial Hymenoptera (Ratnieks et al. 2006; Wenseleers and Ratnieks 2006a).

Colony kin-structure does not predict the extent of worker policing and the proportion of adult worker-produced males in every case (Kikuta and Tsuji 1999; Foster et al. 2000; Hartmann et al. 2003; Pirk et al. 2003). However, comparative studies show that, across the eusocial Hymenoptera as a whole, policing and the frequency of adult worker-produced males covary with colony kin-structure as inclusive fitness theory predicts (Table 2.5). In cases where workers police eggs when they are not more related to queen-produced offspring, it has been suggested that policing is maintained because it keeps down the colony-level productivity costs of worker reproduction (Ratnieks 1988; Hartmann et al. 2003; Ratnieks and Wenseleers 2008; Ohtsuki and Tsuji 2009). But costs of worker reproduction have proved hard to measure (Section 5.4). Therefore, the relative contributions of productivity and relatedness effects to worker policing remain uncertain. It is possible that in some cases worker policing originally evolved for reasons unconnected with colony kin-structure, such as colony-level productivity benefits, and was later modulated by variations in colony kin-structure (Ohtsuki and Tsuji 2009). Better knowledge of the costs of worker reproduction and of the phylogenetic history of policing would go far in clarifying this issue.

One reason why worker policing by egg-eating, which is essentially a form of infanticide, may be particularly frequent in eusocial Hymenoptera is that each act of policing is likely to be of low cost to the policer (Ratnieks and Reeve 1992). Workers have to interact with eggs as part of their normal duties, eggs do not defend themselves and—at least in large colonies—cannot all be guarded, and little investment is embodied in an insect egg (plus what there is gets recycled by ingestion). However, for effective policing, there must be some mechanism for workers to minimize recognition errors, and destroy worker-laid but not queen-laid eggs (Ratnieks and Reeve 1992; Nonacs 1993). In the ant *Camponotus floridanus*, a species in which worker policing occurs in the absence of workers being more closely related to queen-produced males, the mechanism by which workers discriminate against worker-laid eggs has been elucidated (Endler et al. 2004). Queen-laid and worker-laid eggs bear different blends of chemicals on their surfaces, allowing workers to tell one type from the other. In addition, the chemical profiles of queen-laid and worker-laid eggs match those on the cuticles of adult queens and workers, respectively (Fig. 5.4). The suggestion is that queens label their eggs with a caste-specific chemical signature to ensure their eggs are not mistakenly eaten. Moreover, the ability to produce the component chemicals is believed to be physiologically linked to the queens' greater fecundity, so preventing laying workers from faking the queens' egg-signature in order to cheat the system (Endler et al. 2004).

In sum, the coercive behaviours of dominance, punishment, and policing can each promote broad-sense cooperation within eusocial societies, but do so in a variety of ways. One thing they have in common is that they decrease the benefits and increase the costs of acting selfishly in social cheats (Ratnieks and Wenseleers 2008), so discouraging the occurrence of cheating either over the time-scale of the life of the eusocial society or over evolutionary time. Moreover, since they represent social actions in their own right, dominance, punishment, and policing must evolve according

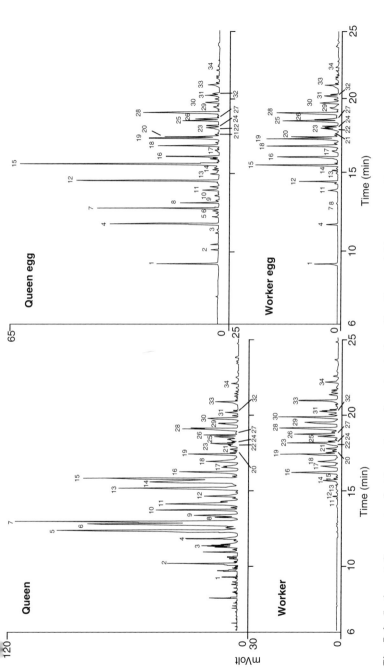

Fig. 5.4 In the ant *Camponotus floridanus*, the surface chemical profile of queen-laid eggs resembles that of adult queens and the surface chemical profile of worker-laid eggs resembles that of adult workers. Each peak represents a different hydrocarbon, with position on the horizontal axis (retention time in GC/MS analysis) reflecting molecular weight and height of the peak representing amount. Numbers identify particular chemicals in the original analysis. Reproduced from Endler et al. (2004) with kind permission of the authors and publisher. Copyright (2004) National Academy of Sciences, USA.

to the same rules, based on inclusive fitness theory, as social actions in general (Section 3.2).

Enforced fairness in interspecific mutualisms

The temptation to cheat is greater in interspecific mutualisms than in groups of relatives (other things being equal), because, in interspecific mutualisms, relatedness cannot act as a brake on the level of damage inflicted on one's partner (Section 3.2) Hence it is not surprising that mutualisms are prone to breakdown through cheating or that coercive means of preventing cheating have evolved (Sachs et al. 2004; Douglas 2008).

Coercive prevention of cheating in interspecific mutualisms occurs through a variety of mechanisms (Herre 1999; Sachs et al. 2004). First, mutualists may try to impose vertical transmission on their partners, so guaranteeing a large degree of shared reproductive fate (Section 4.2). For example, fungus-growing ants actively prevent the formation of fruiting structures (mushrooms) of their mutualistic fungi, with the result that fungal interests remain more aligned with those of their host ants (Mueller 2002). Second, mutualists may take steps to remain with a single partner over their lifetime. Such partner fidelity again increases the level of shared reproductive fate (Section 4.2). Connected with this, and analogously with the effect of nuclear genomes on organellar genetic diversity, mutualists would benefit by imposing genetic uniformity on their partners, so minimizing costly internal conflict within partners (Frank 1996c, 1997). Evidence that this occurs comes from a recent study of fungus-growing termites (Aanen et al. 2009). The work showed that, although the termites may bring in to their nests several strains of their mutualistic fungus at the start of colonial life, they create conditions for one strain to outcompete the others, so ensuring genetic uniformity of the symbiont over the remainder of the lifetime of each nest. Since most termites do not transmit their fungus vertically, this suggests that ensuring genetic uniformity among symbionts can override the instability inherent in horizontal transmission of symbionts (Aanen et al. 2009). Third, mutualists may pick only those partners that they assess will act cooperatively and reject those that they assess will not act this way, i.e. partner choice (Bull and Rice 1991; Sachs et al. 2004). A nice example comes from a mutualism involving the cleaner fish, *Labroides dimidiatus*. A variety of fish (clients) solicit this cleaner fish to remove parasites and dead tissue from them, but some cleaners cheat by nipping off bits of healthy tissue from their clients. Bshary (2002) found that client fish preferentially select cleaner fish that they have observed not to cheat when cleaning other clients. Finally, analogously to coercion within eusocial societies, mutualists may impose costs (sanctions) on cheating partners (West et al. 2002b; Douglas 2008; Kiers and Denison 2008). A common form of mutualism between ants and plants involves plants providing ants with food and dwellings (hollows called domatia) in return for protection from herbivory. Edwards et al. (2006) found that the ant-plant *Cordia nodosa* sheds domatia on stems with simulated herbivory, suggesting that the plants retaliate against partners that they assess not to be serving their interests.

5.6 Predicting the outcome of the limitation of exploitation

The limitation of exploitation of social groups by cheats is essential for social group maintenance. The processes considered in this chapter all demonstrate how such limitation can evolve by natural selection. In other words, as has to be the case, it can come about in the absence of top-down imposition. External cheats are controlled by selection on group members for the self-interested exclusion of would-be intruders. Internal cheats are controlled either by selection on themselves leading to self-limitation or by selection on other group members for the self-interested prevention of cheating within the group. However, in the case of internal cheats, the exact form and outcome of the limitation of cheating is not always predictable. The effect of self-limitation is, in theory, predictable from the effects of an increased frequency of cheats, or increased per capita costs of cheating, on the growth, survivorship, or productivity of the social group. Similarly, adding coercive policing to such a system allows one to predict the fate of cheats as a function of the costs of policing (e.g. Frank 1995; Wenseleers et al. 2003, 2004a). But, as far as the evolution of coercive control of within-group cheating in general is concerned, it is harder to predict how evolution will proceed. This is because the evolutionary outcome depends to a large extent on what party possesses most practical power, which tends in turn to depend on detailed features of specific systems (Ratnieks and Reeve 1992; Hurst et al. 1996; Beekman and Ratnieks 2003; Beekman et al. 2003; Ratnieks et al. 2006). Such a lack of a predictable outcome applies to conflicts at all levels, including those within organisms, eusocial societies, and interspecific mutualisms.

Nonetheless, despite potential idiosyncrasies affecting exactly how coercion may suppress cheating, some general principles do present themselves. For example, as mentioned in the previous section, dominance through physical aggression seems easier to achieve in smaller social groups, for the obvious reason that a dominant then has fewer potential competitors to control (Wilson 1971; Ratnieks 1988; Alexander et al. 1991; Bourke 1999). This in turn implies that, in small social groups, it is easier for the interests of a single group member to prevail over those of the collective. As a corollary, it should be easier for the interests of the collective to prevail as group size rises, since then it becomes less likely that a single group member can be physically dominant (Ratnieks 1988; Bourke 1999). This leads to the prediction that punishment and policing, which often serve a collective interest and are not necessarily performed by physically dominant group members, are more likely to occur as group size rises (Sections 6.4, 6.5).

In cases such as these, it is important to appreciate in what sense the collective's interests are served or a collective can hold power. Essentially, group members may be united in a 'community of interest' (Cosmides and Tooby 1981) within a social group if each one of them shares the same inclusive fitness optimum. Each acts for itself but, because many others share the same standpoint, the result is a conditional unification of interests (Dawkins 1982; Bourke and Franks 1995). Members of such a community can then collectively wield greater power because there are more group members from which mutants selected to perform the same, self-interested coercive action can be drawn. This argument was the one presented by Leigh (1977) when proposing that a 'parliament of genes' limits meiotic drive, as outlined in the previous section. Hurst et al. (1996) further discuss the concept of power in genetic conflicts. Parallels with how power affects other forms of within-group coercion occurring at different hierarchical levels, such as policing, were pointed out by Leigh (1991) and Pomiankowski (1999).

Another general principle governing the outcome of coercive limitation of within-group cheats concerns the control of information within the social group (Beekman and Ratnieks 2003; Beekman et al. 2003; Ratnieks et al. 2006). As described in the previous section, Haig and Grafen (1991) proposed that genes for recombination could have been selected to break up the linkage of loci involved in meiotic drive. In a sense, recombination would be effective in this because it prevents a drive gene from discriminating between itself and its allele (Ridley 2000), so depriving the drive gene of the information it requires to operate. Keller (1997) and Reeve (1998b) have considered how, analogously, active scrambling of recognition cues could be a general means by which members of a social group prevent cheating. Evidence exists that eusocial societies are selected to increase their levels of internal genetic variation to provide resistance against parasites and improve colony efficiency (Section 5.2). But this has the effect of increasing potential conflict between different genetic lineages within the society. One way in which such conflict could be prevented from becoming actual would be by the mixing of recognition cues among group members. For example, in eusocial Hymenoptera, larvae of different patrilines or matrilines might be made chemically indistinguishable to prevent nepotistic rearing of particular lineages (Keller 1997; Reeve 1998b; Ratnieks et al. 2006). Consistent with this, some homogenization of cuticular chemical profiles has been found in adult worker Honey bees (Arnold et al. 2000). Monnin and Ratnieks (2001) pointed out that egg-laying workers in eusocial Hymenoptera could deliberately mix up the colony's eggs to prevent their eggs being recognized and policed. In this case, cue-scrambling would facilitate cheating. Therefore, as with other within-group manipulative behaviours, whether active manipulation of information occurs and whose interests it serves will depend on the party that has control of it.

Finally, if they temporarily share overlapping interests, different relatedness classes within social groups may unite in coalitions against cheats. A good example comes from the ant *Dinoponera quadriceps* discussed in the previous section, in which the alpha female and low-ranking workers share an interest in punishing a beta female that mounts a reproductive challenge to the alpha (Monnin et al. 2002). Egg-marking by the queens of eusocial Hymenoptera may also represent the queen's way of enlisting the support of policing workers, since destruction of eggs that are not the queen's serves the queen's interest (Ratnieks 1995; Endler et al. 2004). In short, communities of interest, themselves conditional on sharing inclusive-fitness optima, may be selected to unite in conditional coalitions of interest. But whether this will happen in any one case is again likely to depend on its specific features.

5.7 Summary

1. Social group maintenance occurs through the limitation of exploitation by both external and internal elements. In this context, external elements include parasites and pathogens, and internal elements are would-be social cheats.

2. Limitation of exploitation by external elements relies on recognition of self versus non-self, followed by rejection of non-self. These processes occur to an extent in some interspecific mutualisms. They are almost universal in multicellular organisms (immune systems) and in eusocial societies (nestmate or groupmate recognition). Recognition systems are often highly effective without being 100% perfect. Pressure from parasites can select for greater levels of genetic variation within social groups, whereas selection for reduced potential conflict within groups acts in the opposite direction. Hence levels of within-group genetic variation may be the product of balancing selection.

3. Limitation of exploitation by social cheats can take place by several means. One is self-limitation through negative frequency-dependent selection, in which the reproductive success of a social cheat falls as it grows more frequent, often through an increased frequency of cheating leading to reductions in group-level productivity. Examples include cheating social bacteria and reproductive workers in eusocial Hymenoptera.

4. Another means of limitation of exploitation by social cheats is self-limitation by excessive per capita costs inflicted by cheats on the group. Evidence for limits to the costs inflicted by cheats comes from eusocial Hymenoptera with reproductive workers and from interspecific mutualisms with potentially cheating partners. Transmissible cancers in which cells are the infectious agent represent internal exploiters of multicellular organisms that have become external ones, and, in one case (canine transmissible venereal tumour), they also appear limited in their effects on hosts by the need to avoid excessive costs.

5. Limitation of exploitation by social cheats may occur through coercion by other group members. Coercion of various forms is found at all hierarchical levels, and includes enforced uniparental inheritance of mitochondria, enforced suppression of meiotic drive, nuclear suppression of cytoplasmic male sterility, dominance, punishment, and policing in eusocial societies, and sanctions against defecting partners in interspecific mutualisms.

6. The conclusion is that the limitation of exploitation of social groups by cheats in its various forms evolves by natural selection. The exact evolutionary outcome is predictable in some cases but not others. Specifically, both the occurrence and the effects of coercive prevention of internal cheating are not always predictable, because they depend on which parties within a social group have the most power, which may vary on a case-by-case basis.

6

Social group transformation

6.1 The size-complexity hypothesis for social group transformation

Many examples of social groups exist that do not readily fit the definition of an individual, since they lack one or more of the qualities associated with individuality. These qualities include interdependence and coordination of parts, cooperation in achieving common goals, and attenuation of internal conflict (Section 1.3). For example, a nest of squabbling paper wasps seems less individualistic than a hive of honey bees, just as a scratch militia lacks the unity of a well-drilled army platoon. Hence, for the evolution of individuality to reach completion, some process must turn stable social groups into more cohesive, integrated entities worthy of being regarded as individuals. This process is social group transformation.

While it has not been overlooked, social group transformation has not received the level of theoretical and empirical study that has been directed at the formation and maintenance of social groups (Section 1.3). The present chapter aims to redress this imbalance. The importance of social group transformation is not hard to see. First, as the crowning step in the evolution of individuality, social group transformation represents an integral part of any major evolutionary transition. Second, a sequence of major evolutionary transitions cannot occur unless units emerging at one level become sufficiently integrated to serve as the subunits that enter the next level, which requires social group transformation (Fig. 1.4). Therefore, understanding the biological hierarchy in its entirety requires an understanding of social group transformation. Overall, a fully expanded view of social evolution needs to include the key phenomena and principles that social group transformation entails.

This chapter describes the phenomena of social group transformation and then explores a hypothesis that, it is proposed, permits the underlying principles to be picked out. To begin with, the chapter describes the suites of traits (or syndromes) that characterize social groups that have not undergone social group transformation and those that characterize groups that have. Respectively, such groups are labelled 'simple' and 'complex'. Bonner (1988, 1993, 2003b, 2004, 2006) described what he termed a size-complexity rule—an association between cell number and organismal complexity—in multicellular organisms. Along with other authors, he highlighted a similar association between worker number and social complexity in eusocial societies (e.g. Wilson 1971; Alexander et al. 1991; Bonner 1993, 2004; Bourke 1999). All these authors argued that the association between size and complexity is causal. Following Bonner, I term the idea that an increase in the size of social groups causes complexity, whether in multicellular organisms or eusocial societies, the size-complexity hypothesis. The bulk of the chapter develops an extended version of the size-complexity hypothesis that, I propose, accounts for the transformation of simple social groups into complex ones. Note that, throughout, group size means the number of component subunits, i.e. the number of cells within a multicellular organism or the number of members within a eusocial society.

In essence, the extended version of the size-complexity hypothesis proposes the following (Fig. 6.1). First, external ecological and evolutionary drivers tend to favour increased size in social groups (Section 6.3). Second, increased size selects for the syndrome of social traits that (independently of size itself) collectively define social complexity (Section 6.4). Third, via positive feedback (i.e. through self-reinforcing social evolution), these traits promote further increases in group size (Section 6.5). The result is the transformation of small, simple social groups with one suite of traits, which collectively amount to a lesser degree of individuality, into large, complex social groups with another, which collectively confer greater individuality (Section 6.6). The hypothesis focuses on social groups of relatives, namely multicellular organisms and eusocial societies. However, the final section of the chapter briefly considers what social group transformation involves in the case of interspecific mutualisms. The basic approach resembles the one taken in Bourke (1999). It differs in that the chapter applies the extended size-complexity hypothesis to multicellular organisms as well as to eusocial societies, and incorporates, where appropriate, more recent theoretical findings and data (e.g. Jeon and Choe 2003; Reeve and Jeanne 2003; Wenseleers et al. 2004a; Fjerdingstad and Crozier 2006; Michod 2006; Boomsma 2009).

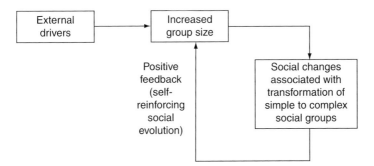

Fig. 6.1 Schematic diagram of the extended version of the size-complexity hypothesis proposed in this chapter to underlie social group transformation in multicellular organisms and eusocial societies.

6.2 Simple versus complex social groups

Simplicity and complexity in social groups

A number of authors have noted the existence of syndromes of simplicity and complexity in social organization. In the case of multicellular organisms, they include Bonner (1988, 1993, 2003b, 2004, 2006), McShea (1996), Bell and Mooers (1997), Ridley (2000), Grosberg and Strathmann (2007), and Herron and Michod (2008). In the case of eusocial societies, they include Wilson (1971), Alexander et al. (1991), Bonner (1988, 1993, 2004), Bourke (1999), Anderson and McShea (2001), Strassmann and Queller (2007), and Boomsma (2009). Several have also pointed out correspondences in the features defining these syndromes across multicellular organisms and eusocial societies. What these authors have collectively highlighted is that, within both multicellular organisms and eusocial societies, it is possible to divide species into those in which social groups are simple and those in which social groups are complex, based on a shared set of criteria, such as the degree of reproductive and non-reproductive division of labour and overall size (Tables 6.1, 6.2). One might quibble with the use of 'simple' and 'complex' in this context. However, the labels attached to these syndromes are less important than the fact that an evolutionary explanation for them appears to be required, which necessitates calling them something. In addition, using a pair of terms to define a set of consistent differences does not preclude the existence of intermediates (Anderson and McShea 2001). There is a continuum of complexity in the organization of social groups, and 'simple' and 'complex' describe its extremes.

Table 6.1 Features associated with simplicity and complexity in multicellular organisms (for source references see Section 6.2)

	Simplicity	Complexity
Reproductive division of labour:		
Degree of dimorphism between germline and somatic cells	Low	High
Reproductive potential of somatic cells[1]	High	Low
Segregation of germline[2]	Absent	Present
Timing in development of divergence of germline and somatic cell lineages	Late	Early[3]
Non-reproductive division of labour:		
Number of somatic cell-types	Low	High
Size (number of component members):		
Number of cells	Low	High

[1] Hence, somatic cells in the simplest multicellular organisms are totipotent, whereas those in the most complex ones are terminally differentiated.

[2] Segregation of the germline refers to whether or not there is a single lineage of germline cells that is confined to one part of the body.

[3] An early-diverging germline is here hypothesized, on average, to be a feature of large, complex multicellular organisms, but evidence for this is currently equivocal (Section 6.2).

Table 6.2 Features associated with simplicity and complexity in eusocial societies (for source references see Section 6.2)

	Simplicity	Complexity
Reproductive division of labour:		
Degree of dimorphism between reproductive (queens) and non-reproductive forms (helpers or workers)[1]	Low	High
Reproductive potential of helpers or workers[2]	High	Low
Segregation of developing reproductive forms[3]	Absent	Present
Timing in development of immatures of divergence of reproductive and non-reproductive forms[4]	Late	Early
Non-reproductive division of labour:		
Number of morphological (physical) castes among helpers or workers, i.e. degree of worker polymorphism	Low	High
Size (number of component members):		
Number of helpers or workers	Low	High

[1] Encompasses both difference in size and in morphology; in principle these can vary independently, but usually size differences between reproductives and non-reproductives are accompanied by morphological differences.

[2] High reproductive potential implies that adult helpers or workers are totipotent, whereas low reproductive potential, in its most extreme form, implies a lack of functional reproductive organs and hence irreversible sterility.

[3] The formal analogue of segregation of the germline in eusocial societies could involve both the existence of a morphological reproductive caste and the physical confinement of developing reproductive forms of the new generation to one part of the nest (such that developing worker-destined young are less likely to 'aspire' to development as reproductives because of their unsuitable physical location). The latter has not been systematically investigated, but certainly occurs in many complex eusocial Hymenoptera, which segregate sexual brood from worker brood. For example, ants segregate brood by size (Carlin 1988) and at least in some species by sex and caste (Jemielity and Keller 2003). The Honey bee separates its brood by sex and caste (Winston 1987) whereas some stingless bees (*Melipona*) do not (Bourke and Ratnieks 1999).

[4] The formal analogue of early divergence of the germline in eusocial societies would involve foundress females laying a batch of eggs early in their reproductive lives that were then destined to develop (slowly) into the new generation of reproductive adults. As far as is known, this does not happen in any eusocial insect. Instead, queens initially lay worker-yielding eggs, and produce reproductive-yielding eggs later in their lives (Oster and Wilson 1978). However, as the table notes, during the development of each colony member there is variation in the relative timing of divergence into reproductive or non-reproductive forms (caste determination).

In brief, simple multicellular organisms are those with a low degree of reproductive division of labour, a low degree of non-reproductive division of labour, and (it is hypothesized) a relatively low number of group members (cells). Examples of simple multicellular organisms include the cellular slime moulds (Fig. 3.1) and the smaller species of volvocine algae (Fig. 4.1). Complex multicellular organisms are those with a high degree of reproductive division of labour, a high degree of non-reproductive division of labour, and (it is hypothesized)

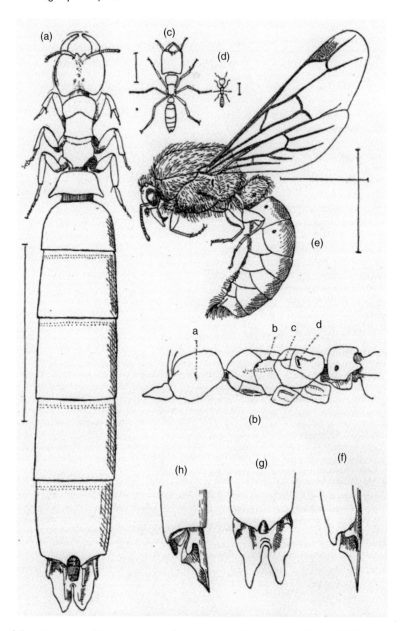

Fig. 6.2 A eusocial Hymenopteran, the army ant *Dorylus helvolus*, exhibiting complex eusociality, characterized by extreme queen–worker dimorphism and by worker polymorphism. (a) queen (top view); (b) queen (side view of head and thorax); (c) major worker; (d) minor worker; (e) male. Reproduced from Wheeler (1910) with kind permission of the publisher. Copyright © 1910 Columbia University Press.

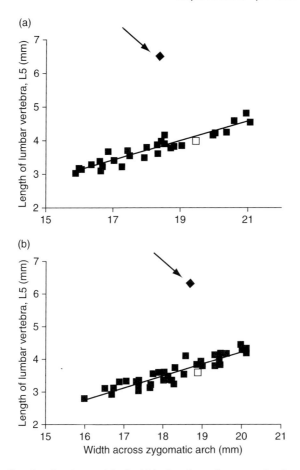

Fig. 6.3 Breeding females (queens) in the Naked mole-rat have greatly elongated vertebrae relative to their body size. Elongation of vertebrae occurs after a female's first pregnancy as a queen and facilitates the bearing of the large litters typical of Naked mole-rats (up to 28 young per litter). Shown are width across zygomatic arch, a measure of body size, plotted against vertebral length (lumbar vertebra 5) in two colonies of Naked mole-rats (a and b). Black squares: non-breeding females and males; white square: breeding male; black diamond (arrowed): breeding female. Reproduced from O'Riain et al. (2000) with kind permission of the authors and publisher. Copyright (2000) National Academy of Sciences, USA.

a relatively high number of group members (Table 6.1). Examples of complex multicellular organisms include animals, plants, and fungi. Simple and complex eusocial societies show corresponding syndromes of traits with respect to reproductive division of labour, non-reproductive division of labour, and number of group members (Table 6.2). Examples of simple eusocial societies include those of allodapine and halictid bees and, under a broad definition of eusociality, the societies of cooperatively-breeding vertebrates. Examples of complex eusocial societies include those of many ant and termite species (Fig. 6.2). The most complex society among eusocial vertebrates occurs in the Naked mole-rat, which shows a form of caste dimorphism between reproductive and non-reproductive adults (Fig. 6.3). The analogue of complex eusociality is also found in the colonial marine invertebrates (Fig. 1.3).

Simple and complex multicellular organisms or eusocial societies may differ from one another in other ways as well (Buss 1987; Bourke 1999; Anderson and McShea 2001). For example, complex multicellular organisms exhibit elaborate processes of development (Buss 1987; Maynard Smith and Szathmáry 1995). Complex eusocial societies undergo a coordinated set of changes as they grow, analogous in some respects to organismal development (Wilson 1985). But, because the changes in the degree of reproductive and non-reproductive division of labour are the ones most tightly linked to the emergence of individuality, it is these traits that the size-complexity hypothesis concentrates upon.

Evidence for size-associated syndromes of simplicity and complexity in multicellular organisms

The size-complexity hypothesis predicts that an increase in the size of social groups causes increases in the degree of reproductive and non-reproductive division of labour. Hence one expects to find positive associations between these traits. Existing evidence for the relevant syndromes and associations is of variable quality. Nonetheless, such evidence exists, and comes from both the study of particular cases and comparative surveys. Consider first the evidence from multicellular organisms. Volvocine algae contain approximately 10 to 10^5 cells (Fig. 4.1). Species with few cells have a single cell-type, with all cells being capable of reproduction, although some cells have a lower probability of dividing than others (Section 4.2). By contrast, species with many cells have two cell-types, namely morphologically distinct reproductive (germline) and somatic cells (Kirk 1998; Bonner 2003b; Herron and Michod 2008). Hence, in the volvocine algae, larger size is associated with the evolution of a morphologically distinct germline and soma.

Across eukaryotes more generally, however, the relationship between the size of multicellular organisms, the presence or absence of a germline, and the timing of divergence of the germline, if present, is not straightforward (Pál and Szathmáry 2000). Within animals, there is a very crude relationship between cell number and mode of germline formation, in that the two phyla with the largest mean cell numbers always exhibit a differentiated germline (Table 6.3). But some representatives of these two phyla have a late-diverging germline, and other phyla have an early-diverging germline despite their low mean cell number (Table 6.3). Within phyla in which there is variation in the relative timing of germline divergence, it appears there are not enough data to determine whether or not size is a correlate of this variation.

It also needs noting that the phylum-level averages for cell number in Table 6.3 are based on relatively few species per phylum and that the table omits many phyla (Buss 1987; Bell and Mooers 1997). Hence the data on the relationship of cell number and mode of germline formation are very incomplete and need improving. Overall, in multicellular organisms, an association between size and a morphologically distinct germline is reasonably well supported, but an association between size and an early-diverging germline is tentative.

Multicellular organisms with relatively few cells have few types of somatic cell. In the prokaryotes, filaments of multicellular cyanobacteria such as *Anabaena* contain of the order of 10^5 cells (Bonner 1988), which is not a big number by the standards of larger complex multicellular organisms with as many as 10^{14} cells (Section 3.3). *Anabaena* has just two cell-types in total (Table 4.4). One of them is a specialist nitrogen-fixing cell-type, the heterocyst (Kaiser 2001; Golden and Yoon 2003), which is incapable of further cell division (Willensdorfer 2008). Hence this prokaryote shows a division into reproductive cells and one type of terminally differentiated somatic cell. The slugs and fruiting bodies of cellular slime moulds contain up to 2×10^6 cells (Bonner 2009). Their number of cell types varies from two to four, including the terminally differentiated cells of the stalk of the fruiting body, and it is the species with more cells that tend to have the higher numbers of cell types (Bonner 2003b; Schaap et al. 2006). No simple multicellular organism consists of just two cells that differ morphologically from one another (Bell and Mooers 1997).

Table 6.3 Mode of germline formation in relation to total cell number in animals

Phylum	Ranked mean log.$_{10}$ number of cells[1]	Mode of germline formation[2]
Mesozoa	2.29	Early
Placozoa	2.50	No germline
Nematoda	2.80	Early
Gastrotricha	3.48	Early
Mollusca	3.65	Early, late
Ctenophora	4.00	Early
Porifera	4.00	No germline
Entoprocta	4.08	No germline
Rotifera	4.33	Early
Annelida	5.58	No germline, early, late
Platyhelminthes	6.40	No germline, early, late
Cnidaria	6.85	No germline
Arthropoda	10.50	Early, late
Chordata	11.95	Early, late

[1] Data from Bell and Mooers (1997); names of phyla have been left as in the original.
[2] Data from Buss (1987). No germline = Buss's 'somatic embryogenesis', i.e. no differentiated germline; late = Buss's 'epigenetic embryogenesis', i.e. germline diverges relatively late in development; early = Buss's 'preformistic embryogenesis', i.e. germline diverges relatively early in development. More than one entry per phylum means that taxa within the phylum vary in their mode of germline formation.

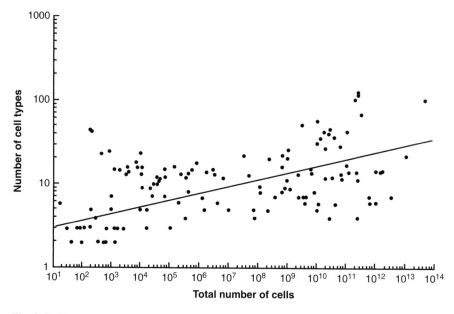

Fig. 6.4 Larger multicellular organisms have more cell types. Plot of log cell number against log number of cell types (mostly somatic) in a wide range of organisms, e.g. cellular slime moulds, algae, fungi, plants, and animals. Data from Bell and Mooers (1997). Reproduced from Bonner (2004) with kind permission of the author.

A comparative survey also shows that, across the whole range of cell number within eukaryotes, organisms with higher cell numbers have more somatic cell-types (Fig. 6.4). For example, vertebrates have over 10^{11} cells and over one hundred cell-types, most of which are somatic (Bonner 1988; Bell and Mooers 1997).

Evidence for size-associated syndromes of simplicity and complexity in eusocial societies

Almost all analyses of the correlates of social complexity in eusocial societies have been carried out on social insects. This is unsurprising, since, among vertebrates, only the Naked mole-rat approaches a high degree of social complexity (Fig. 6.3). However, incipient caste dimorphism between reproductives and non-reproductives in Naked mole-rats is consistent with the size-complexity hypothesis, since these rodents have the largest vertebrate societies (Bourke 1999; O'Riain et al. 2000). Pursuing this logic, one can conclude that vertebrate societies fail to become as complex as invertebrate ones because they fail to grow as populous (Bourke 1999). If ever a Pygmy naked mole-rat were found, with colonies containing thousands rather than hundreds of members, its form of sociality would be expected to be more complex. In social insects, comparative analyses of varying levels of sophistication have shown that, as predicted, the degree of dimorphism between reproductive and non-reproductive forms is positively associated with colony size in several taxa of eusocial bees and wasps (Bourke 1999) and in ants (Fjerdingstad and Crozier 2006). Among other eusocial invertebrates, a similar relationship occurs in eusocial shrimps (Tóth and Duffy 2008). The size-complexity hypothesis predicts that, in colonial marine invertebrates, the degree of dimorphism between reproductive and non-reproductive zooids should increase with zooid number (as should the number of different types of non-reproductive zooid); but relevant data appear absent. Note that, from now on, when speaking of eusocial societies, I will refer to reproductive forms as queens and non-reproductive, helper forms as workers. This admittedly insect-centred terminology is simply to avoid the clumsiness of the 'reproductive versus non-reproductive' formulation.

In social insects, loss of reproductive potential by workers occurs in several steps (Ratnieks 1988; Bourke 1999; Khila and Abouheif 2010). There may be loss of the ability to disperse effectively, so reducing the probability of successful reproduction outside the natal group. This is exemplified by the disappearance of wings in worker termites and ants (workers of bees and wasps retain their wings). There may be loss of ability to mate. This is exemplified by loss of the sperm receptacle by workers in many species of eusocial Hymenoptera. These workers cannot produce daughters sexually, but, if they retain their ovaries, haplodiploidy allows them to produce sons parthenogenetically. Finally, there may be a loss of fecundity through the ovaries becoming smaller, culminating in some species in the total loss of functional ovaries in adults. Such irreversible worker sterility represents the analogue of terminal differentiation in the somatic cells of multicellular organisms

No study has examined how colony size affects the occurrence of each of these steps in loss of workers' reproductive potential. At a broad scale, consistent with the size-complexity hypothesis, total worker sterility is found only in major taxa of social insects with large colonies, such as ants, stingless bees, and termites (Bourke 1999). However, in a phylogenetically-controlled comparative analysis across 50 species of eusocial Hymenoptera in which workers could produce sons, Hammond and Keller (2004) found no significant association between colony size and the percentage of adult worker-produced males. This did not support the size-complexity hypothesis (Table 6.2). But this study omitted species with very large colonies (10^6 to 10^7 workers), such as the leaf-cutter ant *Atta* or the army ants *Dorylus* and *Eciton*, in which worker reproductive potential appears to be particularly low (e.g. Dijkstra and Boomsma 2006; Kronauer et al. 2006a, 2010). Helanterä and Sundström (2007) found no association between colony size and degree of worker male-production across nine species of *Formica* ants, which again did not support the size-complexity hypothesis. This study included highly polydomous species (those whose colonies occupy several interconnected nests) in which colony boundaries and colony size are hard to define. The concept of social complexity defined by a syndrome of interconnected traits applies principally to multicolonial eusocial species (i.e. ones living in separate, mutually hostile colonies), since these are the ones in which societies are sufficiently distinct genetically to be candidates for consideration as individuals. The conclusion is that, in social insects, the claim of the size-complexity hypothesis that large colony size is associated with reduced worker reproductive potential currently has only equivocal support.

In social insects with morphologically distinct queens and workers, the timing of divergence (caste determination) between the developmental pathways leading to each phenotype varies widely, with divergence occurring at every point between the egg and the pupal stages (Wheeler 1986; O'Donnell 1998). There is the hint of a positive relationship between early divergence and colony size (Wheeler 1986), but there has been no systematic investigation of this relationship across a large number of species.

In ants, the number of worker castes in even the most complex societies (fewer than ten) never approaches the number of somatic cell-types in the most complex multicellular organisms. Nonetheless, it is the species with very large colonies (e.g. *Atta*, *Dorylus*, and *Eciton*) that the highest levels of worker polymorphism are found (Hölldobler and Wilson 1990). At the other end of the scale of colony size, just as no simple multicellular organism consists of two cells that differ morphologically from one another, no eusocial society consists of a single queen and a single, morphologically distinct worker. A positive association between the level of worker polymorphism and colony size in ants (Fig. 6.5) was supported in across-species comparative surveys by Bonner (1993) and Anderson and McShea (2001). In their later phylogenetically-controlled comparative analysis of 35 ant species, Fjerdingstad and Crozier (2006) found only a marginally significant positive association between the level of worker polymorphism and colony size. Furthermore, there was no significant association between these two variables once the level of queen–worker size dimorphism was statistically controlled. However, again, Fjerdingstad and Crozier's (2006) dataset did not include species with a colony size of 10^6 or more, i.e. those with the most polymorphic workers.

In sum, though there is some support for the existence of size-associated syndromes of simplicity and complexity in social groups among both multicellular organisms and eusocial societies, more comprehensive comparative analyses are required to test rigorously all the predicted associations with size. Nonetheless, with the aim of developing and investigating the size-complexity hypothesis as fully as possible, in the remainder of the chapter I assume that these associations exist and explore likely explanations and consequences.

Number of independent evolutions of complexity in social groups

Defining social group transformation as a distinct step in a major evolutionary transition implies that the origin of simple social groups and the origin of complex social groups are separate events. Several authors have recognized this point (though not always in these terms) in the contexts of both multicellularity and eusociality

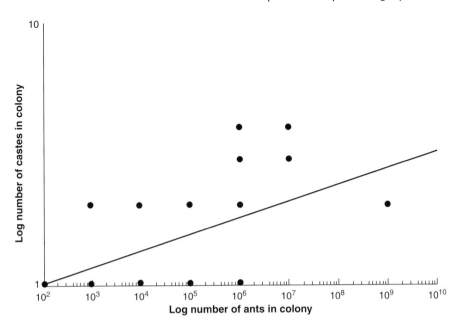

Fig. 6.5 Larger ant societies have a greater level of worker polymorphism. Plot of log number of workers against log number of physical worker castes. Reproduced from Bonner (1993) with kind permission of the author and the Indian Academy of Sciences.

(Buss 1987; Bourke 1999; Lachmann et al. 2003; Szathmáry and Wolpert 2003; Grosberg and Strathmann 2007). To expand it, consider that when a social group first arises (social group formation), it must be simple. Complex social groups cannot spring into being fully formed, as this would be too improbable (Bourke 1999). They must evolve from simple social groups. We have already examined the number of independent origins of simple social groups (Box 1.1). Hence, the question now arises as to how many times complex social groups have arisen from their simple predecessors.

This question has not been studied systematically, but an increasing number of phylogenetic analyses allow it to be answered in some lineages (Table 6.4). One such analysis, Herron and Michod's (2008) study of the volvocine algae, has mapped out in rich detail the sequence of steps taken in the path from solitary living, to simple group life (here, simple multicellularity), to the beginnings of complex group life (here, complex multicellularity). It found three independent origins of one facet of multicellular complexity, namely terminal differentiation of somatic cells (Table 6.4). Indeed, the basic message from relevant phylogenetic analyses within various taxa is that, following social group formation, complex social groups have evolved multiple times independently (Table 6.4). However, within single taxa, it is hard at present to discern any consistent relationship between the number of origins of simplicity and complexity. For example, within eukaryotes, simple multicellularity evolved at least 16 times independently (Box 1.1). Complex multicellularity in eukaryotes is confined to animals, plants, fungi, red algae, and brown algae (Box 1.1). Hence it has evolved at least five times independently, though the total may be higher since complex multicellularity is believed to have evolved more than once in the fungi (Table 6.4). From this it follows that many multicellular lineages in eukaryotes have retained a simple organization (Fig. 1.1). By contrast, in termites, eusociality evolved once (Table 1.4), which is less than the number (three) of later, independent origins of complex eusociality (Table 6.4).

One point that is apparent from considering origins of simplicity and complexity within lineages is that they can be widely separated in time. Between the first cell around 3500 mya and the appearance of simple multicellularity (in prokaryotes) there seems to have been a relatively short interval of a few hundred million years or less (Box 1.1). But there was a much longer wait for the appearance of complex

Table 6.4 Selected examples of independent origins of complex social groups, based on phylogenetic analyses

Taxon (trait denoting complexity)	Number of independent origins of complex social groups	References
Multicellular organisms:		
Fungi (complex multicellularity = with differentiation of tissues, present independently in Ascomycota and Basidiomycota)	2	Stajich et al. (2009)
Volvocine algae (complex multicellularity = with terminally differentiated somatic cells)	3	Herron and Michod (2008); Sachs (2008)
Eusocial societies:		
Ants (complex eusociality = with worker polymorphism)	3[1]	Brady et al. (2006a); Moreau et al. (2006)
Swarm-founding neotropical wasps, i.e. Epiponini (complex eusociality = with morphologically distinct queen)	>1	Noll and Wenzel (2008)
Termites (complex eusociality = with irreversibly wingless, sterile workers)	3	Thorne (1997); Inward et al. (2007b)

[1] Not formally analysed, but the distribution of ant subfamilies with worker polymorphism (Oster and Wilson 1978) within recent ant phylogenies suggests at least three independent origins of worker polymorphism.

multicellularity, which, being found only in eukaryotes, must have arisen some time after the origin of eukaryotes around 2000 mya (Box 1.1). Hence, as Ridley (2000) put it, 'Complex life looks evolutionarily difficult'. On a lesser scale, a similar delay is evident in the transformation to complexity in eusocial societies. Ants and termites both arose by the early Cretaceous, at least 115 mya and 140 mya, respectively (Table 1.4). But, as Grimaldi and Engel (2005) pointed out, their sparse fossil records suggest that they were rare at this time, implying they existed as small-colony (simple) eusocial forms. According to these authors, the two taxa 'exploded in biomass and diversity' only in the Eocene, roughly 50 mya, when species with large colonies appeared. Hence, in ants and termites, at least 65 million years elapsed before the transformation from simple to complex eusociality took place (Grimaldi and Engel 2005). Reasons for the delayed evolution of complexity, in the case of either multicellularity or eusociality, are not well understood. A likely one is that complexity requires an incremental build-up of many separate traits (e.g. Herron and Michod 2008). In the case of multicellularity, as next discussed, another could involve a connection with the evolution of sex.

Complexity, sexual reproduction, and genetic variation

Simple multicellularity occurs in prokaryotes and hence can evolve in the absence of sexual reproduction as found in eukaryotes. But, as already mentioned, complex multicellularity occurs only in eukaryotes. Because sexual reproduction appears to be the ancestral condition in eukaryotes (Box 1.1), it follows that complex multicellularity arose only in lineages with sexual reproduction (though some complex multicellular organisms have secondarily lost sexual reproduction; Section 4.3). Furthermore, facultative sex is found commonly in unicellular eukaryotes but far less often in multicellular ones (Dacks and Roger 1999). Hence there is a strong connection between complex multicellularity and sex (Ruvinsky 1997; Ridley 2000). This raises the issue of whether complex multicellularity requires sex for its long-term persistence. Hamilton (1982), as part of his parasite theory for the maintenance of sex (Section 5.2), suggested that large multicellular organisms require sex to maintain resistance over evolutionary time against parasites that, being small and having shorter generations, evolve faster. More broadly, Ridley (2000) argued that complex multicellularity entails larger genomes and that, as genome size increases, sex becomes necessary to prevent the degeneration of adaptations through the accumulation of harmful mutations. So, powerful arguments find that sex is indeed necessary for the evolution of complex multicellularity. The implication is that complex multicellularity was delayed in its appearance in evolutionary time because sex had to evolve first. But the link between sex and complex multicellularity could be fortuitous (Bonner 1998). Since the eukaryotic cell and eukaryotic sex each appear to have arisen just once (Box 1.1), before the multiple origins of complex multicellularity, we lack a set of multiple, independent lineages that would allow us to test for an obligate link between complex multicellularity and sex. At present, therefore, only the strength of the arguments that complex multicellularity requires sexual reproduction to persist serves to make the case that evolution could not have proceeded otherwise.

Notwithstanding the uncertainty surrounding the link of complex multicellularity and sex, a possible parallel exists with the case of eusociality. Complex eusociality in the Hymenoptera seems to be associated with a high level of polyandry in queens, which increases within-group levels of genetic variation relative to those found under ancestral monandry (Section 5.2). In general, a positive association exists between colony size and degree of polyandry in eusocial Hymenoptera (Cole 1983; Boomsma and Ratnieks 1996). The largest, most complex societies of ants (*Atta*, *Dorylus*, and *Eciton*) are characterized by moderate to high levels of polyandry (Kronauer et al. 2004, 2006b; Helmkampf et al. 2008). Studies have also detected a genetic component to worker polymorphism in these or closely related species (Hughes et al. 2003; Jaffé et al. 2007). Therefore, as well as enhancing resistance against parasites and increasing colony efficiency (Section 5.2), the additional genetic variation introduced into colonies of these species by polyandry may facilitate the evolution of worker polymorphism. In this sense, as with multicellularity, complexity in eusocial societies may partly depend on processes that maintain within-group genetic variation. However, the entire matter of how complexity and genetic variation are connected is underexplored at present and requires further investigation.

6.3 External drivers leading to greater size in social groups

Social groups will not initially increase in size spontaneously, so the size-complexity hypothesis asserts that, for the evolution of size-associated complexity to begin, something external to simple groups must cause them to become larger (e.g. Bonner 1988; Bourke 1999). The extended size-complexity hypothesis proposes two classes of external drivers that potentially have this effect, namely short-term ecological drivers and long-term evolutionary ones.

Short-term ecological drivers of greater size in social groups

Ecological factors that promote the formation of social groups (Section 4.3) are also likely to bring about increases in group size in established social groups. For multicellular organisms, therefore, the external drivers of greater group (body) size will include any that put a premium on benefits from increases in mobility and dispersal, in resilience to environmental stresses, in capacity for nutrient uptake and storage, or in protection against predators (Table 4.3). To this list can be added, in sexual organisms, any that confer benefits of larger body size through increased mating success under sexual selection (Bonner 1988). For eusocial societies, relevant drivers are those yielding benefits of larger group (colony) size from increases in resilience to environmental stresses, in capacity to monopolize high-quality sites, and in protection against competitors, parasites, and predators (Table 4.3). Evidence for natural selection on body size in wild populations of multicellular organisms occurring via at least some of the proposed benefits was collated by Endler (1986). Direct demonstrations of advantages of larger size in eusocial societies, and the factors underlying them, are rarer (Bourke 1999). One example comes from a field study of the Red imported fire ant (*Solenopsis invicta*). Adams and Tschinkel (2001) showed that, in a monogynous population of this species, annual survivorship of colonies was higher as colony size increased. Furthermore, experimental removal of colonies led to remaining colonies expanding their foraging territories. These results strongly suggested that selection for success in intraspecific competition favoured larger colonies. Modelling and data on encounters between colonies also suggest that intraspecific competition drives large colony size in the African army ants, *Dorylus* (Boswell et al. 2001).

Long-term evolutionary drivers of greater size in social groups

Bonner (1988) pointed out that, at any moment in evolutionary time, the potential niche represented by the size class immediately above the currently largest organism within a given lineage is empty. So, as he put it, 'there is always room at the top'. This means that, over evolutionary time, maximum body size within a lineage might creep upwards as successive taxa discover this empty niche and fill it. The same argument applied to colony size in eusocial societies would contribute to an upward bias in colony size over evolutionary time (Bourke 1999). The limits to such size increases would be set by physical limits to body size in multicellular organisms and by decreasing returns to scale, in terms of colony efficiency, in eusocial societies (e.g. Michener 1964).

Two additional evolutionary processes might add to any tendency for body or colony size to increase over evolutionary time. The first is that the attainment of greater size in one lineage might catalyse the attainment of greater size in a competing lineage, through an evolutionary arms race (e.g. Dawkins and Krebs 1979). As described in the previous section, ants and termites evolved large colony size at roughly the same geological time, in the Eocene (Grimaldi and Engel 2005). This may be more than coincidence. To the extent that these two taxa compete with one another, the appearance of larger colonies in one taxon, which would therefore have represented superior competitors, may have led to selection for greater colony size in the other. The same effect might also have operated between different lineages within each taxon. Wilson and Hölldobler (2005) similarly invoked the competitive superiority of complex eusociality in the attainment of ecological dominance by ants and termites. An analogous process could have promoted the evolution of ever-greater body size in multicellular organisms. For example, in land plants, the occurrence of multiple, independent origins of large woody forms (trees) is easy to understand as the consequence of an arms race within and between species for competitive access to the critical resource represented by light (Dawkins 2009).

A second means by which size might tend to increase over evolutionary time within lineages concerns the issue of whether social evolution can ever go into reverse. Several authors have discussed the likelihood that taxa, caught near the origin of eusociality, could slip back and forth across the border between solitary-living and eusociality (Michener 1990; Gadagkar 1997; Wcislo and Danforth 1997). Evidence for such reversals in social evolution comes from phylogenetic analyses of halictid bees (Fig. 1.2). However, these bees form relatively simple societies. In the ponerine ants, there is evidence that species with morphologically distinct queens and workers have given rise to species in which all females are morphologically alike (Peeters 1993). If so, these would represent cases of moderately complex eusocial societies giving rise to simpler ones (Bourke 1999; Boomsma 2009). By contrast, what has never been observed are cases of complex eusocial societies producing solitary descendants (Gadagkar 1997; Wilson and Hölldobler 2005). Instead, eusocial lineages appear to pass, in Wilson and Hölldobler's (2005) phrase, a 'point of no return'. If such a point exists, beyond it the only options would be further evolution along a social trajectory, degeneration to social parasitism, or extinction. The same argument could apply to multicellular organisms. Hence, by this logic, a complex multicellular organism should be unlikely to give rise to free-living, unicellular descendants. The basic reason for expecting multicellular organisms or eusocial societies not to revert to unicellular or solitary life once they have attained complexity is that the sequence of steps required to accomplish this is too improbable. Chenoweth et al. (2007) queried the existence of a point of no return by pointing out that, following the origin of eusociality in the allodapine bees at least 40 mya (Table 1.4), this taxon has maintained a simple level of eusociality without any known reversions to solitary living. They therefore argued that social complexity cannot be the only factor discouraging reversions to solitary living in eusocial lineages. This could be correct, but social complexity may nonetheless make such reversions less likely.

If ecological niches for societies with greater colony size are always available, and if greater colony size in any one lineage promotes selection for greater colony size in competitors, and if greater colony size (via increased social complexity) makes reversals to solitary living less likely, the result would be a trend for eusocial societies to increase in size and complexity over evolutionary time. A parallel argument applies to multicellular organisms. Maynard Smith and Szathmáry (1995) proposed that 'contingent irreversibility' is a characteristic feature of any major evolutionary transition, since, once the transition has occurred, formerly independent entities become less capable of reproducing alone. The current discussion is essentially an expansion of this proposal. Irreversibility would be strengthened by the self-reinforcing element of social group transformation to be fully discussed later in the chapter (Section 6.5). In sum, social evolution in its broadest sense may not be an escalator but it may have a ratchet. There is no selective motor automatically propelling a lineage towards the next major transition (Section 1.2), but ascent to a higher level in the biological hierarchy may nonetheless make descent to a lower level less likely than a further ascent.

6.4 Effect of increasing size of the social group on group complexity

The size-complexity hypothesis predicts that, once greater group size has evolved as a result of external drivers, there is a consistent set of consequences as regards social evolution (e.g. Bonner 1988; Alexander et al. 1991; Bourke 1999). Specifically, there is selection for the syndrome of traits defining complexity (Tables 6.1, 6.2). This section explores in detail the reasons why this might be so, examining each principal set of traits first in multicellular organisms and then in eusocial societies.

Reproductive division of labour within multicellular organisms

In the case of small multicellular organisms that develop from a single propagule, there is no difficulty in understanding why some cells should specialize on reproduction and others on somatic functions. Being clones of one another, the cells have entirely coincident evolutionary interests, so inclusive fitness theory predicts that cells would sacrifice their reproduction to aid other cells to reproduce, even if this yielded just a slight gain in total productivity (Section 3.3). The question that now poses itself concerns why increasing cell number should strengthen the reproductive division of labour among cells, bringing about a segregated, early-diverging germline and terminal differentiation of somatic cells (Table 6.1). Several inclusive-fitness models have tackled this question. I divide them into those that propose a role for increased size *per se* and those based on the increase in the number of somatic mutations that accompanies greater size. General discussions of the evolution of germline segregation are provided by Buss (1987), Maynard Smith and Szathmáry (1995), Pál and Szathmáry (2000), Michod and Roze (2001), and Grosberg and Strathmann (2007).

The main example of the first type of model is the fitness trade-off model of Michod and colleagues (Michod 2006, 2007; Michod et al. 2006). Consider a set of single cells on the point of forming a multicellular group. Assume that, for each cell, there is a negative covariance (trade-off) between fecundity and viability. Then, the model showed, if the cells grouped, the fitness (per cell) of the group would exceed the average fitness of the cells considered singly. If, as was assumed, the cells were genetically identical, this effect would translate into a group-level benefit promoting grouping. (In a sense, therefore, such an effect represents a special type of synergistic gain in total productivity of the sort that Hamilton's rule will always predict to favour grouping among clonal units.) Assume furthermore that the negative association between cell fecundity and viability is convex (i.e. a small increase in fecundity brings about a disproportionately large decrease in viability, and vice versa). The model showed that cells in a multicellular group would then be selected to specialize either on being fecund (reproductive function) or on being viable (somatic function). In addition, the strength of this effect would increase as cell number rose. This would happen if rising group size entailed an increasing cost of reproduction (proportionately greater initial investment in reproduction) to the group, since such a cost would detract from overall viability and so translate into increased convexity in the relationship between fecundity and viability for all cells. These predictions were consistent with what is seen in volvocine algae (Michod 2006, 2007; Michod et al. 2006), in which larger species have, on average, greater germline-soma differentiation (Section 6.2). Therefore, although the fitness trade-off model makes some fairly specific assumptions (trade-off between fecundity and viability in unicells, greater costs of reproduction in larger groups), it accounts for cells increasingly specializing on germline or somatic functions as group size increases.

Buss (1983, 1987) stimulated the creation of the second type of model by drawing attention to the potential role of somatic mutations in strengthening the reproductive division of labour within multicellular organisms. His argument was that a segregated, early-diverging germline evolved in order to minimize the risk that selfish cell-lineages, having arisen by somatic mutation, would achieve onward transmission and hence disrupt the integrity of early multicellular organisms. However, as

---

Box 6.1 The concept of the virtual dominant

Reeve and Jeanne (2003) considered whether members of social groups might cooperate in aiding one particular group member to reproduce, even in the absence of physical dominance. As they pointed out, their model has an important bearing on the evolution of a segregated germline of morphologically distinct reproductive cells in multicellular organisms, as well as on the evolution of a morphological, reproductive caste in eusocial societies. However, the model has not been widely recognized in either context. This box presents the conclusions of the model in a much simplified algebraic form.

Let the following terms be defined:

π_i = the amount of physical power held by a focal group member i,

r_{ij} = the relatedness of the jth group member within the group to the (potential) offspring of a focal group member i,

n = group size.

The model finds that the dominant within the group can be defined as the member whose value of $\Sigma(\pi_i . r_{ij})$ is a maximum (the summation being performed over all group members). In other words, a monopoly of reproduction (complete reproductive skew) in favour of this group member would be evolutionarily stable. This is because this member combines the power to enforce its interests with offspring to which other group members are maximally related.

Now assume that no single group member monopolizes physical power within the group, i.e. that there is 'majority rule' (Reeve and Jeanne 2003). Hence:

$$\pi_i = 1/n,$$

i.e. power is distributed equally among all members. This allows one to define as a 'virtual dominant' the group member with the maximum value of:

$$\Sigma(\pi_i \cdot r_{ij}) = \Sigma((1/n) \cdot r_{ij}) = (\Sigma r_{ij})/n.$$

The 'virtual dominant' is the group member whose monopoly of reproduction all other group members are predicted to favour, even though no-one wields sole physical power. In words, the virtual dominant is the group member to whose offspring the other group members have the greatest mean relatedness (Reeve and Jeanne 2003). Other group members are therefore selected to promote the virtual dominant's reproduction because this allows them to maximize their inclusive fitness.

For example, in a monogynous, subsocial colony of insects headed by a monandrous mother (queen), the queen is the virtual dominant simply on account of the kin structure of the colony. This is because other group members (i.e. workers, which—under subsociality—are also offspring of the mother) are more closely related to her reproductive offspring (their siblings, the new queens and males) than they are to the potential offspring of any other group member (i.e. of any other worker).

In diploids, this comes about because the relatedness of the workers (of both sexes) to the offspring of the queen is sibling–sibling relatedness or 0.5, whereas their relatedness to any offspring of another worker would be the product of sibling–sibling relatedness (0.5) and parent–offspring relatedness (0.5), i.e. 0.25.

In haplodiploids, the relatedness of the workers (females only) to the offspring of the queen is the average of sister–sister relatedness (0.75) and sister–brother relatedness (0.25), i.e. 0.5. (Averaging the relatednesses assumes that workers cannot discriminate between female and male siblings.) But workers' relatedness to any offspring of another worker would be the product of sister–sister relatedness (0.75) and mother–offspring relatedness (0.5), i.e. 0.375 (Fig. 2.2; Table 2.2).

Hence in both subsocial diploids and haplodiploids, workers' average relatedness to the queen's offspring always exceeds workers' relatedness to the offspring of another worker. Note that any one worker is as closely related to its own offspring ($r = 0.5$) as to those of the queen (Section 4.2), but all other workers still remain more closely related to offspring of the queen. Workers, therefore, share a coincidence of fitness interests with respect to rearing the queen's offspring but not with respect to rearing workers' offspring, including their own. So, if (as assumed) no worker has a monopoly of power, the queen emerges as the virtual dominant.

This conclusion needs qualifying in eusocial haplodiploids if workers can discriminate between female and male eggs laid by the queen and selectively replace the queen's male eggs with their own. Such discrimination is possible, for example, in the Honey bee, in which males and females develop on separate combs within the nest (Winston 1987). If discrimination occurs, a monandrous queen would be the virtual dominant with respect to female production but not with respect to male production (Ratnieks 1988; Bourke 1999; Reeve and Jeanne 2003), since workers are more closely related to the queens' daughters ($r = 0.75$) than to any daughters produced by another worker ($r = 0.375$), but less closely related to queen's sons ($r = 0.25$) than to any sons produced by another worker ($r = 0.375$). This fact has interesting implications for the evolution of worker reproduction in the eusocial Hymenoptera (Sections 2.3, 5.4, 5.5), but does not affect the basic concept of virtual dominance.

previously discussed, Buss's (1987) idea realistically only applies to large multicellular organisms, since in small ones the number of somatic mutations would be negligibly low (assuming a unitary propagule). In small multicellular organisms, clonality would be maintained and would alone guarantee within-organism harmony (Section 3.3).

The obvious corollary is that somatic mutations are an increasing threat to the evolution of individuality as cell number rises. This point can be highlighted by considering how many somatic mutations occur in the lifetime of a large, multi-cellular organism. Frank (2010b) estimated this number in humans. Adult humans contain an estimated 10^{13} to 10^{14} cells (Section 3.3). Each cell division increases cell number by one. That is, one cell division produces 2 cells, a second cell division produces 3 cells, a third one 4 cells, and so on. Hence, if cell death is ignored, the number of cell divisions required to produce N cells is $(N-1)$. So the number of cell divisions required to produce an adult human is at least 10^{13} to 10^{14}. The estimated rate of somatic mutations in humans is 10^{-7} to 10^{-6} mutations per gene per cell division (Araten et al. 2005). Therefore, during development to adulthood, each gene in the soma can be expected to mutate 10^6 to 10^8 times (Frank 2010b). Note that this does not mean that every somatic cell is expected to house a mutant copy of a given gene. The mutant genes will be spread across the 10^{13} to 10^{14} cells of the soma as a whole, and, in general, the frequency of a particular mutation across the somatic cells will depend on how early in development it arises (Frank 2010b). In addition, not all these mutations need be cancerous. But this estimate vividly illustrates how greatly somatic mutations might threaten the integrity of large organisms. Indeed, for this reason, Nunney (1999b) proposed that risk of cancers sets an upper limit on the size and longevity of multicellular organisms.

Reeve and Jeanne (2003) modelled the influence of somatic mutations on the evolution of the germline by invoking the concept of the 'virtual dominant' (Box 6.1). In a social group where no single group member monopolizes physical power, the virtual dominant is defined as the group member to whose offspring the other group members have the greatest mean relatedness. The virtual dominant is therefore predicted to have a stable reproductive monopoly despite its lack of power (Box 6.1). In multicellular organisms, germline cells typically divide far fewer times than somatic ones. For example, in humans, it is estimated that of the order of 10 to 1000 cell divisions within the germline separate the zygote from the gamete (Burt and Trivers 2006). Why germline cells divide less often than somatic ones is unclear, but an important consequence is that they are expected to accumulate far fewer mutations than somatic cells. If so, as Reeve and Jeanne (2003) pointed out, somatic cells would have greater average relatedness to germline cells than they would to other somatic cells. Hence, provided cells in the germline accumulate fewer mutations, the germline would be the virtual dominant in the colony of cells making up the multicellular organism (Reeve and Jeanne 2003). Moreover, it would be all the more likely to be so as organism size increases. The reason is that, because the number of mutations within cell lineages rises exponentially with the number of cell divisions, the difference between the number of accumulated mutations in germline and somatic lineages grows as total cell number increases. In this way, Reeve and Jeanne's (2003) concept of virtual dominance provides a very general explanation for both the existence of a separate germline of cells and the strengthening of the reproductive division of labour among cells as cell number rises. Moreover, it supports earlier, allelic models by Michod and colleagues. These also invoked the threat from somatic mutations and again found a segregated, early-diverging germline to be facilitated by increasing organism size (Section 3.3).

Table 6.5 Alternative hypotheses for the existence of a segregated germline within multicellular organisms

Hypothesis (References)
1. Cells containing many and/or virulent selfish genetic elements opt to be somatic cells as an act of altruism on behalf of fellow cells with fewer and/or less virulent selfish genetic elements, which form the prospective germline (Hurst 1990).
2. There is between-organism selection to reduce either the mutation rate or the rate of inherited epigenetic errors in cells producing gametes, with this being most easily achieved within a specialized cell lineage (Maynard Smith and Szathmáry 1995; Michod 1996; Pál and Szathmáry 2000).
3. In a lineage of gamete-producing cells that has undergone the minimum of divisions, it would be easier to 'reset' genes to a totipotent state, as is required since gametic genomes have to give rise to fully-developed organisms each generation (Maynard Smith and Szathmáry 1995; Pál and Szathmáry 2000).
4. Selfish genetic elements favour the germline-soma distinction if the replication of such elements within cell lineages that fail to produce gametes reduces overall gamete production (Johnson 2008).

Evidence that segregation of the germline arises via the need to suppress selfish mutant lineages of somatic cells comes from comparing animals with plants and fungi. Plants and fungi lack a differentiated, segregated germline, with the result that numerous cell lineages within any one organism can form gametes or spores (Buss 1987). To this we owe such pleasant sights as a Cherry tree covered in blossom or a fairy ring of mushrooms in a field. The situation in plants and fungi is consistent with suppression of selfish cell-lineages not having evolved because mutant somatic cells, being unable to travel around internally, would be limited in their negative impact on the organism as a whole (Section 5.4). As alternatives to the idea that a segregated germline evolved in animals to suppress selfish cell-lineages, several other hypotheses have been proposed (Table 6.5). Although the state of knowledge seems insufficient to rule any out, none offers an explanation for why the relative timing of divergence of the germline varies within and across animal phyla (Table 6.3). In sum, several models exist that propose cogent reasons for why larger multicellular organisms might evolve a segregated, early-diverging germline. But the association of size with early divergence of the germline is tentative and alternative hypotheses for segregation of the germline exist.

Non-reproductive division of labour within multicellular organisms

There are two basic reasons why, within multicellular organisms, increasing cell number should promote a greater number of types of somatic cell. The first assumes that a morphological division of labour among somatic cells, i.e. the occurrence of differentiated cells each suited to a particular function, is beneficial to the organism as a whole. This assumption follows from the basic principle of division of labour dating back at least to Adam Smith (Bonner 1993, 2006), namely that a multi-step task can be carried out more efficiently if it can be performed by sets of agents specializing on each task. But when cell number is low, an organism with several, specialist somatic cell-types is vulnerable to the loss of just a few cells knocking out its ability to perform a key function. So a small organism cannot achieve a division of labour among its somatic cells without endangering its robustness to accidental damage. By contrast, a large organism is buffered against accidental loss of function, since, if damage occurred, there would still be many specialist cells of each type left behind. Therefore, only large, multicellular organisms can enjoy the full benefit of a division of labour among somatic cells. Likewise, in human manufacturing, small companies need employees each able to take on every essential task if co-workers go sick. Only large companies can benefit by having sets of employees each able to perform just one task (Sudd and Franks 1987). In multicellular organisms, another means by which large cell numbers promote non-reproductive division of labour was suggested by Willensdorfer (2008) using models involving digital (computer-based) organisms. He showed that large organisms evolved specialized, somatic cells more easily than small ones because large organisms were more tolerant of mutations that temporarily reduced the performance of specific cells but eventually led to higher organismal fitness.

The second basic reason for high cell number promoting a greater number of types of somatic cell concerns the physics of increasing body size (Bonner 1993). As is well known, large organisms have a smaller ratio of surface area to volume than smaller organisms. Hence nutrients and oxygen can no longer be taken in through passive diffusion as in small organisms. They must be actively brought into the body by means of specialized structures, such as mouths, guts, gills, and lungs. Other structures are needed to expel waste, and still others to support the organism as a whole. Specialized structures need to be built from specialized tissues. This would again lead to larger organisms requiring more somatic cell-types (Bonner 1993).

Reproductive division of labour within eusocial societies

From the perspective of the size-complexity hypothesis, eusocial societies (with some exceptions) differ from multicellular organisms in three important ways. They are rarely clonal, their 'somatic' subunits (helpers or workers) are usually all first-generation offspring of the founding propagule (pair), and there are far fewer workers in even the largest eusocial societies than there are cells in the largest multicellular organisms (Section 3.3). Therefore, large size in eusocial societies is not associated with a rising number of 'somatic mutations' (here, mutant lineages of workers) as in multicellular organisms (Section 3.3). Despite these differences, the size-complexity hypothesis predicts that large colony size induces in eusocial societies a set of social changes similar to those induced by high cell number in multicellular organisms (Tables 6.1, 6.2). Furthermore, inclusive fitness considerations again lie at the root of these predictions.

If large group size is ecologically advantageous to a given eusocial society, queens must be capable of producing numerous offspring. This provides a reason why large eusocial societies should be headed by big, fecund queens, which would be most readily achieved by these queens having a specialized morphology. But why should larger colonies typically also have workers characterized by a morphology increasingly specialized for helping and by diminishing reproductive potential (Alexander et al. 1991; Bourke 1999)? In short, why should queen–worker dimorphism, and the degree to which workers lose reproductive potential, increase with colony size (Table 6.2)? There are several possible reasons, all likely to be interconnected.

Since it represents the modal ancestral situation (Section 4.2), consider the case of a subsocial society headed by one, singly-mated queen. Alexander et al. (1991) suggested that, as colony size increases, any one member of the worker generation would have a smaller and smaller chance of becoming a replacement queen, that is, as an adult, of becoming the colony's new queen should the original queen die. For, in a colony with N workers, the average chance of any single worker replacing the queen falls as $1/N$. Therefore, with rising colony size, workers should more readily abandon an adult morphology fitting them to become replacement queens and, as a corollary, more readily adopt a morphology suited to working, so leading to greater queen–worker dimorphism (Alexander et al. 1991; Bourke 1999; Jeon and Choe 2003). This effect would be compounded by larger colony size selecting, as discussed below, for a greater degree of polymorphism among workers. Worker morphologies suited to specialized worker tasks, such as defence or foraging, are less likely simultaneously to be suited to reproduction, especially reproduction as a highly fecund queen. This is particularly true if specialized worker actions involve an element of suicidal altruism, as many do (Wilson 1971; Tofilski et al. 2008). Such a factor would again make workers in large societies less likely to retain an adult morphology fitting them to function as queens.

Adding to these effects is a shift in power away from single group members as colony size increases (Wilson 1971; Ratnieks 1988; Bourke 1999). Power shifts either towards the workers as a collective, or (virtually) to the queen in her capacity as the virtual dominant (Reeve and Jeanne 2003), or to both parties. For example, increasing colony size leads to power over caste fate shifting away from each developing immature itself to the workers as a collective. Adult workers are more able to control the food supply to immatures at larger colony sizes. This means that adult workers can constrain immatures to develop into workers, even against the immatures' interests, by restricting the amount of food they receive (Bourke and Ratnieks 1999; Wenseleers et al. 2003, 2004b).

Another situation in which increasing colony size shifts power away from single group members concerns aggressive dominance. As colony size increases, it becomes less and less feasible for a single group member to monopolize reproduction by physical aggression (Section 5.6). This in turn means that a focal adult worker making the evolutionary decision whether or not to have offspring undergoes a shift of perspective. In larger colonies, the average worker-produced offspring would be not its own, but the less closely related offspring of other workers. This makes rearing the queen's offspring a relatively more profitable choice at larger colony sizes. In Bourke (1999) I argued that, other things equal, this facilitates selection for policing by workers of one another's reproduction in larger colonies, since such policing results in workers rearing the queen's offspring. The point can also be made in terms of Reeve and Jeanne's (2003) virtual dominance theory. When power is evenly distributed, Reeve and Jeanne's (2003) theory predicts the queen of a subsocial society, whether diploid or haplodiploid, to be the virtual dominant (Box 6.1). In other words, in comparison with their relatedness to their own offspring, workers are more closely related to the queen's offspring than to other workers' offspring (leaving aside for the moment the complication in the eusocial Hymenoptera that the queen is not the virtual dominant with respect to male production; Box 6.1). So, if the choice were between rearing the queen's offspring or those of another worker, workers would choose the former option. And, since increasing colony size smooths out the distribution of power, this is the choice that workers in larger colonies face.

The preceding arguments apply in principle to eusocial diploids or haplodiploids. In the latter, they also need to be qualified in two respects. To begin with, because haplodiploidy allows non-mating females to produce offspring (sons), combining reproductive and helping morphologies is easier in eusocial Hymenoptera than in other taxa (Bourke 1999). Therefore, male-producing workers can retain comparatively high reproductive potential even when the degree of queen–worker dimorphism is high. Nonetheless, a trade-off between the traits required for effective reproduction and those required for effective working seems hard to escape entirely. The other qualification also stems from the quirks of haplodiploidy. If workers under one, singly-mated queen can effectively replace the queen's male progeny with their own, the queen is the virtual dominant with respect to female production but not with respect to male production (Box 6.1). This is likely to be why Hymenopteran workers so commonly give up their ability to mate (lose their sperm receptacle), so preventing them from producing female offspring, but retain their ovaries, so permitting them to continue to produce male offspring (Ratnieks 1988; Bourke 1999). However, since the queen lacks virtual dominance with respect to male production, policing by workers able to produce their male offspring in place of the queen's would not be favoured at large colony sizes unless other factors intervened, since workers would be more closely related to other workers' male offspring than to the queen's. Note, though, that if larger colony size in the eusocial Hymenoptera is associated with polyandry, as appears to be the case (Section 6.2), a positive association between colony size and reduced worker reproduction would be predicted, since polyandry by itself promotes worker policing (Sections 5.4, 5.5).

Formal models that predict associations between colony size and the level of reproduction by workers in eusocial societies offer some support for the predictions of the size-complexity hypothesis that increasing colony size leads to more policing, less worker reproduction, or both (Table 6.6). But some of the models have conclusions that go against these predictions. Such mixed results probably stem, in part, from the wide variety of modelling approaches and assumptions (Table 6.6). For

Table 6.6 Models predicting covariation between colony size and degree of reproduction by workers in eusocial societies

Type of inclusive fitness model	Model findings (References)	Comments
Inclusive fitness/ game theory model	In eusocial Hymenoptera, the ESS frequency of egg-laying workers is (a) higher in larger colonies unless (b) policing is more effective in larger colonies (as it appears to be), in which case it is lower (Wenseleers et al. 2004a, 2004b).	Result (a) arose in part because it was assumed that collective worker fecundity would outweigh the queen's fecundity at very large colony sizes. Result (b) supports the size-complexity hypothesis.
Dynamic game model	In eusocial Hymenoptera, worker policing evolves to prevent selfish reproduction by workers during colony growth (ergonomic phase), predicting that, within species, smaller (i.e. growing) colonies should show more policing (Ohtsuki and Tsuji 2009).	Within-species result does not support size-complexity hypothesis, but there was no explicit prediction as regards between-species comparisons (Ohtsuki and Tsuji 2009).
Reproductive skew model	In eusocial societies, total worker sterility through self-restraint is favoured by subsociality, functional monogyny, and large colony size (Jeon and Choe 2003).	Result supports size-complexity hypothesis. The result arose because the conditions identified favoured workers remaining in the colony to gain indirect benefits rather than reproducing either inside (through nest inheritance) or outside the colony. The model did not include policing or explicitly tackle the issue that, in eusocial Hymenoptera, workers lacking a sperm receptacle cannot found their own colonies externally and, within the society, are (if there is queen monandry) more closely related to workers' male offspring than to the queen's (Box 6.1).
Reproductive skew model	In eusocial societies, members of eusocial colonies are selected to invest more in within-group cooperation as group size increases, provided that larger groups occur in resource-rich patches with a greater number of competing groups (Reeve and Hölldobler 2007).	Result supports size-complexity hypothesis.

example, some of the models explicitly concern the eusocial Hymenoptera, and so address the point that the queen can be the virtual dominant with respect to female but not male progeny. Others are more general. Some permit policing, others do not. None of the models systematically derives conditions for each possible step in the loss of workers' reproductive potential (Section 6.2). Hence the strength of formal theoretical support for the size-complexity hypothesis in the present context remains unclear.

Of special significance is the question of when, in the transformation of eusocial societies, total worker sterility should evolve. In general terms, inclusive fitness theory explains total sterility as the result of selection for extreme altruism (Section 2.2). The issue of interest is what precise combination of circumstances brings about such extreme altruism and, for current purposes, whether these circumstances are as expected from the size-complexity hypothesis. In proximate terms, total worker sterility could come about through workers' voluntarily refraining from all reproduction (self-restraint) or through an effect of the queen on caste determination (a form of maternal effect). In other words, workers could either voluntarily fail to develop or activate their ovaries, or they could be forced into so doing via an epigenetic influence of the queen. Recent work by Khila and Abouheif (2008, 2010) has begun to elucidate the proximate basis of workers' loss of reproductive function in ants. These authors showed that workers' ovaries become defective through a series of developmental failures; for example, through messenger RNAs and proteins required for proper oocyte development in insects failing to be correctly localized. This suggested that genes have been selected that block key steps in the development of workers' reproductive organs, i.e. that there has been selection for self-restraint (Khila and Abouheif 2008, 2010).

Under what selective conditions would complete workers' self-restraint be expected? One set would arise if a small amount of worker reproduction induced a large cost to colony productivity (Wenseleers et al. 2004a). This is equivalent to saying that workers' self-restraint would be favoured if it brought about large productivity benefits. But, although costs of worker reproduction are hard to measure, there is little indication of disproportionately high costs in eusocial Hymenoptera (Sections 5.4, 5.5). With regard to the size-complexity hypothesis, there are also no data on how costs of worker reproduction vary with colony size. Jeon and Choe (2003) explicitly modelled the evolution of total worker sterility in eusocial societies and found that, consistent with the size-complexity hypothesis, it was favoured by large colony size, in part because of falling worker benefits from replacing the queen in large colonies (Table 6.6). But their model did not explicitly address selection on workers under one, singly-mated Hymenopteran queen to retain just their ability to produce sons.

Another route by which complete workers' self-restraint might evolve is as a response to policing. In eusocial Hymenoptera, the model of Wenseleers et al. (2004a, 2004b) found that, above a threshold of effectiveness in policing, the ESS frequency of workers laying male eggs would be zero. This implied that there would then be selection for complete acquiescence to non-reproduction by workers. To the extent that policing is more effective in larger colonies (Table 6.6), this would predict larger colonies to have a lower frequency of reproductive workers. If highly effective policing led to complete workers' self-restraint, policing would be rendered unnecessary, since worker would no longer lay eggs. A fine balance might then be maintained between selection for reinvasion of worker reproduction on the one hand and selection for reinvasion of policing on the other (Ratnieks 1988). Khila and Abouheif (2008, 2010) interpreted their results on the mechanisms leading to loss of reproductive function in worker ants as uniquely supporting selection for the productivity benefits of workers' self-restraint. However, the finding of Wenseleers et al. (2004a, 2004b) that policing can promote self-restraint shows that this need not be the case. Moreover, the two routes to self-restraint are not mutually exclusive. Policing might lead to workers' self-restraint both because it makes worker reproduction unprofitable (Wenseleers et al. 2004a, 2004b) and because it brings about productivity benefits at the colony level (Ratnieks 1988; Ratnieks and Wenseleers 2008).

The idea that an effect of the queen on caste determination contributes to worker sterility is not far-fetched, since it is known that caste fate in some ants is influenced by the hormonal content of eggs (Wheeler 1991; Schwander et al. 2008). So, conceivably, queens could influence workers from their earliest development such that, as adults, workers lacked functional ovaries. This would represent a complete loss by workers of the ability to control their own reproductive fate. Whether queens have such an ability is unknown, and still less is it known if it covaries with colony size. If queens can induce worker sterility this way, it is probably a derived feature, since queens in early eusocial societies would be unlikely to have acquired such an extreme form of parental manipulation (Bourke and Franks 1995; Ratnieks and Wenseleers 2008). Moreover, even in complex societies, it seems unlikely that it would be evolutionarily stable in the face of strong selection for workers to oppose it, that is, unless workers were already at least partly acquiescent (Keller and Nonacs 1993; Bourke and Franks 1995).

In sum, there is a set of linked reasons for why greater colony size in eusocial societies should become associated with greater queen–worker dimorphism and the loss of reproductive potential among the workers. Some complications arise in the eusocial Hymenoptera because of haplodiploidy and its associated relatedness asymmetries. Uncertainty still surrounds the relative roles of worker policing, self-restraint, and queen influence on caste in the total loss of reproductive ability by adult workers. Models that integrate the different approaches taken to date, along with better data on the frequency of reproductive workers, relative queen and worker fecundities, presence or absence of functional worker ovaries, the effectiveness of worker policing, the costs of worker reproduction, modes of caste determination, colony size, colony kin structures, and the phylogenetic history of all these traits, are still required.

Finally, if indeed larger colony size promotes greater queen–worker dimorphism, one would also expect it to be associated with caste determination occurring at an earlier stage during the development of immatures (Table 6.2). This is because earlier divergence would allow more distinct adult phenotypes to be generated (Wheeler 1986; Bourke 1999). However, as mentioned already (Section 6.2), it remains unknown if divergence is indeed generally earlier in species with larger colonies.

Non-reproductive division of labour within eusocial societies

If worker polymorphism improves colony performance, larger eusocial societies are expected to have more types (physical castes) of workers than smaller societies because only larger societies with polymorphic workers would be buffered by size against accidental loss of key worker functions (Sudd and Franks 1987; Wheeler 1991). Essentially the same argument was used above to explain why large multicellular organisms should have more somatic cell-types than small organisms. The other reason proposed to account for larger multicellular organisms having more somatic cell-types was their reduced surface area-to-volume ratio. Physical factors of this sort do not operate with quite the same force in eusocial societies, since (colonial marine invertebrates apart) members of these societies are not stuck together in a single physical mass. However, by analogy, large eusocial societies are likely to have greater logistical problems than small ones in bringing food to the interior of the nest and disposing of waste generated there. This might create a greater variety of worker tasks in larger than smaller eusocial societies, so again encouraging a greater level of worker polymorphism. For example, in fungus-growing ants with large colonies, the smallest worker caste is believed to be adapted to working in the fine interstices of the nest centre (Hölldobler and Wilson 2009) and there is a notably complex system for exporting waste (Hart and Ratnieks 2002). Further explorations of the similarities between multicellular organisms and eusocial societies in how group-level processes scale with size may be found in Shik (2008) and Hou et al. (2010).

6.5 Self-reinforcing social evolution in social group transformation

The final part of the extended size-complexity hypothesis advocated in this chapter proposes that the social traits brought about by increased group size, which were described in the previous section, themselves promote further increases in group size. This creates a positive-feedback loop that reinforces the evolution of large group size, social complexity, and individuality (Fig. 6.1). This section explores the main ways in which such a process might occur.

Positive feedback favouring large group size in multicellular organisms

In multicellular organisms, the main type of positive feedback promoting ever-larger group size stems simply from the likelihood that the segregated germline and multiple somatic cell-types associated with increased cell number, if they are selected at all, would enhance performance of the organism as a whole. Enhanced performance would allow organisms to grow yet larger (if their ecology continued to favour this and physical limits allowed), so reinforcing germline segregation and a greater degree of non-reproductive division of labour.

Positive feedback favouring large group size in eusocial societies

Some of the processes of positive feedback contributing to increasing group size in eusocial societies are similar to those operating in multicellular organisms. As discussed in Bourke (1999), if the degree of dimorphism between queens and workers increases, and the reproductive potential of workers concomitantly falls, the implication is that queens become better at reproducing (more fecund) and workers become better at working (more productive). Hence queens furnish workers with more siblings to rear, and workers are (up to a limit) increasingly capable of rearing them. The result would be further increases in colony size, driven evolutionarily by associated increases in the number of new reproductives produced. Indeed, such a feedback effect linked to colony size emerged as one of the outcomes of Jeon and Choe's (2003) model of worker sterility (Table 6.6).

Several additional processes might contribute to this effect. One is that, as colony size increases and power over caste fate and worker reproduction shifts away from single group members (Section 6.4), costly conflicts become resolved, which is to say they become less costly (Ratnieks et al. 2006). An example comes from control of caste fate in bees. In *Melipona* stingless bees, conflict over caste fate results in a costly cull of virgin queens each generation (Section 5.5). In trigonine stingless bees and honey bees, colony size is greater and so is the degree of queen–worker dimorphism. Furthermore, queens are reared in large cells and workers in small cells. This means that workers can control the number of new queens by regulating the number of large cells that they build and that, within small cells, developing immatures (female larvae) are constrained to be workers. (In some trigonine stingless bees, female larvae in small cells may also develop as dwarf queens, but these achieve reproductive success only under limited circumstances (Bourke and Ratnieks 1999; Ribeiro et al. 2006).) Hence, unlike the case in *Melipona*, female larvae have lost control over their own caste fate to the workers, who favour far fewer new queens being produced. The result is that virgin queens are not overproduced and culled, so saving the costs of this wasteful practice (Bourke and Ratnieks 1999). This presumably helps account for trigonine stingless bees and honey bees having larger, more productive colonies.

A second process is more speculative and stems from long-standing suggestions regarding the best response to loss of fecundity in workers (West-Eberhard 1981; Craig 1983). Say workers begin to lose fecundity for any of the reasons (policing, self-restraint, queen influence) discussed in the previous section. Any fall in workers' expected fecundity makes helping raise relatives comparatively more attractive to workers, essentially because the cost term in their Hamilton's rule would fall. In other words, there would be a decrease in the difference between the number of offspring that workers could have if helping and the number they could have if not helping (Section 2.2). This would make workers more likely to accumulate adaptations for working more efficiently, so increasing the benefit term in their Hamilton's rule. To the extent that working and reproducing are incompatible, this would lead to a further fall in worker fecundity, thence to a further increase in worker efficiency, and so on in a process of positive feedback. The previous section suggested that, if queens in complex eusocial species are able to manipulate the fecundity of workers through effects on worker-destined eggs, this would require an element of worker acquiescence. The process just described might provide this element (cf. West-Eberhard 1981; Craig 1983). If an increase in colony size brings about an initial drop in worker fecundity, its continued, accelerating fall would also facilitate an evolutionary trajectory towards societies with large colonies and low-fecundity, high-efficiency workers.

A final way in which positive feedback contributes to increased colony size in eusocial societies involves a divergence in the lifespans of queens and workers. Alexander et al. (1991) showed that, as queen fecundity rose and workers' reproductive potential fell within eusocial societies, selection would tend to increase the lifespans of queens relative to those of workers. The evidence is that this indeed occurs (Alexander et al. 1991; Keller and Genoud 1997; Bourke 1999, 2007). Coupled with large colony size not being conducive to replacement of the queen by adult workers (Section 6.4), extended lifespan in queens has the important social effect of making would-be workers more ready to commit to a worker role (Alexander et al. 1991; Bourke 1999). This is because it increases the predictability of the colony's kin structure. In other words, if the colony's kin structure is initially favourable to a commitment to working, and if queens live a long time and are unlikely to be replaced, the colony's kin structure is more likely to continue to be favourable in this respect. This would allow developing immatures to commit to acquiring irreversible worker status more readily.

Boomsma (2007, 2009) proposed that the absence of remating by reproductives heading eusocial societies would likewise contribute to this effect. He defined obligate eusociality as occurring when worker caste is irreversibly determined early in the development of immatures. So, in the present terminology, Boomsma's obligate eusociality is complex eusociality. Boomsma (2007, 2009) argued that the transformation to obligate eusociality only occurs once eusocial societies have passed through a stage in which each colony is headed by a single foundress female paired for life with a single founding male (lifetime monogamy). This would create initial relatedness conditions maximally conducive to helping (Section 4.2). Alongside long-lived queens, lack of queen replacement, and the rigid exclusion of would-be external exploiters (Section 5.2), it would also help ensure that colony kin-structure was stable for the duration of an average worker's lifetime. This would again promote workers' irreversible commitment to helping. Put another way, under these conditions the queen would more readily represent the virtual dominant (Reeve and Jeanne 2003). Boomsma's (2007, 2009) argument, although not logically requiring complex eusociality to be confined to large colonies, therefore complements the size-complexity hypothesis to a high degree. Moreover, the phylogenetic evidence of Hughes et al. (2008) that worker's loss of totipotency begins before the evolution of high levels of queen polyandry within lineages of eusocial Hymenoptera supports Boomsma's (2007, 2009) argument.

Lastly, consider how positive feedback might lead to increased colony size in eusocial societies with respect to non-reproductive division of labour. One way in which this could happen is simply through worker polymorphism, which is favoured in larger colonies (Section 6.4), leading to increased worker efficiency and hence to the potential for yet more increases in colony size. A second way invokes the link between colony size and self-organization. Recall that self-organization occurs when interacting subunits, through following a set of simple, local rules, generate higher-level order (Section 1.5). In eusocial societies, the element of self-organization in the performance of worker tasks is potentially advantageous because of the robustness and flexibility of the resulting systems (Bourke and Franks 1995; Camazine et al. 2001). In the present context, the key point is that self-organization, being reliant on multiple interactions between subunits, generally requires a minimum group size (Bonabeau et al. 1997; Anderson and McShea 2001). Such a reliance has been supported by experiments showing that effective pheromonal foraging trails only form by self-organization in ants above a threshold colony size (Beekman et al. 2001). In addition, models of self-organization show that increased colony size alone can promote the specialization of workers on particular tasks (Gautrais et al. 2002; Jeanson et al. 2007). The occurrence of such specialization in large colonies would reinforce the tendency of worker polymorphism to evolve in them. Therefore, once large colony size has been reached, self-organization of worker behaviour can take effect. If this were advantageous for any reason, the result would be enhanced colony productivity and hence the potential for further increases in colony size.

6.6 The size-complexity hypothesis: conclusions

The size-complexity hypothesis proposes that, within social groups of relatives, the transformation of simple groups into complex groups is driven primarily by increases in group size, i.e. in the number of subunits making up the group. The hypothesis is a multi-faceted one, drawing on insights of many authors and on many different applications of inclusive fitness theory. It is relevant to social group transformation in both multicellular organisms and eusocial societies. The hypothesis is provisional. Although it helps account, in a manner partly supported by theory, for several patterns seen in the social evolution of multicellular organisms and eusocial societies, uncertainties, inconsistencies, and, above all, gaps in the data and conceptual underpinnings remain. In short, the size-complexity hypothesis awaits full validation or rejection by further modelling and data collection.

The core of the size-complexity hypothesis is that, within social groups, increasing group size causes an increase in the degree of both the reproductive division of labour (increased germline–soma differentiation or queen–worker dimorphism) and the non-reproductive division of labour (more somatic cell-types or worker polymorphism). The effect is to enhance the interdependence of the reproductive and non-reproductive members of the social group, such that each party becomes less and less capable of gaining inclusive fitness in a solitary state and more and more reliant on gaining inclusive fitness as part of the social group. This, then, is how the size-complexity hypothesis accounts for the emergence of individuality as the end-product of social group transformation.

The size-complexity hypothesis suggests a role for self-reinforcing social evolution in social group transformation. That is, once external factors select for greater size of social groups, a consistent set of social traits evolves that themselves promote greater group size. A particular type of self-reinforcing social evolution occurs with respect to the control of social conflicts within the group. Specifically, the resolution of one type of social conflict may create conditions for the evolution of new social traits. This is found in several contexts. For example, the imposition of uniparental inheritance on organellar symbionts in sexual eukaryotes creates conditions for sex-ratio distortion by organelles (e.g. cytoplasmic male sterility), which in turn creates conditions for the suppression of such distorters (Section 5.5). In social group transformation, prime examples of social resolution of conflicts leading to further social consequences are worker policing of caste fate and worker policing of one another's reproduction. These potentially lead, respectively, to the loss of the ability of immatures to develop selfishly as queens and to workers' acquiescence to reductions in their reproductive potential (Section 6.5). The result would be that previously costly social conflicts are avoided and workers' commitment to worker functions can increase. A parallel argument proposes that early segregation of the germline within multicellular organisms avoids the costs of within-group conflict between selfish cell-lineages and reinforces the commitment of somatic cells to somatic functions (Section 6.4). Hence, paradoxically, as Michod and Roze (1997) wrote, 'Conflict leads, through the evolution of adaptations that reduce it, to greater individuality and harmony for the organism'. For 'organism' here one could read 'social group', and one might add the rider that the 'adaptations' are better regarded as adaptations of the genes responsible for the conflict-reducing behaviour rather than as group-level adaptations of the collective as a single entity (Section 2.5). But, these comments aside, the point is a fundamental one. It gives a new twist to a suggestion of earlier authors (Hamilton 1972; West-Eberhard 1981) that when we examine a complex social group we frequently see, like tourists watching a ceremonial changing of the guard, features that makes sense only as the products of a more turbulent past.

This chapter has applied the size-complexity hypothesis exclusively to fraternal transitions. Does the hypothesis also apply to egalitarian ones? The answer is not exactly. Take interspecific mutualisms as a representative example. There is no obvious association between an increase in the size of an interspecific mutualism and its complexity. Indeed, in defining the number of components within an interspecific mutualism, the number of partners (species) within it seems the more salient measure of its size than the number of component cells or organisms. The lack of applicability of the size-complexity hypothesis to social groups of non-relatives is not surprising.

Since egalitarian transitions do not result in a reproductive division of labour (Section 3.1), one would not expect to see the same kind of growing interdependence between reproductive and non-reproductive partners as is found within social groups of relatives.

Nonetheless, there are important commonalities underlying the consolidation of the evolution of individuality in each type of transition. The most obvious correlate of social complexity in interspecific mutualisms is the degree to which the partners are physically bound together. The eukaryotic cell, which encloses its organellar partners, and symbiotic organisms such as lichens, which incorporate fungal and algal cells side-by-side in a single structure, are more complex and individualistic mutualisms than, say, an ant-plant and its partner ants. The evolution of individuality relies ultimately on the different parties making up a social group achieving an ever-closer coincidence of fitness interests (Sections 3.1, 4.2). As just discussed, the processes invoked by the size-complexity hypothesis lead to the members within a social group made of relatives becoming more interdependent. This is true with respect to both reproductive and somatic function. The result is an increase in the degree to which group members achieve a coincidence of fitness interests via shared genes. In interspecific mutualisms, a stronger level of physical association also leads to partners becoming more interdependent with respect to both reproductive and somatic function. In this case, the result is an increase in the degree to which partners achieve a coincidence of fitness interests via shared reproductive fate. So, in both cases, social group transformation consolidates the evolution of individuality by aligning more closely the inclusive fitness interests of group members.

6.7 Summary

1. The size-complexity hypothesis proposes that, in multicellular organisms and eusocial societies, social group transformation occurs when the number of members per group increases. As group size increases, groups transform from simple, less individualistic ones to complex, more individualistic ones.

2. Simple multicellular organisms exhibit poorly separated germline and somatic cell-lineages, totipotent somatic cells, few somatic cell-types, and low cell numbers. Complex ones exhibit a segregated, early-diverging germline, terminally differentiated somatic cells, many somatic cell-types, and high cell numbers. Simple eusocial societies exhibit low queen–worker dimorphism, workers with high reproductive potential, a monomorphic, generalist worker caste, and low worker numbers. Complex ones exhibit high queen–worker dimorphism, workers with low reproductive potential, a polymorphic worker caste, and high worker numbers. Hence, social complexity is associated with a greater degree of reproductive and non-reproductive division of labour. Evidence for these syndromes of traits exists in both multicellular organisms and eusocial societies, although more is needed. Complex social groups evolve from simple ones and, within taxa, may do so multiple times independently. Some evidence suggests that, for long-term

stability, complex social groups require features that maintain within-group genetic variation.

3. Once formed, social groups are selected to increase in size as a result of a variety of external drivers. These include short-term ecological drivers (e.g. intraspecific or interspecific competition for space and resources) and long-term evolutionary ones (e.g. arms races between lineages resulting in parallel size increases over evolutionary time). Trends to larger size could be reinforced by evolutionary reversals in size and complexity being difficult to achieve.

4. If external drivers cause social groups to become larger, groups are then selected to become more complex. In multicellular organisms, larger size leads to an increasing number of somatic mutations. This causes any lineage of more slowly-dividing cells (germline) to become the virtual dominant, where, in a group lacking physical dominance, the virtual dominant is defined as the member to whose offspring the other group members have the greatest mean relatedness. So, larger size promotes the germline–soma distinction. It also promotes the evolution of more somatic cell-types, through permitting enhanced efficiency without compromising robustness to accidental damage. In eusocial societies, larger colony size requires highly fecund queens and, with workers becoming less likely to act as replacement queens, reduces workers' reproductive potential. It is argued that larger colony size also leads to the emergence of a virtual dominant, since power becomes more evenly distributed as colony size increases. Since the queen is the virtual dominant of a subsocial, eusocial society, larger colony size leads to reproduction becoming more concentrated in the queen. For these reasons, larger colony size induces greater queen–worker dimorphism and loss of workers' reproductive potential. Some formal models similarly find that large colony size promotes loss of workers' reproductive potential, but others do not. A complication is that, in eusocial Hymenoptera, a single, monandrous queen is the virtual dominant with respect to female production but not with respect to male production. It remains uncertain what combination of worker policing, self-restraint, and queen influence on caste ultimately brings about total worker sterility, where it occurs. Larger colonies are expected to exhibit worker polymorphism for very similar reasons to those leading to larger multicellular organisms having more somatic cell-types.

5. Social complexity allows social groups to become more productive, and so, in a process of positive feedback, promotes further increases in group size. An important component of this process is the saving of the costs of conflict as power shifts away from single group members to collectives of group members (or to the virtual dominant) as group size increases. In eusocial societies, greater queen–worker dimorphism leads to queen lifespans increasing and worker lifespans decreasing. By making the colony's kin structure more predictable, this favours workers becoming irreversibly committed to helping. Larger colony size also allows self-organization of workers' labour to occur, again enhancing colony productivity and so leading to yet larger colonies.

6. By enhancing reproductive and non-reproductive division of labour within social groups, increased group size creates more interdependence between group members, and hence increases group-level individuality. The size-complexity hypothesis applies primarily to social groups of relatives. In social groups of non-relatives, such as interspecific mutualisms, complexity and individuality increase in proportion to the degree to which partners are physically bound together. In both kinds of social groups, individuality is consolidated by social complexity leading to group members having more closely aligned inclusive fitness interests.

7

Synthesis and conclusions

7.1 The principles of social evolution: a summing-up

Over evolutionary time, individuals have repeatedly undergone major transitions in which they have grouped into larger units that evolve to resemble individuals in their own right. The expanded view of social evolution advocated in this book holds that each major transition arises from the operation of shared principles of social evolution. The basic reason for expecting this is the universal applicability of Hamilton's (1964) inclusive fitness theory. With this background, I sought to address two key unresolved issues (Section 1.5). The first concerned the extent to which common principles operate at each stage of a major transition across all levels in the biological hierarchy, and the strength of the empirical evidence for their operation. The conclusion is, resoundingly, that common principles operate to a very large extent. Each stage in a major transition to individuality (social group formation, maintenance, and transformation) frequently involves the same social evolutionary principles acting in analogous ways at the different hierarchical levels (Table 7.1). In addition, as earlier chapters have explored in detail, the empirical evidence for the operation of these principles is very considerable. It includes behavioural, comparative, ecological, and genetic evidence, and derives from many taxa. These taxa include bacteria, plants, colonial marine invertebrates, eusocial Hymenoptera, termites, other eusocial insects, cooperatively-breeding birds and mammals, and the eusocial Naked mole-rat. Common principles operate to a greater extent within each of the main classes of transition (egalitarian or fraternal). However, principles shared across both classes of transition (e.g. limitation of internal cheating by coercion) are also found.

The second key unresolved issue involved asking what integrative processes occur, within lineages, between one major transition and the next. That is, how does the final step in the evolution of individuality, social group transformation, occur? The main conclusion here is that, at least in social groups of relatives, it is likely to have occurred via the steps defined by the size-complexity hypothesis. This is because, as argued in the previous chapter, the hypothesis provisionally accounts for the evolution of many of the features observed in those social groups of relatives most closely resembling individuals.

These points established, I can repeat with renewed confidence my belief that social evolutionary theory in its expanded form offers an interpretation of the selective basis of life's basic organization and history more profound and satisfying than alternative

Table 7.1 The main principles of social evolution summarized

Principle

General

- Major transitions in evolution, leading to individuality at a new hierarchical level, come about through previously independent subunits coming together to form a new, collective unit, or social group (Section 1.1).
- In each such transition, the evolutionary interests of previously independent subunits must, at least in part, have become subordinated to the evolutionary interests of the social group as a whole. Cooperation (broad sense) must outweigh conflict, but cooperation remains contingent on the continued suppression of conflict and conflict need not be abolished entirely (Sections 1.1, 5.3–5.5, 6.6).
- Social evolution is facilitated in proportion to the coincidence of fitness interests experienced, through sociality, by the component subunits (partners). Such a coincidence may come about through two basic methods, namely shared genes (relatedness) or shared reproductive fate (Sections 3.1, 4.2, 6.6).
- Major transitions involve non-relatives (egalitarian transitions) or relatives (fraternal transitions). Non-relatives (including separate species) may participate only in major transitions based on narrow-sense cooperation, in which each partner retains reproductive ability. Relatives may participate in major transitions based on altruism, in which one partner (the altruist) becomes non-reproductive (Section 3.1).

Social group formation

- Development of a social group from a single propagule (bottleneck) containing few genomes (spore, zygote, mated pair) is most conducive to social group formation among relatives by elevating relatedness and reducing within-group potential conflict (Sections 3.3, 4.2).
- Subsocial formation of a social group (i.e. via offspring associating with parents) is likewise most conducive to social group formation among relatives, through automatically elevating relatedness (Section 4.1).
- Several shared ecological and synergistic factors have promoted social group formation in the origin of multicellularity and eusociality, by increasing either the benefits of grouping or the costs of remaining alone. Others have the same effect but are specific to each transition (Sections 4.3, 4.4).

Social group maintenance

- Social groups are maintained partly by limitation of exploitation by external elements, which relies on the evolution of processes of recognition of self versus non-self followed by rejection of non-self (Section 5.2).
- Social groups are also maintained by limitation of exploitation from internal cheats, through conflict resolution. This evolves through either (a) self-limitation, i.e. effect of self (including self-limiting effects of relatedness on selfishness) or (b) the evolution of coercion, i.e. effect of others (Sections 3.2, 5.3–5.5).
- Coercion is a form of social action and so is itself subject to the rules of inclusive fitness theory; hence coercion cannot bring about altruism in the absence of relatedness (Section 3.2).
- Conflict in fraternal major transitions can be resolved either by relatedness, or coercion, or both, but in egalitarian ones must be resolved by coercion alone, as in many forms of interspecific mutualism (Sections 3.2, 5.5).

(Cont.)

Table 7.1 (*Cont.*)

Social group transformation

- According to the size-complexity hypothesis, an increase in group size (number of subunits) within social groups of relatives brings about a greater degree of reproductive and non-reproductive division of labour. This increases the level of interdependence of group members and so increases individuality at the group level (Sections 6.4–6.6).
- An important component of this process is that an increase in group size within social groups of relatives facilitates one party (germline cells in multicellular organisms, the queen in eusocial societies) emerging as a virtual dominant, i.e. the party to whose offspring the other group members have the greatest mean relatedness, so increasing the degree of reproductive division of labour (Section 6.4).
- Another important component is the saving of the costs of within-group conflict (more effective conflict resolution) as power shifts away from single group members as group size increases, transferring either to collectives of group members or to the virtual dominant, e.g. respectively, workers and the queen in eusocial societies headed by one, singly-mated queen (Section 6.5).
- Any process that contributes to the stability of kin structures within social groups, e.g. in eusocial societies, effective exclusion of non-nestmates, lack of queen replacement, and lifetime monogamy of the founding pair, facilitates altruists' commitment to altruism and hence promotes individuality at the group level (Section 6.5).
- Social evolution can be self-reinforcing, in that the evolution of one set of social traits can create conditions favourable to their maintenance or to the evolution of another set of social traits. This occurs when the evolution of conflict resolution creates conditions for further social change and, in social group transformation, when increased group size promotes further increases in group size. At the largest scale, it is evident from a transition to one hierarchical level permitting a later transition to the next (Section 6.6).

---

frameworks. It conceptually unifies a very large range of different phenomena across all taxa, hierarchical levels, and times. Looking backwards, it explains not only the major transitions to individuality themselves, and hence the biological hierarchy, but also the reasons for which the hierarchy extends; that is, why life grows more complex. Looking forwards, it predicts that newly-discovered cases of the phenomena it covers should conform to the principles on which it is based (Table 7.1).

It is also worth repeating what the expanded view of social evolution does not imply. Although it explains the hierarchy of life with shared evolutionary principles, it does not imply that all levels in the hierarchy are equivalent. The gene remains as a distinct entity—the replicator—throughout the hierarchy. Cells and organisms are not replicators. Natural selection at all levels concerns changes in gene frequency, not changes in cell or organism frequency. Adaptations serve the interests of the genes responsible for them. They serve the interests of units at higher levels in the hierarchy incidentally, to the extent that genes achieve a coincidence of fitness interests at a given level (Section 2.5). In addition, the view of social evolution espoused in this book does not imply that major transitions are inevitable. As suggested by life's first two billion years, the long prokaryotic aeons, major transitions occur if the selective conditions at the time they evolve are conducive to them. Hence, social groups that do not exhibit individualistic levels of organization, such as the many different kinds of non-eusocial societies in vertebrates and invertebrates (Wilson 1971, 1975; Costa 2006), are not somehow incomplete. They are not pausing on an unavoidable trajectory towards individuality. Instead, they presumably represent fully viable social structures in their current selective environments. Nonetheless, social evolution may exhibit biases such that, within multiple lineages over evolutionary time, the ladder of life grows extra rungs (Sections 6.3, 6.5).

7.2 Open questions in the study of social evolution

The overall approach of this book has been one of unabashed system-building. If the system advocated is valid, such an approach has the potential pay-off of helping achieve an intellectually satisfying unification of previously disconnected ideas. Through this, it can go on to help explain a large array of otherwise disconnected facts. If the system is not valid, the risk is that one ends up trying to force all relevant phenomena into it to make everything fit, as if cramming a set of ill-chosen items into a suitcase of the wrong size and shape. I naturally hold the system to be valid and the fit to be good. But this is not to declare that everything is included and neatly sorted, that nothing is sticking out awkwardly, or that future developments will not force us to reconsider the direction of travel. Like every idea in science, the interpretation of social evolution advocated in this book draws its validity from its theoretical cogency, its scope, the strength of its empirical base, its predictive power, and nothing else. With this in mind, I have suggested a list of issues in the field of social evolution and the major transitions that remain open (Table 7.2). The list is not intended to be exhaustive. Some of the issues have long been recognized as still needing resolution, others involve newly-posed questions that are only beginning to be investigated. Some issues have been raised in the preceding chapters, others have not but would nonetheless reward investigation. Some might lead to findings that reinforce the approach taken here, others might compel us to reassess it. The study of social evolution in its broadest sense has yielded a large set of novel and exciting results over recent years. There is every reason to expect that it will carry on doing so.

7.3 The next major transition

The next major transition, if it occurs, could take several forms. Among the egalitarian transitions, the next step could consist of two different interspecific mutualisms coming together to form a single, symbiotic organism composed of four or more species. The fusion of two different kinds of lichen would be an example. Among the fraternal transitions, the next step could consist of two or more societies fusing to form a society of societies. Stearns's (2007) analysis of the later stages of human social evolution proposes that such a transition is already underway in our own species. In this section, we end our exploration of the major transitions by considering whether the unicolonial ants also represent such a transition in progress.

Table 7.2 Twenty open questions in social evolution

Question

General

- Within sets of related lineages whose phylogenies are known, does inclusive fitness theory allow us to identify factors that correctly predict why some lineages undergo major transitions whereas others do not (Section 4.5)?
- Which of the following affects social traits more strongly: variation in relatedness or variation in the non-genetic parameters in Hamilton's rule, e.g. benefit and cost (Korb and Heinze 2008a)?
- Will knowledge of the detailed genetic basis of social evolution require us to alter our concepts of how it occurs (Robinson et al. 2005; Keller 2009)?
- Does social evolution influence genome evolution (Gadau et al. 2000; Wilfert et al. 2007)?
- What is the role of maternal effects, and, more generally, what is the role of indirect genetic effects in social evolution, and does considering these effects provide insights and explain phenomena in ways that add to insights and explanations derived from inclusive fitness theory (Linksvayer and Wade 2005; Bijma and Wade 2008; Russell and Lummaa 2009)?

Social group formation

- How and why did genes (replicating molecules) first aggregate into cells and form genomes (Maynard Smith and Szathmáry 1995)?
- What was the ecological basis of the origin of sexual reproduction (Section 4.3)?
- In the origin of sexual reproduction, what was the proximate and ultimate basis for the evolution of meiosis (Maynard Smith and Szathmáry 1995)?
- Can the origin of recombination be explained with inclusive fitness theory (Haig and Grafen 1991)?
- In general, how useful is an inclusive-fitness perspective on the evolution of sexual reproduction and associated phenomena (Bourke 2009)?
- Are there ecological factors that consistently facilitate the evolution of interspecific mutualisms (Section 4.3)?
- Does social group formation in the origin of multicellularity satisfy Hamilton's rule quantitatively, i.e. when all terms (genetic and non-genetic) in Hamilton's rule are accurately measured (Grosberg and Strathmann 2007)?
- When more cases than have previously been investigated are researched, does social group formation in the origin of eusociality satisfy Hamilton's rule quantitatively (Bourke 1997)?
- Is the haplodiploidy hypothesis for the origin of eusociality correct (Hamilton 1964; Bourke and Franks 1995; Bourke 1997)? Note that the question at issue here is not, 'Are relatedness asymmetries as found in haplodiploids essential for the origin of eusociality?'; clearly the answer to this is no. Instead, the question is, 'Do relatedness asymmetries make haplodiploid lineages more likely to evolve eusocial members than non-haplodiploid lineages?'.

Social group maintenance

- In social evolution in cooperatively breeding vertebrates and small-colony eusocial insects, what is the relative importance of direct and indirect fitness (Cant and Field 2001; Clutton-Brock 2002; Griffin and West 2003)?
- What are the conditions under which within-group kin discrimination evolves, and do they operate differently in different taxa (Komdeur et al. 2008)?

Table 7.2   (*Cont.*)

Question

• In eusocial societies, how is selection for increased genetic variability, e.g. for defence against parasites, balanced against selection for genetic similarity, to minimize potential conflict (Sections 5.2, 6.2)?
• In eusocial societies, how does group productivity covary with the frequency of worker reproduction and do costs of worker reproduction help explain its distribution (Section 5.4)?

Social group transformation

• In multicellular organisms and eusocial societies, are all the different elements of the size-complexity hypothesis supported (Chapter 6)?
• In eusocial societies, what are the roles of policing, self-restraint, and mode of caste determination in the evolution of extreme reproductive altruism, e.g. total worker sterility (Sections 6.4)?

Unicolonial ants are ones in which there has been a breakdown in the usual barriers of nestmate recognition (Section 5.2). As a result, populations no longer exist as sets of distinct colonies whose workers never mix (multicoloniality), but as far larger groupings consisting of an indefinite number of nests that cannot readily be sorted into separate colonies. Higher-levels of grouping can, in some cases, be detected at the level of the 'supercolony'. For example, the unicolonial Argentine ant (*Linepithema humile*) forms a supercolony that stretches for 6000 km around the Atlantic and Mediterranean coasts of south-western Europe and is estimated to contain millions of nests and billions of workers (Giraud et al. 2002). Within the supercolony, workers are not mutually hostile, but they are hostile to workers of a smaller supercolony embedded within the main one (Fig. 7.1).

The evolution of unicoloniality in ants has occurred several times independently. In all cases, it is believed to represent a relatively recent phylogenetic development (Bourke and Franks 1995; Helanterä et al. 2009). Supporting this conclusion is the

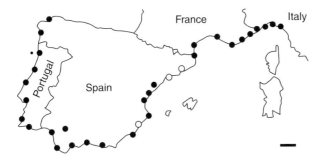

Fig. 7.1   Distribution of a supercolony of Argentine ants, *Linepithema humile* (black circles). Aggression tests established that workers within the supercolony were not aggressive to one another, but they were aggressive to workers of a smaller supercolony (open circles) embedded within the main one. Scale bar is 100 km. Reproduced from Giraud et al. (2002) with kind permission of the authors and publisher. Copyright (2002) National Academy of Sciences, USA.

fact that there are no taxa consisting entirely of unicolonial species, no 'ancient unicolonials'. Passera (1994) noted a tendency for unicolonial ants to have completely sterile workers, suggesting that unicolonial evolution is facilitated by the attainment of a certain degree of social complexity, specifically the loss of workers' totipotency. All these features suggest that unicoloniality is a derived and, to date, evolutionarily transient phenomenon. The existence of unicoloniality raises a host of fascinating ecological, evolutionary, and behavioural questions (Hölldobler and Wilson 1977; Bourke and Franks 1995; Helanterä et al. 2009). One concerns how worker altruism is maintained in the face of very low levels of relatedness (Bourke and Franks 1995; Helanterä et al. 2009). The basic answer appears to be that, through the suite of other traits associated with unicoloniality, such as high queen and worker numbers per nest and the ability of nests to reproduce rapidly by splitting into parts, unicolonial species purchase short-term ecological success at the expense of long-term evolutionary stability (Bourke and Franks 1995; Helanterä et al. 2009). This also helps explain why several unicolonial ants are notoriously successful invasive species. So damaging are some of them that they threaten the faunal and floral biodiversity of their non-native ranges (Chapman and Bourke 2001; Holway et al. 2002).

Here, however, the issue is whether, as has been suggested (McShea and Changizi 2003), unicoloniality in ants can be regarded as one of the next major transitions. The answer is a qualified yes. On the one hand, unicolonial populations represent higher-level groupings of multiple eusocial societies and in this sense are the next step in the hierarchy of fraternal transitions. The fact that unicolonial populations arise from the serial splitting of a smaller number of founder colonies, rather than the fusion of pre-existing colonies (Helanterä et al. 2009), can, in present terms (Sections 3.3, 4.1), be restated by saying that they develop subsocially from a low number of propagules. In addition, evidence that habitat saturation can be one of the ecological drivers of unicoloniality (Ross and Keller 1995) suggests a similarity between eusociality and unicoloniality in the factors promoting social group formation (Section 4.3). On the other hand, unicolonial populations, or supercolonies, have not become fully integrated collectives worthy of being regarded as individuals in their own right. This would occur if, for example, some nests tended to concentrate on reproduction (e.g. nests of queens alone) and others on helping (e.g. nests of workers alone), or if some worker-dominated nests concentrated on some tasks (e.g. collecting food) and others on different tasks (e.g. defence). In short, unicolonial populations lack a reproductive and a non-reproductive division of labour at the level of the grouped subunits, i.e. their constituent nests. This leads to the conclusion that, in the major transition that they are undergoing, unicolonial ants have entered the stage of social group formation without having passed through social group transformation. There is no reason why they should necessarily complete this final stage. Their apparent long-term instability would need to be overcome, and various other ecological and physical constraints might prevent them from doing so. But, if they ever do, a time-traveller to the distant future could anticipate with relish an encounter with another astounding product of social evolution.

7.4 Summary

1. A major conclusion of this book is that each stage in a major transition involves common principles of social evolution acting in analogous ways at the different hierarchical levels. Evidence supporting this conclusion comes from numerous taxa and from several levels within both the egalitarian and the fraternal transitions. Another major conclusion is that, in the evolution of multicellularity and eusociality, the size-complexity hypothesis provides a potentially powerful explanation for how and why social group transformation occurs.
2. There are many open questions remaining in the study of social evolution and the major transitions. They concern each stage in a major transition, and transitions at every level.
3. The evolution of unicoloniality in ants represents the evolution, via a subsocial pathway, of societies of eusocial societies, and in this sense represents the next step in the hierarchy of fraternal transitions. However, unicolonial ants are at the stage of social group formation, and might never reach the stage of social group transformation.

References

Aanen, D. K., Debets, A. J. M., de Visser, J. A. G. M., and Hoekstra, R. F. (2008). The social evolution of somatic fusion. *BioEssays*, **30**, 1193–203.

Aanen, D. K., de Fine Licht, H. H., Debets, A. J. M., Kerstes, N. A. G., Hoekstra, R. F., and Boomsma, J. J. (2009). High symbiont relatedness stabilizes mutualistic cooperation in fungus-growing termites. *Science*, **326**, 1103–106.

Abbot, P. (2009). On the evolution of dispersal and altruism in aphids. *Evolution*, **63**, 2687–96.

Abbot, P., Withgott, J. H., and Moran, N. A. (2001). Genetic conflict and conditional altruism in social aphid colonies. *Proceedings of the National Academy of Sciences, USA*, **98**, 12068–71.

Abedin, M. and King, N. (2008). The premetazoan ancestry of cadherins. *Science*, **319**, 946–48.

Adams, E. S. and Tschinkel, W. R. (2001). Mechanisms of population regulation in the fire ant *Solenopsis invicta*: an experimental study. *Journal of Animal Ecology*, **70**, 355–69.

Addicott, J. F. (1981). Stability properties of 2-species models of mutualism: simulation studies. *Oecologia*, **49**, 42–49.

Agnarsson, I., Avilés, L., Coddington, J. A., and Maddison, W. P. (2006). Sociality in theridiid spiders: repeated origins of an evolutionary dead end. *Evolution*, **60**, 2342–51.

Agnarsson, I., Maddison, W. P., and Avilés, L. (2007). The phylogeny of the social *Anelosimus* spiders (Araneae: Theridiidae) inferred from six molecular loci and morphology. *Molecular Phylogenetics and Evolution*, **43**, 833–51.

Agrawal, A. F. (2006). Evolution of sex: why do organisms shuffle their genotypes? *Current Biology*, **16**, R696–704.

Akino, T., Yamamura, K., Wakamura, S., and Yamaoka, R. (2004). Direct behavioral evidence for hydrocarbons as nestmate recognition cues in *Formica japonica* (Hymenoptera: Formicidae). *Applied Entomology and Zoology*, **39**, 381–87.

Alexander, R. D. (1974). The evolution of social behavior. *Annual Review of Ecology and Systematics*, **5**, 325–83.

Alexander, R. D. and Borgia, G. (1978). Group selection, altruism, and the levels of organization of life. *Annual Review of Ecology and Systematics*, **9**, 449–74.

Alexander, R. D., Noonan, K. M., and Crespi, B. J. (1991). The evolution of eusociality. In P. W. Sherman, J. U. M. Jarvis, and R. D. Alexander, eds. *The biology of the naked mole-rat*, pp. 3–44. Princeton University Press, Princeton.

Alonso, W. J. and Schuck-Paim, C. (2002). Sex-ratio conflicts, kin selection, and the evolution of altruism. *Proceedings of the National Academy of Sciences, USA*, **99**, 6843–47.

Amdam, G. V., Csondes, A., Fondrk, M. K., and Page, R. E. (2006). Complex social behaviour derived from maternal reproductive traits. *Nature*, **439**, 76–78.

Amoah-Buahin, E., Bone, N., and Armstrong, J. (2005). Hyphal growth in the fission yeast *Schizosaccharomyces pombe*. *Eukaryotic Cell*, **4**, 1287–97.

Anderson, C. and McShea, D. W. (2001). Individual versus social complexity, with particular reference to ant colonies. *Biological Reviews*, **76**, 211–37.

Araten, D. J., Golde, D. W., Zhang, R. H. et al. (2005). A quantitative measurement of the human somatic mutation rate. *Cancer Research*, **65**, 8111–17.

Archibald, J. M. (2006). Endosymbiosis: double-take on plastid origins. *Current Biology*, **16**, R690–92.

Archibald, J. M. (2009). The puzzle of plastid evolution. *Current Biology*, **19**, R81–88.

Archibald, J. M. and Keeling, P. J. (2002). Recycled plastids: a 'green movement' in eukaryotic evolution. *Trends in Genetics*, **18**, 577–84.

Arnold, G., Quenet, B., and Masson, C. (2000). Influence of social environment on genetically based subfamily signature in the honeybee. *Journal of Chemical Ecology*, **26**, 2321–33.

Avilés, L. (1997). Causes and consequences of cooperation and permanent-sociality in spiders. In J. C. Choe and B. J. Crespi, eds. *Social behavior in insects and arachnids*, pp. 476–98. Cambridge University Press, Cambridge.

Axelrod, R. and Hamilton, W. D. (1981). The evolution of cooperation. *Science*, **211**, 1390–96.

Axelrod, R., Axelrod, D. E., and Pienta, K. J. (2006). Evolution of cooperation among tumor cells. *Proceedings of the National Academy of Sciences, USA*, **103**, 13474–79.

Ayre, D. J. and Grosberg, R. K. (2005). Behind anemone lines: factors affecting division of labour in the social cnidarian *Anthopleura elegantissima*. *Animal Behaviour*, **70**, 97–110.

Baer, B. and Schmid-Hempel, P. (2001). Unexpected consequences of polyandry for parasitism and fitness in the bumblebee, *Bombus terrestris*. *Evolution*, **55**, 1639–43.

Baglione, V., Canestrari, D., Marcos, J. M., and Ekman, J. (2006). Experimentally increased food resources in the natal territory promote offspring philopatry and helping in cooperatively breeding carrion crows. *Proceedings of the Royal Society B*, **273**, 1529–35.

Baldauf, S. L. (2003). The deep roots of eukaryotes. *Science*, **300**, 1703–706.

Banfield, W. G., Woke, P. A., MacKay, C. M., and Cooper, H. L. (1965). Mosquito transmission of reticulum cell sarcoma of hamsters. *Science*, **148**, 1239–40.

Bantinaki, E., Kassen, R., Knight, C. G., Robinson, Z., Spiers, A. J., and Rainey, P. B. (2007). Adaptive divergence in experimental populations of *Pseudomonas fluorescens*. III. Mutational origins of wrinkly spreader diversity. *Genetics*, **176**, 441–53.

Barr, C. M., Neiman, M., and Taylor, D. R. (2005). Inheritance and recombination of mitochondrial genomes in plants, fungi and animals. *New Phytologist*, **168**, 39–50.

Baudry, E., Kryger, P., Allsopp, M. et al. (2004). Whole-genome scan in thelytokous-laying workers of the Cape honeybee (*Apis mellifera capensis*): central fusion, reduced recombination rates and centromere mapping using half-tetrad analysis. *Genetics*, **167**, 243–52.

Beekman, M. and Oldroyd, B. P. (2008). When workers disunite: intraspecific parasitism by eusocial bees. *Annual Review of Entomology*, **53**, 19–37.

Beekman, M. and Ratnieks, F. L. W. (2003). Power over reproduction in social Hymenoptera. *Philosophical Transactions of the Royal Society of London B*, **358**, 1741–53.

Beekman, M., Sumpter, D. J. T., and Ratnieks, F. L. W. (2001). Phase transition between disordered and ordered foraging in Pharaoh's ants. *Proceedings of the National Academy of Sciences, USA*, **98**, 9703–706.

Beekman, M., Komdeur, J., and Ratnieks, F. L. W. (2003). Reproductive conflicts in social animals: who has power? *Trends in Ecology and Evolution*, **18**, 277–82.

Bell, G. (1982). *The masterpiece of nature: the evolution and genetics of sexuality*. University of California Press, Berkeley.

Bell, G. (1985). The origin and early evolution of germ cells as illustrated by the Volvocales. In H. O. Halvorson and A. Monroy, eds. *The origin and evolution of sex*, pp. 221–56. Alan R. Liss, New York.

Bell, G. and Mooers, A. O. (1997). Size and complexity among multicellular organisms. *Biological Journal of the Linnean Society*, **60**, 345–63.

Bennett, P. M. and Owens, I. P. F. (2002). *Evolutionary ecology of birds: life histories, mating systems, and extinction*. Oxford University Press, Oxford.

Bentolila, S., Alfonso, A. A., and Hanson, M. R. (2002). A pentatricopeptide repeat-containing gene restores fertility to cytoplasmic male-sterile plants. *Proceedings of the National Academy of Sciences, USA*, **99**, 10887–92.

Bergmüller, R., Heg, D., and Taborsky, M. (2005). Helpers in a cooperatively breeding cichlid stay and pay or disperse and breed, depending on ecological constraints. *Proceedings of the Royal Society B*, **272**, 325–31.

Bergmüller, R., Johnstone, R. A., Russell, A. F., and Bshary, R. (2007a). Integrating cooperative breeding into theoretical concepts of cooperation. *Behavioural Processes*, **76**, 61–72.

Bergmüller, R., Russell, A. F., Johnstone, R. A., and Bshary, R. (2007b). On the further integration of cooperative breeding and cooperation theory. *Behavioural Processes*, **76**, 170–81.

Bergstrom, C. T., Bronstein, J. L., Bshary, R. et al. (2003). Interspecific mutualism: puzzles and predictions. In P. Hammerstein, ed. *Genetic and cultural evolution of cooperation*, pp. 241–56. The MIT Press, Cambridge MA.

Bhattacharya, D., Archibald, J. M., Weber, A. P. M., and Reyes-Prieto, A. (2007). How do endosymbionts become organelles? Understanding early events in plastid evolution. *BioEssays*, **29**, 1239–46.

Bijma, P. and Aanen, D. K. (2010). Assortment, Hamilton's rule and multilevel selection. *Proceedings of the Royal Society B*, **277**, 673–75.

Bijma, P. and Wade, M. J. (2008). The joint effects of kin, multilevel selection and indirect genetic effects on response to genetic selection. *Journal of Evolutionary Biology*, **21**, 1175–88.

Birky, C. W. (1995). Uniparental inheritance of mitochondrial and chloroplast genes: mechanisms and evolution. *Proceedings of the National Academy of Sciences, USA*, **92**, 11331–38.

Birky, C. W. (2001). The inheritance of genes in mitochondria and chloroplasts: laws, mechanisms, and models. *Annual Review of Genetics*, **35**, 125–48.

Birky, C. W. (2008). Uniparental inheritance of organelle genes. *Current Biology*, **18**, R692–95.

Birmingham, A. L., Hoover, S. E., Winston, M. L., and Ydenberg, R. C. (2004). Drifting bumble bee (Hymenoptera: Apidae) workers in commercial greenhouses may be social parasites. *Canadian Journal of Zoology*, **82**, 1843–53.

Blackstone, N. W. (1995). A units-of-evolution perspective on the endosymbiont theory of the origin of the mitochondrion. *Evolution*, **49**, 785–96.

Bonabeau, E., Theraulaz, G., Deneubourg, J.-L., Aron, S., and Camazine, S. (1997). Self-organization in social insects. *Trends in Ecology and Evolution*, **12**, 188–93.

Bone, Q. and Trueman, E. R. (1983). Jet propulsion in salps (Tunicata: Thaliacea). *Journal of Zoology*, **201**, 481–506.

Bonner, J. T. (1969). Hormones in social amoebae and mammals. *Scientific American*, **220**, 78–91.

Bonner, J. T. (1974). *On development: the biology of form*. Harvard University Press, Cambridge MA.

Bonner, J. T. (1988). *The evolution of complexity by means of natural selection*. Princeton University Press, Princeton.

Bonner, J. T. (1993). Dividing the labour in cells and societies. *Current Science*, **64**, 459–66.

Bonner, J. T. (1998). The origins of multicellularity. *Integrative Biology*, **1**, 27–36.

Bonner, J. T. (2003a). Evolution of development in the cellular slime molds. *Evolution & Development*, **5**, 305–13.

Bonner, J. T. (2003b). On the origin of differentiation. *Journal of Biosciences*, **28**, 523–28.

Bonner, J. T. (2004). The size-complexity rule. *Evolution*, **58**, 1883–90.

Bonner, J. T. (2006). *Why size matters: from bacteria to blue whales*. Princeton University Press, Princeton.

Bonner, J. T. (2009). *The social amoebae: the biology of cellular slime molds*. Princeton University Press, Princeton.

Boomsma, J. J. (2007). Kin selection versus sexual selection: why the ends do not meet. *Current Biology*, **17**, R673–83.

Boomsma, J. J. (2009). Lifetime monogamy and the evolution of eusociality. *Philosophical Transactions of the Royal Society B*, **364**, 3191–207.

Boomsma, J. J. and Franks, N. R. (2006). Social insects: from selfish genes to self organisation and beyond. *Trends in Ecology and Evolution*, **21**, 303–308.

Boomsma, J. J. and Grafen, A. (1991). Colony-level sex ratio selection in the eusocial Hymenoptera. *Journal of Evolutionary Biology*, **4**, 383–407.

Boomsma, J. J. and Ratnieks, F. L. W. (1996). Paternity in eusocial Hymenoptera. *Philosophical Transactions of the Royal Society of London B*, **351**, 947–75.

Boomsma, J. J., Nielsen, J., Sundström, L. et al. (2003). Informational constraints on optimal sex allocation in ants. *Proceedings of the National Academy of Sciences, USA*, **100**, 8799–804.

Boraas, M. E., Seale, D. B., and Boxhorn, J. E. (1998). Phagotrophy by a flagellate selects for colonial prey: a possible origin of multicellularity. *Evolutionary Ecology*, **12**, 153–64.

Boswell, G. P., Franks, N. R., and Britton, N. F. (2001). Arms races and the evolution of big fierce societies. *Proceedings of the Royal Society of London B*, **268**, 1723–30.

Boucher, D. H., James, S., and Keeler, K. H. (1982). The ecology of mutualism. *Annual Review of Ecology and Systematics*, **13**, 315–47.

Bourke, A. F. G. (1988). Worker reproduction in the higher eusocial Hymenoptera. *Quarterly Review of Biology*, **63**, 291–311.

Bourke, A. F. G. (1997). Sociality and kin selection in insects. In J. R. Krebs and N. B. Davies, eds. *Behavioural ecology: an evolutionary approach*, 4th edn, pp. 203–27. Blackwell Science Ltd, Oxford.

Bourke, A. F. G. (1999). Colony size, social complexity and reproductive conflict in social insects. *Journal of Evolutionary Biology*, **12**, 245–57.

Bourke, A. F. G. (2002). Genetics of social behaviour in fire ants. *Trends in Genetics*, **18**, 221–23.

Bourke, A. F. G. (2005). Genetics, relatedness and social behaviour in insect societies. In M. D. E. Fellowes, G. J. Holloway, and J. Rolff, eds. *Insect evolutionary ecology*, pp. 1–30. CABI Publishing, Wallingford.

Bourke, A. F. G. (2007). Kin selection and the evolutionary theory of aging. *Annual Review of Ecology, Evolution, and Systematics*, **38**, 103–28.

Bourke, A. F. G. (2009). The kin structure of sexual interactions. *Biology Letters*, **5**, 689–92.

Bourke, A. F. G. and Franks, N. R. (1995). *Social evolution in ants*. Princeton University Press, Princeton.

Bourke, A. F. G. and Ratnieks, F. L. W. (1999). Kin conflict over caste determination in social Hymenoptera. *Behavioral Ecology and Sociobiology*, **46**, 287–97.

Boyd, R. and Richerson, P. J. (2009). Culture and the evolution of human cooperation. *Philosophical Transactions of the Royal Society B*, **364**, 3281–88.

Boyle, R. A. and Lenton, T. M. (2006). Fluctuation in the physical environment as a mechanism for reinforcing evolutionary transitions. *Journal of Theoretical Biology*, **242**, 832–43.

Brady, S. G., Schulz, T. R., Fisher, B. L., and Ward, P. S. (2006a). Evaluating alternative hypotheses for the early evolution and diversification of ants. *Proceedings of the National Academy of Sciences, USA*, **103**, 18172–77.

Brady, S. G., Sipes, S., Pearson, A., and Danforth, B. N. (2006b). Recent and simultaneous origins of eusociality in halictid bees. *Proceedings of the Royal Society B*, **273**, 1643–49.

Branda, S. S., Gonzalez-Pastor, J. E., Ben-Yehuda, S., Losick, R., and Kolter, R. (2001). Fruiting body formation by *Bacillus subtilis*. *Proceedings of the National Academy of Sciences, USA*, **98**, 11621–26.

Brandt, M., Foitzik, S., Fischer-Blass, B., and Heinze, J. (2005). The coevolutionary dynamics of obligate ant social parasite systems - between prudence and antagonism. *Biological Reviews*, **80**, 251–67.

Breed, M. D. and Bennett, B. (1987). Kin recognition in highly eusocial insects. In D. J. C. Fletcher and C. D. Michener, eds. *Kin recognition in animals*, pp. 243–85. John Wiley and Sons, Chichester.

Brockmann, H. J. (1997). Cooperative breeding in wasps and vertebrates: the role of ecological constraints. In J. C. Choe and B. J. Crespi, eds. *The evolution of social behavior in insects and arachnids*, pp. 347–71. Cambridge University Press, Cambridge.

Brocks, J. J., Logan, G. A., Buick, R., and Summons, R. E. (1999). Archean molecular fossils and the early rise of eukaryotes. *Science*, **285**, 1033–36.

Bronstein, J. L. (1994). Conditional outcomes in mutualistic interactions. *Trends in Ecology and Evolution*, **9**, 214–17.

Bshary, R. (2002). Biting cleaner fish use altruism to deceive image-scoring client reef fish. *Proceedings of the Royal Society of London B*, **269**, 2087–93.

Bshary, R. and Bergmüller, R. (2008). Distinguishing four fundamental approaches to the evolution of helping. *Journal of Evolutionary Biology*, **21**, 405–20.

Bull, J. J. and Rice, W. R. (1991). Distinguishing mechanisms for the evolution of co-operation. *Journal of Theoretical Biology*, **149**, 63–74.

Burki, F., Shalchian-Tabrizi, K., and Pawlowski, J. (2008). Phylogenomics reveals a new 'megagroup' including most photosynthetic eukaryotes. *Biology Letters*, **4**, 366–69.

Burland, T. M., Bennett, N. C., Jarvis, J. U. M., and Faulkes, C. G. (2002). Eusociality in African mole-rats: new insights from patterns of genetic relatedness in the Damaraland mole-rat (*Cryptomys damarensis*). *Proceedings of the Royal Society of London B*, **269**, 1025–30.

Burt, A. and Trivers, R. (2006). *Genes in conflict: the biology of selfish genetic elements*. Belknap Press of Harvard University Press, Cambridge MA.

Buss, L. W. (1983). Evolution, development, and the units of selection. *Proceedings of the National Academy of Sciences, USA*, **80**, 1387–91.

Buss, L. W. (1987). *The evolution of individuality*. Princeton University Press, Princeton.

Buss, L. W. (1999). Slime molds, ascidians, and the utility of evolutionary theory. *Proceedings of the National Academy of Sciences, USA*, **96**, 8801–803.

Buston, P. M. and Zink, A. G. (2009). Reproductive skew and the evolution of conflict resolution: a synthesis of transactional and tug-of-war models. *Behavioral Ecology*, **20**, 672–84.

Butterfield, N. J. (2000). *Bangiomorpha pubescens* n. gen., n. sp.: implications for the evolution of sex, multicellularity, and the Mesoproterozoic/Neoproterozoic radiation of eukaryotes. *Paleobiology*, **26**, 386–404.

Butterfield, N. J. (2009). Modes of pre-Ediacaran multicellularity. *Precambrian Research*, **173**, 201–11.

Cairns-Smith, A. G. (1985). *Seven clues to the origin of life*. Cambridge University Press, Cambridge.

Camazine, S., Deneubourg, J.-L., Franks, N. R., Sneyd, J., Theraulaz, G., and Bonabeau, E. (2001). *Self-organization in biological systems*. Princeton University Press, Princeton.

Cameron, S. A. and Mardulyn, P. (2001). Multiple molecular data sets suggest independent origins of highly eusocial behavior in bees (Hymenoptera: Apinae). *Systematic Biology*, **50**, 194–214.

Canestrari, D., Marcos, J. M., and Baglione, V. (2009). Cooperative breeding in carrion crows reduces the rate of brood parasitism by great spotted cuckoos. *Animal Behaviour*, **77**, 1337–44.

Cant, M. A. and Field, J. (2001). Helping effort and future fitness in cooperative animal societies. *Proceedings of the Royal Society of London B*, **268**, 1959–64.

Cant, M. A. and Johnstone, R. A. (2009). How threats influence the evolutionary resolution of within-group conflict. *American Naturalist*, **173**, 759–71.

Cao, L., Shitara, H., Sugimoto, M., Hayashi, J.-I., Abe, K., and Yonekawa, H. (2009). New evidence confirms that the mitochondrial bottleneck is generated without reduction of mitochondrial DNA content in early primordial germ cells of mice. *PLoS Genetics*, **5** (12), e1000756.

Carlin, N. F. (1988). Species, kin and other forms of recognition in the brood discrimination behavior of ants. In J. C. Trager, ed. *Advances in myrmecology*, pp. 267–95. E.J. Brill, Leiden.

Carr, M., Leadbeater, B. S. C., Hassan, R., Nelson, M., and Baldauf, S. L. (2008). Molecular phylogeny of choanoflagellates, the sister group to Metazoa. *Proceedings of the National Academy of Sciences, USA*, **105**, 16641–46.

Carroll, S. B. (2001). Chance and necessity: the evolution of morphological complexity and diversity. *Nature*, **409**, 1102–109.

Cavalier-Smith, T. (2006). Cell evolution and earth history: stasis and revolution. *Philosophical Transactions of the Royal Society B*, **361**, 969–1006.

Châline, N., Martin, S. J., and Ratnieks, F. L. W. (2005). Absence of nepotism toward imprisoned young queens during swarming in the honey bee. *Behavioral Ecology*, **16**, 403–409.

Chapman, T. W. (2003). An inclusive fitness-based exploration of the origin of soldiers: the roles of sex ratio, inbreeding, and soldier reproduction. *Journal of Insect Behavior*, **16**, 481–501.

Chapman, R. E. and Bourke, A. F. G. (2001). The influence of sociality on the conservation biology of social insects. *Ecology Letters*, **4**, 650–62.

Chapman, T. W., Crespi, B. J., and Perry, S. P. (2008). The evolutionary ecology of eusociality in Australian gall thrips: a 'model clades' approach. In J. Korb and J. Heinze, eds. *Ecology of social evolution*, pp. 57–83. Springer-Verlag, Berlin.

Chapman, N. C., Makinson, J., Beekman, M., and Oldroyd, B. P. (2009). Honeybee, *Apis mellifera*, guards use adaptive acceptance thresholds to limit worker reproductive parasitism. *Animal Behaviour*, **78**, 1205–11.

Charlat, S., Hornett, E. A., Fullard, J. H. et al. (2007). Extraordinary flux in sex ratio. *Science*, **317**, 214.

Charlesworth, B. (1978). Some models of the evolution of altruistic behaviour between siblings. *Journal of Theoretical Biology*, **72**, 297–319.

Charnov, E. L. (1978). Evolution of eusocial behavior: offspring choice or parental parasitism? *Journal of Theoretical Biology*, **75**, 451–65.

Charnov, E. L. (1982). *The theory of sex allocation*. Princeton University Press, Princeton.

Chenoweth, L. B., Tierney, S. M., Smith, J. A., Cooper, S. J. B., and Schwarz, M. P. (2007). Social complexity in bees is not sufficient to explain lack of reversions to solitary living over long time scales. *BMC Evolutionary Biology*, **7**, 246.

Choe, J. C. (1988). Worker reproduction and social evolution in ants (Hymenoptera: Formicidae). In J. C. Trager, ed. *Advances in myrmecology*, pp. 163–87. E. J. Brill, Leiden.

Clutton-Brock, T. (2002). Breeding together: kin selection and mutualism in cooperative vertebrates. *Science*, **296**, 69–72.

Clutton-Brock, T. (2009a). Cooperation between non-kin in animal societies. *Nature*, **462**, 51–57.

Clutton-Brock, T. (2009b). Structure and function in mammalian societies. *Philosophical Transactions of the Royal Society B*, **364**, 3229–42.

Clutton-Brock, T. H. and Parker, G. A. (1995). Punishment in animal societies. *Nature*, **373**, 209–16.

Cockburn, A. (1998). Evolution of helping behavior in cooperatively breeding birds. *Annual Review of Ecology and Systematics*, **29**, 141–77.

Coggin, J. H., Oakes, J. E., Huebner, R. J., and Gilden, R. (1981). Unusual filterable oncogenic agent isolated from horizontally transmitted Syrian hamster lymphomas. *Nature*, **290**, 336–38.

Cole, B. J. (1983). Multiple mating and the evolution of social behavior in the Hymenoptera. *Behavioral Ecology and Sociobiology*, **12**, 191–201.

Cole, B. J. (1986). The social behavior of *Leptothorax allardycei* (Hymenoptera, Formicidae): time budgets and the evolution of worker reproduction. *Behavioral Ecology and Sociobiology*, **18**, 165–73.

Colman, A. M. (2006). The puzzle of cooperation. *Nature*, **440**, 744–45.

Cornwallis, C. K., West, S. A., and Griffin, A. S. (2009). Routes to indirect fitness in cooperatively breeding vertebrates: kin discrimination and limited dispersal. *Journal of Evolutionary Biology*, **22**, 2445–57.

Cosmides, L. M. and Tooby, J. (1981). Cytoplasmic inheritance and intragenomic conflict. *Journal of Theoretical Biology*, **89**, 83–129.

Costa, J. T. (2006). *The other insect societies*. Belknap Press of Harvard University Press, Cambridge MA.

Cousens, R., Dytham, C., and Law, R. (2008). *Dispersal in plants: a population perspective*. Oxford University Press, Oxford.

Craig, R. (1983). Subfertility and the evolution of eusociality by kin selection. *Journal of Theoretical Biology*, **100**, 379–97.

Creel, S. and Macdonald, D. (1995). Sociality, group size, and reproductive suppression among carnivores. *Advances in the Study of Behavior*, **24**, 203–57.

Cremer, S. and Sixt, M. (2009). Analogies in the evolution of individual and social immunity. *Philosophical Transactions of the Royal Society B*, **364**, 129–42.

Cremer, S., Armitage, S. A. O., and Schmid-Hempel, P. (2007). Social immunity. *Current Biology*, **17**, R693–702.

Crespi, B. J. (1994). Three conditions for the evolution of eusociality: are they sufficient? *Insectes Sociaux*, **41**, 395–400.

Crespi, B. J. (2001). The evolution of social behavior in microorganisms. *Trends in Ecology and Evolution*, **16**, 178–83.

Crespi, B. J. (2007). Comparative evolutionary ecology of social and sexual systems: water-breathing insects come of age. In J. E. Duffy and M. Thiel, eds. *Evolutionary ecology of social and sexual systems: crustaceans as model organisms*, pp. 442–60. Oxford University Press, New York.

Crespi, B. J. and Choe, J. C. (1997). Explanation and evolution of social systems. In J. C. Choe and B. J. Crespi, eds. *The evolution of social behavior in insects and arachnids*, pp. 499–524. Cambridge University Press, Cambridge.

Crespi, B. and Summers, K. (2005). Evolutionary biology of cancer. *Trends in Ecology and Evolution*, **20**, 545–52.

Crespi, B. J. and Yanega, D. (1995). The definition of eusociality. *Behavioral Ecology*, **6**, 109–15.

Crozier, R. H. (1992). The genetic evolution of flexible strategies. *American Naturalist*, **139**, 218–23.

Crozier, R. H. (2008). Advanced eusociality, kin selection and male haploidy. *Australian Journal of Entomology*, **47**, 2–8.

Crozier, R. H. and Pamilo, P. (1996). *Evolution of social insect colonies: sex allocation and kin selection*. Oxford University Press, Oxford.

Cruz, Y. P. (1981). A sterile defender morph in a polyembryonic hymenopterous parasite. *Nature*, **294**, 446–47.

Currie, C. R. and Stuart, A. E. (2001). Weeding and grooming of pathogens in agriculture by ants. *Proceedings of the Royal Society of London B*, **268**, 1033–39.

Currie, C. R., Scott, J. A., Summerbell, R. C., and Malloch, D. (1999). Fungus-growing ants use antibiotic-producing bacteria to control garden parasites. *Nature*, **398**, 701–704.

Currie, C. R., Poulsen, M., Mendenhall, J., Boomsma, J. J., and Billen, J. (2006). Coevolved crypts and exocrine glands support mutualistic bacteria in fungus-growing ants. *Science*, **311**, 81–83.

Dacks, J. and Roger, A. J. (1999). The first sexual lineage and the relevance of facultative sex. *Journal of Molecular Evolution*, **48**, 779–83.

Darwin, C. (1859). *On the origin of species*. John Murray, London.

Davies, N. B. (2000). *Cuckoos, cowbirds and other cheats*. T & AD Poyser, London.

Dawkins, R. (1976). *The selfish gene*. Oxford University Press, Oxford.

Dawkins, R. (1979). Twelve misunderstandings of kin selection. *Zeitschrift für Tierpsychologie*, **51**, 184–200.

Dawkins, R. (1982). *The extended phenotype*. W. H. Freeman, Oxford.

Dawkins, R. (1986). *The blind watchmaker*. Longman, Harlow.

Dawkins, R. (1989). *The selfish gene*, 2nd edn. Oxford University Press, Oxford.

Dawkins, R. (1990). Parasites, desiderata lists and the paradox of the organism. *Parasitology*, **100**, S63–73.

Dawkins, R. (2004). *The ancestor's tale*. Weidenfeld & Nicolson, London.

Dawkins, R. (2009). *The greatest show on earth: the evidence for evolution*. Bantam Press, London.

Dawkins, R. and Krebs, J. R. (1979). Arms races between and within species. *Proceedings of the Royal Society of London B*, **205**, 489–511.

Deisboeck, T. S. and Couzin, I. D. (2009). Collective behavior in cancer cell populations. *BioEssays*, **31**, 190–97.

D'Ettorre, P. and Lenoir, A. (2010). Nestmate recognition. In L. Lach, C. L. Parr, and K. L. Abbott, eds. *Ant ecology*, pp. 194–209. Oxford University Press, Oxford.

Dickinson, J. L. and Hatchwell, B. J. (2004). Fitness consequences of helping. In W. D. Koenig and J. L. Dickinson, eds. *Ecology and evolution of cooperative breeding in birds*, pp. 48–66. Cambridge University Press, Cambridge.

Diggle, S. P., Griffin, A. S., Campbell, G. S., and West, S. A. (2007). Cooperation and conflict in quorum-sensing bacterial populations. *Nature*, **450**, 411–14.

Dijkstra, M. B. and Boomsma, J. J. (2006). Are workers of *Atta* leafcutter ants capable of reproduction? *Insectes Sociaux*, **53**, 136–40.

Dijkstra, M. B. and Boomsma, J. J. (2007). The economy of worker reproduction in *Acromyrmex* leafcutter ants. *Animal Behaviour*, **74**, 519–29.

Dingli, D. and Nowak, M. A. (2006). Infectious tumour cells. *Nature*, **443**, 35–36.

Dobata, S., Sasaki, T., Mori, H., Hasegawa, E., Shimada, M., and Tsuji, K. (2009). Cheater genotypes in the parthenogenetic ant *Pristomyrmex punctatus*. *Proceedings of the Royal Society B*, **276**, 567–74.

Dodgson, J., Avula, H., Hoe, K.-L. et al. (2009). Functional genomics of adhesion, invasion, and mycelial formation in *Schizosaccharomyces pombe*. *Eukaryotic Cell*, **8**, 1298–306.

Doebeli, M. and Knowlton, N. (1998). The evolution of interspecific mutualisms. *Proceedings of the National Academy of Sciences, USA*, **95**, 8676–80.

Donaldson, Z. R. and Young, L. J. (2008). Oxytocin, vasopressin, and the neurogenetics of sociality. *Science*, **322**, 900–904.

Douglas, A. E. (2008). Conflict, cheats and the persistence of symbioses. *New Phytologist*, **177**, 849–58.

Douzery, E. J. P., Snell, E. A., Bapteste, E., Delsuc, F., and Philippe, H. (2004). The timing of eukaryotic evolution: does a relaxed molecular clock reconcile proteins and fossils? *Proceedings of the National Academy of Sciences, USA*, **101**, 15386–91.

Duffy, J. E. (2007). Ecology and evolution of eusociality in sponge-dwelling shrimp. In J. E. Duffy and M. Thiel, eds. *Evolutionary ecology of social and sexual systems: crustaceans as model organisms*, pp. 387–409. Oxford University Press, New York.

Duffy, J. E. and Macdonald, K. S. (2010). Kin structure, ecology and the evolution of social organization in shrimp: a comparative analysis. *Proceedings of the Royal Society B*, **277**, 575–84.

Duffy, J. E., Morrison, C. L., and Ríos, R. (2000). Multiple origins of eusociality among sponge-dwelling shrimps (*Synalpheus*). *Evolution*, **54**, 503–16.

Dugatkin, L. A. (1997). *Cooperation among animals: an evolutionary perspective*. Oxford University Press, New York.

Dugatkin, L. A. (2002). Animal cooperation among unrelated individuals. *Naturwissenschaften*, **89**, 533–41.

Dugatkin, L. A. and Reeve, H. K. (1994). Behavioral ecology and levels of selection: dissolving the group selection controversy. *Advances in the Study of Behavior*, **23**, 101–33.

Duncan, L., Nishii, I., Harryman, A. et al. (2007). The *VARL* gene family and the evolutionary origins of the master cell-type regulatory gene, *regA*, in *Volvox carteri*. *Journal of Molecular Evolution*, **65**, 1–11.

Dunn, C. (2009). Siphonophores. *Current Biology*, **19**, R233–34.

Dunn, C. W. and Wagner, G. P. (2006). The evolution of colony-level development in the Siphonophora (Cnidaria: Hydrozoa). *Development Genes and Evolution*, **216**, 743–54.

Eberhard, W. G. (1980). Evolutionary consequences of intracellular organelle competition. *Quarterly Review of Biology*, **55**, 231–49.

Edwards, S. V. and Hedrick, P. W. (1998). Evolution and ecology of MHC molecules: from genomics to sexual selection. *Trends in Ecology and Evolution*, **13**, 305–11.

Edwards, D. P., Hassall, M., Sutherland, W. J., and Yu, D. W. (2006). Selection for protection in an ant-plant mutualism: host sanctions, host modularity, and the principal-agent game. *Proceedings of the Royal Society B*, **273**, 595–602.

Eickwort, G. C., Eickwort, J. M., Gordon, J., and Eickwort, M. A. (1996). Solitary behavior in a high-altitude population of the social bee *Halictus rubicundus* (Hymenoptera: Halictidae). *Behavioral Ecology and Sociobiology*, **38**, 227–33.

Ellis, L. (1995). Dominance and reproductive success among nonhuman animals: a cross-species comparison. *Ethology and Sociobiology*, **16**, 257–333.

Embley, T. M. and Martin, W. (2006). Eukaryotic evolution, changes and challenges. *Nature*, **440**, 623–30.

Embley, T. M., Van der Giezen, M., Horner, D. S., Dyal, P. L., and Foster, P. (2003). Mitochondria and hydrogenosomes are two forms of the same fundamental organelle. *Philosophical Transactions of the Royal Society of London B*, **358**, 191–203.

Emlen, S. T. (1982a). The evolution of helping. I. An ecological constraints model. *American Naturalist*, **119**, 29–39.

Emlen, S. T. (1982b). The evolution of helping. II. The role of behavioral conflict. *American Naturalist*, **119**, 40–53.

Emlen, S. T. (1991). Evolution of cooperative breeding in birds and mammals. In J. R. Krebs and N. B. Davies, eds. *Behavioural ecology: an evolutionary approach*, 3rd edn, pp. 301–37. Blackwell Scientific Publications, Oxford.

Emlen, S. T. (1995). An evolutionary theory of the family. *Proceedings of the National Academy of Sciences, USA*, **92**, 8092–99.

Emlen, S. T. (1997). Predicting family dynamics in social vertebrates. In J. R. Krebs and N. B. Davies, eds. *Behavioural ecology: an evolutionary approach*, 4th edn, pp. 228–53. Blackwell Science Ltd, Oxford.

Endler, J. A. (1986). *Natural selection in the wild*. Princeton University Press, Princeton.

Endler, A., Liebig, J., Schmitt, T. et al. (2004). Surface hydrocarbons of queen eggs regulate worker reproduction in a social insect. *Proceedings of the National Academy of Sciences, USA*, **101**, 2945–50.

Engelstädter, J. and Hurst, G. D. D. (2009). The ecology and evolution of microbes that manipulate host reproduction. *Annual Review of Ecology, Evolution, and Systematics*, **40**, 127–49.

Farnworth, E. G. and Golley, F. B., eds. (1974). *Fragile ecosystems: evaluation of research and applications in the neotropics*. Springer-Verlag, New York.

Faulkes, C. G., Bennett, N. C., Bruford, M. W., O'Brien, H. P., Aguilar, G. H., and Jarvis, J. U. M. (1997). Ecological constraints drive social evolution in the African mole-rats. *Proceedings of the Royal Society of London B*, **264**, 1619–27.

Fehr, E. and Fischbacher, U. (2003). The nature of human altruism. *Nature*, **425**, 785–91.

Fehr, E. and Gächter, S. (2002). Altruistic punishment in humans. *Nature*, **415**, 137–40.

Field, J. (1992). Intraspecific parasitism as an alternative reproductive tactic in nest-building wasps and bees. *Biological Reviews*, **67**, 79–126.

Field, J. (2008). The ecology and evolution of helping in hover wasps (Hymenoptera: Stenogastrinae). In J. Korb and J. Heinze, eds. *Ecology of social evolution*, pp. 85–107. Springer-Verlag, Berlin.

Field, J., Foster, W., Shreeves, G., and Sumner, S. (1998). Ecological constraints on independent nesting in facultatively eusocial hover wasps. *Proceedings of the Royal Society of London B*, **265**, 973–77.

Field, J., Shreeves, G., Sumner, S., and Casiraghi, M. (2000). Insurance-based advantage to helpers in a tropical hover wasp. *Nature*, **404**, 869–71.

Fink, S., Excoffier, L., and Heckel, G. (2006). Mammalian monogamy is not controlled by a single gene. *Proceedings of the National Academy of Sciences, USA*, **103**, 10956–60.

Fjerdingstad, E. J. and Crozier, R. H. (2006). The evolution of worker caste diversity in social insects. *American Naturalist*, **167**, 390–400.

Flärdh, K. and Buttner, M. J. (2009). *Streptomyces* morphogenetics: dissecting differentiation in a filamentous bacterium. *Nature Reviews Microbiology*, **7**, 36–49.

Fletcher, J. A. and Doebeli, M. (2009). A simple and general explanation for the evolution of altruism. *Proceedings of the Royal Society B*, **276**, 13–19.

Fletcher, J. A. and Zwick, M. (2006). Unifying the theories of inclusive fitness and reciprocal altruism. *American Naturalist*, **168**, 252–62.

Floreano, D. and Keller, L. (2010). Evolution of adaptive behaviour in robots by means of Darwinian selection. *PLoS Biology*, **8**(1), e1000292.

Foitzik, S. and Heinze, J. (2001). Microgeographic genetic structure and intraspecific parasitism in the ant *Leptothorax nylanderi*. *Ecological Entomology*, **26**, 449–56.

Foster, K. R. (2004). Diminishing returns in social evolution: the not-so-tragic commons. *Journal of Evolutionary Biology*, **17**, 1058–72.

Foster, K. R. and Wenseleers, T. (2006). A general model for the evolution of mutualisms. *Journal of Evolutionary Biology*, **19**, 1283–93.

Foster, K. R., Ratnieks, F. L. W., and Raybould, A. F. (2000). Do hornets have zombie workers? *Molecular Ecology*, **9**, 735–42.

Foster, K. R., Fortunato, A., Strassmann, J. E., and Queller, D. C. (2002). The costs and benefits of being a chimera. *Proceedings of the Royal Society of London B*, **269**, 2357–62.

Foster, K. R., Wenseleers, T., and Ratnieks, F. L. W. (2006a). Kin selection is the key to altruism. *Trends in Ecology and Evolution*, **21**, 57–60.

Foster, K. R., Wenseleers, T., Ratnieks, F. L. W., and Queller, D. C. (2006b). There is nothing wrong with inclusive fitness. *Trends in Ecology and Evolution*, **21**, 599–600.

Fournier, D., Estoup, A., Orivel, J. et al. (2005). Clonal reproduction by males and females in the little fire ant. *Nature*, **435**, 1230–34.

Frank, S. A. (1992). A kin selection model for the evolution of virulence. *Proceedings of the Royal Society of London B*, **250**, 195–97.

Frank, S. A. (1994a). Kin selection and virulence in the evolution of protocells and parasites. *Proceedings of the Royal Society of London B*, **258**, 153–61.

Frank, S. A. (1994b). Genetics of mutualism: the evolution of altruism between species. *Journal of Theoretical Biology*, **170**, 393–400.

Frank, S. A. (1995). Mutual policing and repression of competition in the evolution of cooperative groups. *Nature*, **377**, 520–22.

Frank, S. A. (1996a). Models of parasite virulence. *Quarterly Review of Biology*, **71**, 37–78.

Frank, S. A. (1996b). Policing and group cohesion when resources vary. *Animal Behaviour*, **52**, 1163–69.

Frank, S. A. (1996c). Host-symbiont conflict over the mixing of symbiotic lineages. *Proceedings of the Royal Society of London B*, **263**, 339–44.

Frank, S. A. (1997). Models of symbiosis. *American Naturalist*, **150**, S80–99.

Frank, S. A. (1998). *Foundations of social evolution*. Princeton University Press, Princeton.

Frank, S. A. (2003). Repression of competition and the evolution of cooperation. *Evolution*, **57**, 693–705.

Frank, S. A. (2007a). All of life is social. *Current Biology*, **17**, R648–50.

Frank, S. A. (2007b). *Dynamics of cancer: incidence, inheritance, and evolution*. Princeton University Press, Princeton.

Frank, U. (2007c). The evolution of a malignant dog. *Evolution & Development*, **9**, 521–22.

Frank, S. A. (2010a). Demography and the tragedy of the commons. *Journal of Evolutionary Biology*, **23**, 32–39.

Frank, S. A. (2010b). Somatic evolutionary genomics: mutations during development cause highly variable genetic mosaicism with risk of cancer and neurodegeneration. *Proceedings of the National Academy of Sciences, USA*, **107**, 1725–30.

Frank, S. A. and Nowak, M. A. (2004). Problems of somatic mutation and cancer. *BioEssays*, **26**, 291–99.

Fuchs, J., Obst, M., and Sundberg, P. (2009). The first comprehensive molecular phylogeny of Bryozoa (Ectoprocta) based on combined analyses of nuclear and mitochondrial genes. *Molecular Phylogenetics and Evolution*, **52**, 225–33.

Gadagkar, R. (1990). Evolution of eusociality: the advantage of assured fitness returns. *Philosophical Transactions of the Royal Society of London B*, **329**, 17–25.

Gadagkar, R. (1997). Social evolution—has nature ever rewound the tape? *Current Science*, **72**, 950–56.

Gadagkar, R. (2001). *The social biology of* Ropalidia marginata*: towards understanding the evolution of eusociality*. Harvard University Press, Cambridge MA.

Gadau, J. and Hunt, G. J. (2009). Behavioral genetics in social insects. In J. Gadau and J. Fewell, eds. *Organization of insect societies: from genome to sociocomplexity*, pp. 315–34. Harvard University Press, Cambridge MA.

Gadau, J., Page, R. E., Werren, J. H., and Schmid-Hempel, P. (2000). Genome organization and social evolution in Hymenoptera. *Naturwissenschaften*, **87**, 87–89.

Galtier, N., Nabholz, B., Glémin, S., and Hurst, G. D. D. (2009). Mitochondrial DNA as a marker of molecular diversity: a reappraisal. *Molecular Ecology*, **18**, 4541–50.

Gardner, A. and Foster, K. R. (2008). The evolution and ecology of cooperation—history and concepts. In J. Korb and J. Heinze, eds. *Ecology of social evolution*, pp. 1–36. Springer-Verlag, Berlin.

Gardner, A. and Grafen, A. (2009). Capturing the superorganism: a formal theory of group adaptation. *Journal of Evolutionary Biology*, **22**, 659–71.

Gardner, A. and West, S. A. (2006). Spite. *Current Biology*, **16**, R662–64.

Gardner, A. and West, S. A. (2009). Greenbeards. *Evolution*, **64**, 25–38.

Gardner, A., West, S. A., and Barton, N. H. (2007a). The relation between multilocus population genetics and social evolution theory. *American Naturalist*, **169**, 207–26.

Gardner, A., Hardy, I. C. W., Taylor, P. D., and West, S. A. (2007b). Spiteful soldiers and sex ratio conflict in polyembryonic parasitoid wasps. *American Naturalist*, **169**, 519–33.

Gaston, K. J. (2007). Latitudinal gradient in species richness. *Current Biology*, **17**, R574.

Gautrais, J., Theraulaz, G., Deneubourg, J.-L., and Anderson, C. (2002). Emergent polyethism as a consequence of increased colony size in insect societies. *Journal of Theoretical Biology*, **215**, 363–73.

Getz, W. M. (1981). Genetically based kin recognition systems. *Journal of Theoretical Biology*, **92**, 209–26.

Gibbons, D. W. (1986). Brood parasitism and cooperative nesting in the moorhen, *Gallinula chloropus*. *Behavioral Ecology and Sociobiology*, **19**, 221–32.

Gilbert, O. M., Foster, K. R., Mehdiabadi, N. J., Strassmann, J. E., and Queller, D. C. (2007). High relatedness maintains multicellular cooperation in a social amoeba by controlling cheater mutants. *Proceedings of the National Academy of Sciences, USA*, **104**, 8913–17.

Gilbert, O. M., Queller, D. C., and Strassmann, J. E. (2009). Discovery of a large clonal patch of a social amoeba: implications for social evolution. *Molecular Ecology*, **18**, 1273–81.

Gilley, D. C. and Tarpy, D. R. (2005). Three mechanisms of queen elimination in swarming honey bee colonies. *Apidologie*, **36**, 461–74.

Giraud, T., Pedersen, J. S., and Keller, L. (2002). Evolution of supercolonies: the Argentine ants of southern Europe. *Proceedings of the National Academy of Sciences, USA*, **99**, 6075–79.

Giron, D., Harvey, J. A., Johnson, J. A., and Strand, M. R. (2007). Male soldier caste larvae are non-aggressive in the polyembryonic wasp *Copidosoma floridanum*. *Biology Letters*, **3**, 431–34.

Gobin, B., Heinze, J., Strätz, M., and Roces, F. (2003). The energetic cost of reproductive conflicts in the ant *Pachycondyla obscuricornis*. *Journal of Insect Physiology*, **49**, 747–52.

Golden, J. W. and Yoon, H.-S. (2003). Heterocyst development in *Anabaena*. *Current Opinion in Microbiology*, **6**, 557–63.

Goodnight, C. J. (2005). Multilevel selection: the evolution of cooperation in non-kin groups. *Population Ecology*, **47**, 3–12.

Gordon, D. M. (1999). *Ants at work: how an insect society is organized*. The Free Press, New York.

Gotzek, D. and Ross, K. G. (2007). Genetic regulation of colony social organization in fire ants: an integrative overview. *Quarterly Review of Biology*, **82**, 201–26.

Gould, S. J. (2002). *The structure of evolutionary theory*. Belknap Press of Harvard University Press, Cambridge MA.

Grafen, A. (1982). How not to measure inclusive fitness. *Nature*, **298**, 425–26.

Grafen, A. (1984). Natural selection, kin selection and group selection. In J. R. Krebs and N. B. Davies, eds. *Behavioural ecology: an evolutionary approach*, 2nd edn, pp. 62–84. Blackwell Scientific Publications, Oxford.

Grafen, A. (1985). A geometric view of relatedness. In R. Dawkins and M. Ridley, eds. *Oxford surveys in evolutionary biology*, vol. 2, pp. 28–89. Oxford University Press, Oxford.

Grafen, A. (1986). Split sex ratios and the evolutionary origins of eusociality. *Journal of Theoretical Biology*, **122**, 95–121.

Grafen, A. (1991). Modelling in behavioural ecology. In J. R. Krebs and N. B. Davies, eds. *Behavioural ecology: an evolutionary approach*, 3rd edn, pp. 5–31. Blackwell Scientific Publications, Oxford.

Grafen, A. (2006). Optimization of inclusive fitness. *Journal of Theoretical Biology*, **238**, 541–63.

Grafen, A. (2007a). Detecting kin selection at work using inclusive fitness. *Proceedings of the Royal Society B*, **274**, 713–19.

Grafen, A. (2007b). An inclusive fitness analysis of altruism on a cyclical network. *Journal of Evolutionary Biology*, **20**, 2278–83.

Gray, M. W., Burger, G., and Lang, B. F. (1999). Mitochondrial evolution. *Science*, **283**, 1476–81.

Grbić, M., Ode, P. J., and Strand, M. R. (1992). Sibling rivalry and brood sex ratios in polyembryonic wasps. *Nature*, **360**, 254–56.

Griffin, A. S. and West, S. A. (2003). Kin discrimination and the benefit of helping in cooperatively breeding vertebrates. *Science*, **302**, 634–36.

Griffin, A. S., West, S. A., and Buckling, A. (2004). Cooperation and competition in pathogenic bacteria. *Nature*, **430**, 1024–27.

Griffin, A. S., Sheldon, B. C., and West, S. A. (2005). Cooperative breeders adjust offspring sex ratios to produce helpful helpers. *American Naturalist*, **166**, 628–32.

Grimaldi, D. and Engel, M. S. (2005). *Evolution of the insects*. Cambridge University Press, Cambridge.

Grogan, K. E., Chhatre, V. E., and Abbot, P. (2010). The cost of conflict in aphid societies. *Journal of Evolutionary Biology*, **23**, 185–93.

Grosberg, R. K. (1988). The evolution of allorecognition specificity in clonal invertebrates. *Quarterly Review of Biology*, **63**, 377–412.

Grosberg, R. K. and Strathmann, R. R. (1998). One cell, two cell, red cell, blue cell: the persistence of a unicellular stage in multicellular life histories. *Trends in Ecology and Evolution*, **13**, 112–16.

Grosberg, R. K. and Strathmann, R. R. (2007). The evolution of multicellularity: a minor major transition? *Annual Review of Ecology, Evolution, and Systematics*, **38**, 621–54.

Gunnels, C. W., Dubrovskiy, A., and Avalos, A. (2008). Social interactions as an ecological constraint in a eusocial insect. *Animal Behaviour*, **75**, 681–91.

Haig, D. (1992). Intragenomic conflict and the evolution of eusociality. *Journal of Theoretical Biology*, **156**, 401–403.

Haig, D. (1996). Gestational drive and the green-bearded placenta. *Proceedings of the National Academy of Sciences, USA*, **93**, 6547–51.

Haig, D. (1997). The social gene. In J. R. Krebs and N. B. Davies, eds. *Behavioural ecology: an evolutionary approach*, 4th edn, pp. 284–304. Blackwell Science Ltd, Oxford.

Haig, D. (2004). Genomic imprinting and kinship: how good is the evidence? *Annual Review of Genetics*, **38**, 553–85.

Haig, D. and Grafen, A. (1991). Genetic scrambling as a defence against meiotic drive. *Journal of Theoretical Biology*, **153**, 531–58.

Haine, E. R. (2008). Symbiont-mediated protection. *Proceedings of the Royal Society B*, **275**, 353–61.

Hamilton, W. D. (1963). The evolution of altruistic behavior. *American Naturalist*, **97**, 354–56.

Hamilton, W. D. (1964). The genetical evolution of social behaviour I, II. *Journal of Theoretical Biology*, **7**, 1–52.

Hamilton, W. D. (1970). Selfish and spiteful behaviour in an evolutionary model. *Nature*, **228**, 1218–20.

Hamilton, W. D. (1972). Altruism and related phenomena, mainly in social insects. *Annual Review of Ecology and Systematics*, **3**, 193–232.

Hamilton, W. D. (1980). Sex versus non-sex versus parasite. *Oikos*, **35**, 282–90.

Hamilton, W. D. (1982). Pathogens as causes of genetic diversity in their host populations. In R. M. Anderson and R. M. May, eds. *Population biology of infectious diseases*, pp. 269–96. Springer-Verlag, Berlin.

Hamilton, W. D. (1987). Kinship, recognition, disease, and intelligence: constraints of social evolution. In Y. Itô, J. L. Brown, and J. Kikkawa, eds. *Animal societies: theories and facts*, pp. 81–102. Japan Scientific Societies Press, Tokyo.

Hamilton, W. D., Axelrod, R., and Tanese, R. (1990). Sexual reproduction as an adaptation to resist parasites (A Review). *Proceedings of the National Academy of Sciences, USA*, **87**, 3566–73.

Hammerstein, P. (1995). A twofold tragedy unfolds. *Nature*, **377**, 478.

Hammerstein, P., ed. (2003a). *Genetic and cultural evolution of cooperation*. The MIT Press, Cambridge MA.

Hammerstein, P. (2003b). Why is reciprocity so rare in social animals? A Protestant appeal. In P. Hammerstein, ed. *Genetic and cultural evolution of cooperation*, pp. 83–93. The MIT Press, Cambridge MA.

Hammond, R. L. and Keller, L. (2004). Conflict over male parentage in social insects. *PLoS Biology*, **2**(9), e248.

Hammond, R. L., Bruford, M. W., and Bourke, A. F. G. (2006). A test of reproductive skew models in a field population of a multiple-queen ant. *Behavioral Ecology and Sociobiology*, **61**, 265–75.

Hardin, G. (1968). The tragedy of the commons. *Science*, **162**, 1243–48.

Hart, A. G. and Ratnieks, F. L. W. (2002). Waste management in the leaf-cutting ant *Atta colombica*. *Behavioral Ecology*, **13**, 224–31.

Härtel, S., Neumann, P., Raassen, F. S., Moritz, R. F. A., and Hepburn, H. R. (2006). Social parasitism by Cape honeybee workers in colonies of their own subspecies (*Apis mellifera capensis* Esch.). *Insectes Sociaux*, **53**, 183–93.

Hartmann, A., Wantia, J., Torres, J. A., and Heinze, J. (2003). Worker policing without genetic conflicts in a clonal ant. *Proceedings of the National Academy of Sciences, USA*, **100**, 12836–40.

Harvell, C. D. (1994). The evolution of polymorphism in colonial invertebrates and social insects. *Quarterly Review of Biology*, **69**, 155–85.

Hatchwell, B. J. (2009). The evolution of cooperative breeding in birds: kinship, dispersal and life history. *Philosophical Transactions of the Royal Society B*, **364**, 3217–27.

Hawlena, H., Bashey, F., Mendes-Soares, H., and Lively, C. M. (2010). Spiteful interactions in a natural population of the bacterium *Xenorhabdus bovienii*. *American Naturalist*, **175**, 374–81.

Hedges, S. B., Blair, J. E., Venturi, M. L., and Shoe, J. L. (2004). A molecular timescale of eukaryote evolution and the rise of complex multicellular life. *BMC Evolutionary Biology*, **4**, 2.

Heg, D., Bachar, Z., Brouwer, L., and Taborsky, M. (2004). Predation risk is an ecological constraint for helper dispersal in a cooperatively breeding cichlid. *Proceedings of the Royal Society of London B*, **271**, 2367–74.

Heinsohn, R. G. (1991). Kidnapping and reciprocity in cooperatively breeding white-winged choughs. *Animal Behaviour*, **41**, 1097–100.

Helanterä, H. (2007). How to test an inclusive fitness hypothesis - worker reproduction and policing as an example. *Oikos*, **116**, 1782–88.

Helanterä, H. and Bargum, K. (2007). Pedigree relatedness, not greenbeard genes, explains eusociality. *Oikos*, **116**, 217–20.

Helanterä, H. and Ratnieks, F. L. W. (2009). Sex allocation conflict in insect societies: who wins? *Biology Letters*, **5**, 700–704.

Helanterä, H. and Sundström, L. (2007). Worker reproduction in *Formica* ants. *American Naturalist*, **170**, E14–25.

Helanterä, H., Strassmann, J. E., Carrillo, J., and Queller, D. C. (2009). Unicolonial ants: where do they come from, what are they and where are they going? *Trends in Ecology and Evolution*, **24**, 341–49.

Helmkampf, M., Gadau, J., and Feldhaar, H. (2008). Population- and sociogenetic structure of the leaf-cutter ant *Atta colombica* (Formicidae, Myrmicinae). *Insectes Sociaux*, **55**, 434–42.

Henshaw, M. T., Strassmann, J. E., and Queller, D. C. (2000). The independent origin of a queen number bottleneck that promotes cooperation in the African swarm-founding wasp, *Polybioides tabidus*. *Behavioral Ecology and Sociobiology*, **48**, 478–83.

Herre, E. A. (1999). Laws governing species interactions? Encouragement and caution from figs and their associates. In L. Keller, ed. *Levels of selection in evolution*, pp. 209–37. Princeton University Press, Princeton.

Herron, M. D. and Michod, R. E. (2008). Evolution of complexity in the volvocine algae: transitions in individuality through Darwin's eye. *Evolution*, **62**, 436–51.

Herron, M. D., Hackett, J. D., Aylward, F. O., and Michod, R. E. (2009). Triassic origin and early radiation of multicellular volvocine algae. *Proceedings of the National Academy of Sciences, USA*, **106**, 3254–58.

Hillesheim, E., Koeniger, N., and Moritz, R. F. A. (1989). Colony performance in honeybees (*Apis mellifera capensis* Esch.) depends on the proportion of subordinate and dominant workers. *Behavioral Ecology and Sociobiology*, **24**, 291–96.

Himler, A. G., Caldera, E. J., Baer, B. C., Fernández-Marín, H., and Mueller, U. G. (2009). No sex in fungus-farming ants or their crops. *Proceedings of the Royal Society B*, **276**, 2611–16.

Hines, H. M., Hunt, J. H., O'Connor, T. K., Gillespie, J. J., and Cameron, S. A. (2007). Multigene phylogeny reveals eusociality evolved twice in vespid wasps. *Proceedings of the National Academy of Sciences, USA*, **104**, 3295–99.

Hochberg, M. E., Rankin, D. J., and Taborsky, M. (2008). The coevolution of cooperation and dispersal in social groups and its implications for the emergence of multicellularity. *BMC Evolutionary Biology*, **8**, 238.

Hoekstra, R. F. (2003). Power in the genome: who suppresses the outlaw? In P. Hammerstein, ed. *Genetic and cultural evolution of cooperation*, pp. 257–70. The MIT Press, Cambridge MA.

Hoffman, P. F., Kaufman, A. J., Halverson, G. P., and Schrag, D. P. (1998). A Neoproterozoic snowball earth. *Science*, **281**, 1342–46.

Hogendoorn, K. and Leys, R. (1993). The superseded female's dilemma: ultimate and proximate factors that influence guarding behaviour of the carpenter bee *Xylocopa pubescens*. *Behavioral Ecology and Sociobiology*, **33**, 371–81.

Hölldobler, B. and Wilson, E. O. (1977). The number of queens: an important trait in ant evolution. *Naturwissenschaften*, **64**, 8–15.

Hölldobler, B. and Wilson, E. O. (1990). *The ants*. Springer-Verlag, Berlin.

Hölldobler, B. and Wilson, E. O. (2009). *The superorganism: the beauty, elegance, and strangeness of insect societies*. W.W. Norton & Company, New York.

Holway, D. A., Lach, L., Suarez, A. V., Tsutsui, N. D., and Case, T. J. (2002). The causes and consequences of ant invasions. *Annual Review of Ecology and Systematics*, **33**, 181–233.

Honeybee Genome Sequencing Consortium (2006). Insights into social insects from the genome of the honeybee *Apis mellifera*. *Nature*, **443**, 931–49.

Hornett, E. A., Charlat, S., Duplouy, A. M. R. et al. (2006). Evolution of male-killer suppression in a natural population. *PLoS Biology*, **4**(9), e283.

Hou, C., Kaspari, M., Vander Zanden, H. B., and Gillooly, J. F. (2010). Energetic basis of colonial living in social insects. *Proceedings of the National Academy of Sciences, USA*, **107**, 3634–38.

Hughes, W. O. H. and Boomsma, J. J. (2004). Genetic diversity and disease resistance in leaf-cutting ant societies. *Evolution*, **58**, 1251–60.

Hughes, W. O. H. and Boomsma, J. J. (2008). Genetic royal cheats in leaf-cutting ant societies. *Proceedings of the National Academy of Sciences, USA*, **105**, 5150–53.

Hughes, W. O. H., Sumner, S., Van Borm, S., and Boomsma, J. J. (2003). Worker caste polymorphism has a genetic basis in *Acromyrmex* leaf-cutting ants. *Proceedings of the National Academy of Sciences, USA*, **100**, 9394–97.

Hughes, W. O. H., Oldroyd, B. P., Beekman, M., and Ratnieks, F. L. W. (2008). Ancestral monogamy shows kin selection is key to the evolution of eusociality. *Science*, **320**, 1213–16.

Hunt, J. H. (1999). Trait mapping and salience in the evolution of eusocial vespid wasps. *Evolution*, **53**, 225-37.

Hunt, J. H. (2007). *The evolution of social wasps*. Oxford University Press, New York.

Hunt, J. H. and Amdam, G. V. (2005). Bivoltinism as an antecedent to eusociality in the paper wasp genus *Polistes*. *Science*, **308**, 264–67.

Hunt, J. H., Kensinger, B. J., Kossuth, J. A. et al. (2007). A diapause pathway underlies the gyne phenotype in *Polistes* wasps, revealing an evolutionary route to caste-containing insect societies. *Proceedings of the National Academy of Sciences, USA*, **104**, 14020–25.

Hurst, L. D. (1990). Parasite diversity and the evolution of diploidy, multicellularity and anisogamy. *Journal of Theoretical Biology*, **144**, 429–443.

Hurst, L. D. (1991). The incidences and evolution of cytoplasmic male killers. *Proceedings of the Royal Society of London B*, **244**, 91–99.

Hurst, L. D. and Hamilton, W. D. (1992). Cytoplasmic fusion and the nature of sexes. *Proceedings of the Royal Society of London B*, **247**, 189–94.

Hurst, L. D., Atlan, A., and Bengtsson, B. O. (1996). Genetic conflicts. *Quarterly Review of Biology*, **71**, 317–64.

Inward, D., Beccaloni, G., and Eggleton, P. (2007a). Death of an order: a comprehensive molecular phylogenetic study confirms that termites are eusocial cockroaches. *Biology Letters*, **3**, 331–35.

Inward, D. J. G., Vogler, A. P., and Eggleton, P. (2007b). A comprehensive phylogenetic analysis of termites (Isoptera) illuminates key aspects of their evolutionary biology. *Molecular Phylogenetics and Evolution*, **44**, 953–67.

Isoda, T., Ford, A. M., Tomizawa, D. et al. (2009). Immunologically silent cancer clone transmission from mother to offspring. *Proceedings of the National Academy of Sciences, USA*, **106**, 17882–85.

Itô, Y. (1993). *Behaviour and social evolution of wasps*. Oxford University Press, Oxford.

Jaenike, J. (2001). Sex chromosome meiotic drive. *Annual Review of Ecology and Systematics*, **32**, 25–49.

Jaenike, J. (2008). X chromosome drive. *Current Biology*, **18**, R508–11.

Jaffé, R., Kronauer, D. J. C., Kraus, F. B., Boomsma, J. J., and Moritz, R. F. A. (2007). Worker caste determination in the army ant *Eciton burchellii*. *Biology Letters*, **3**, 513–16.

Janeway, C. A. and Medzhitov, R. (2002). Innate immune recognition. *Annual Review of Immunology*, **20**, 197–216.

Janzen, D. H. (1985). The natural history of mutualisms. In D. H. Boucher, ed. *The biology of mutualism: ecology and evolution*, pp. 40–99. Oxford University Press, New York.

Jarvis, J. U. M., O'Riain, M. J., Bennett, N. C., and Sherman, P. W. (1994). Mammalian eusociality: a family affair. *Trends in Ecology and Evolution*, **9**, 47–51.

Jeanne, R. L. (1991). The swarm-founding Polistinae. In K. G. Ross and R. W. Matthews, eds. *The social biology of wasps*, pp. 191–231. Comstock Publishing Associates, Ithaca.

Jeanson, R., Fewell, J. H., Gorelick, R., and Bertram, S. M. (2007). Emergence of increased division of labor as a function of group size. *Behavioral Ecology and Sociobiology*, **62**, 289–98.

Jemielity, S. and Keller, L. (2003). Queen control over reproductive decisions - no sexual deception in the ant *Lasius niger*. *Molecular Ecology*, **12**, 1589–97.

Jeon, J. and Choe, J. C. (2003). Reproductive skew and the origin of sterile castes. *American Naturalist*, **36**, 206–24.

Johns, P. M., Howard, K. J., Breisch, N. L., Rivera, A., and Thorne, B. L. (2009). Nonrelatives inherit colony resources in a primitive termite. *Proceedings of the National Academy of Sciences, USA*, **106**, 17452–56.

Johnson, L. J. (2008). Selfish genetic elements favor the evolution of a distinction between soma and germline. *Evolution*, **62**, 2122–24.

Johnson, P. C. D., Whitfield, J. A., Foster, W. A., and Amos, W. (2002). Clonal mixing in the soldier-producing aphid *Pemphigus spyrothecae* (Hemiptera: Aphididae). *Molecular Ecology*, **11**, 1525–31.

Johnstone, R. A. (2000). Models of reproductive skew: a review and synthesis. *Ethology*, **106**, 5–26.

Jones, M. E., Cockburn, A., Hamede, R. et al. (2008). Life-history change in disease-ravaged Tasmanian devil populations. *Proceedings of the National Academy of Sciences, USA*, **105**, 10023–27.

Kaiser, D. (2001). Building a multicellular organism. *Annual Review of Genetics*, **35**, 103–23.

Karatygin, I. V., Snigirevskaya, N. S., and Vikulin, S. V. (2009). The most ancient terrestrial lichen *Winfrenatia reticulata*: a new find and new interpretation. *Paleontological Journal*, **43**, 107–14.

Keim, C. N., Martins, J. L., Abreu, F. et al. (2004). Multicellular life cycle of magnetotactic prokaryotes. *FEMS Microbiology Letters*, **240**, 203–208.

Keller, L. (1997). Indiscriminate altruism: unduly nice parents and siblings. *Trends in Ecology and Evolution*, **12**, 99–103.

Keller, L., ed. (1999). *Levels of selection in evolution*. Princeton University Press, Princeton.

Keller, L. (2009). Adaptation and the genetics of social behaviour. *Philosophical Transactions of the Royal Society B*, **364**, 3209–16.

Keller, L. and Genoud, M. (1997). Extraordinary lifespans in ants: a test of evolutionary theories of ageing. *Nature*, **389**, 958–60.

Keller, L. and Nonacs, P. (1993). The role of queen pheromones in social insects: queen control or queen signal? *Animal Behaviour*, **45**, 787–94.

Keller, L. and Reeve, H. K. (1994). Partitioning of reproduction in animal societies. *Trends in Ecology and Evolution*, **9**, 98–102.

Keller, L. and Reeve, H. K. (1999). Dynamics of conflicts within insect societies. In L. Keller, ed. *Levels of selection in evolution*, pp. 153–75. Princeton University Press, Princeton.

Keller, L. and Ross, K. G. (1998). Selfish genes: a green beard in the red fire ant. *Nature*, **394**, 573–75.

Kent, D. S. and Simpson, J. A. (1992). Eusociality in the beetle *Austroplatypus incompertus* (Coleoptera: Curculionidae). *Naturwissenschaften*, **79**, 86–87.

Kenyon, C. (1988). The nematode *Caenorhabditis elegans*. *Science*, **240**, 1448–53.

Kerszberg, M. and Wolpert, L. (1998). The origin of metazoa and the egg: a role for cell death. *Journal of Theoretical Biology*, **193**, 535–37.

Kessin, R. H., Gundersen, G. G., Zaydfudim, V., Grimson, M., and Blanton, R. L. (1996). How cellular slime molds evade nematodes. *Proceedings of the National Academy of Sciences, USA*, **93**, 4857–61.

Khila, A. and Abouheif, E. (2008). Reproductive constraint is a developmental mechanism that maintains social harmony in advanced ant societies. *Proceedings of the National Academy of Sciences, USA*, **105**, 17884–89.

Khila, A. and Abouheif, E. (2010). Evaluating the role of reproductive constraints in ant social evolution. *Philosophical Transactions of the Royal Society B*, **365**, 617–30.

Kiers, E. T. and Denison, R. F. (2008). Sanctions, cooperation, and the stability of plant-rhizosphere mutualisms. *Annual Review of Ecology, Evolution, and Systematics*, **39**, 215–36.

Kikuta, N. and Tsuji, K. (1999). Queen and worker policing in the monogynous and monandrous ant, *Diacamma* sp. *Behavioral Ecology and Sociobiology*, **46**, 180–89.

Killingback, T., Bieri, J., and Flatt, T. (2006). Evolution in group-structured populations can resolve the tragedy of the commons. *Proceedings of the Royal Society B*, **273**, 1477–81.

King, N. (2004). The unicellular ancestry of animal development. *Developmental Cell*, **7**, 313–25.

Kirk, D. L. (1998). *Volvox: molecular-genetic origins of multicellularity and cellular differentiation*. Cambridge University Press, Cambridge.

Kirk, D. L. (1999). Evolution of multicellularity in the volvocine algae. *Current Opinion in Plant Biology*, **2**, 496–501.

Kirk, D. L. (2003). Seeking the ultimate and proximate causes of *Volvox* multicellularity and cellular differentiation. *Integrative and Comparative Biology*, **43**, 247–53.

Kirk, D. L. (2004). *Volvox. Current Biology*, **14**, R599–600.

Kirkwood, T. B. L. and Austad, S. N. (2000). Why do we age? *Nature*, **408**, 233–38.

Kleiman, M. and Tannenbaum, E. (2009). Diploidy and the selective advantage for sexual reproduction in unicellular organisms. *Theory in Biosciences*, **128**, 249–85.

Knoll, A. H. (1992). The early evolution of eukaryotes: a geological perspective. *Science*, **256**, 622–27.

Knoll, A. H. (2003). *Life on a young planet: the first three billion years of evolution on earth.* Princeton University Press, Princeton.

Knoll, A. H., Javaux, E. J., Hewitt, D., and Cohen, P. (2006). Eukaryotic organisms in Proterozoic oceans. *Philosophical Transactions of the Royal Society B*, **361**, 1023–38.

Kock, D., Ingram, C. M., Frabotta, L. J., Honeycutt, R. L., and Burda, H. (2006). On the nomenclature of Bathyergidae and *Fukomys* n. gen. (Mammalia: Rodentia). *Zootaxa*, **1142**, 51–55.

Koenig, W. D. and Dickinson, J. L., eds. (2004). *Ecology and evolution of cooperative breeding in birds*. Cambridge University Press, Cambridge.

Koenig, W. D., Pitelka, F. A., Carmen, W. J., Mumme, R. L., and Stanback, M. T. (1992). The evolution of delayed dispersal in cooperative breeders. *Quarterly Review of Biology*, **67**, 111–50.

Kokko, H. and Ekman, J. (2002). Delayed dispersal as a route to breeding: territorial inheritance, safe havens, and ecological constraints. *American Naturalist*, **160**, 468–84.

Komdeur, J. (1992). Importance of habitat saturation and territory quality for evolution of cooperative breeding in the Seychelles warbler. *Nature*, **358**, 493–95.

Komdeur, J., Richardson, D. S., and Hatchwell, B. (2008). Kin-recognition mechanisms in cooperative breeding systems: ecological causes and behavioral consequences of variation. In J. Korb and J. Heinze, eds. *The ecology of social evolution*, pp. 175–93. Springer Verlag, Berlin.

Kopp, R. E., Kirschvink, J. L., Hilburn, I. A., and Nash, C. Z. (2005). The Paleoproterozoic snowball Earth: a climate disaster triggered by the evolution of oxygenic photosynthesis. *Proceedings of the National Academy of Sciences, USA*, **102**, 11131–36.

Korb, J. (2008). The ecology of social evolution in termites. In J. Korb and J. Heinze, eds. *Ecology of social evolution*, pp. 151–74. Springer-Verlag, Berlin.

Korb, J. (2009). Termites: an alternative road to eusociality and the importance of group benefits in social insects. In J. Gadau and J. Fewell, eds. *Organization of insect societies: from genome to sociocomplexity*, pp. 128–47. Harvard University Press, Cambridge MA.

Korb, J. and Heinze, J. (2004). Multilevel selection and social evolution of insect societies. *Naturwissenschaften*, **91**, 291–304.

Korb, J. and Heinze, J., eds. (2008a). *Ecology of social evolution*. Springer-Verlag, Berlin.

Korb, J. and Heinze, J. (2008b). The ecology of social life: a synthesis. In J. Korb and J. Heinze, eds. *Ecology of social evolution*, pp. 245–59. Springer-Verlag, Berlin.

Korb, J., Weil, T., Hoffmann, K., Foster, K. R., and Rehli, M. (2009). A gene necessary for reproductive suppression in termites. *Science*, **324**, 758.

Koufopanou, V. (1994). The evolution of soma in the Volvocales. *American Naturalist*, **143**, 907–31.

Koufopanou, V. and Bell, G. (1993). Soma and germ: an experimental approach using *Volvox*. *Proceedings of the Royal Society of London B*, **254**, 107–13.

Krause, J. and Ruxton, G. D. (2002). *Living in groups*. Oxford University Press, Oxford.

Krieger, M. J. B. and Ross, K. G. (2002). Identification of a major gene regulating complex social behavior. *Science*, **295**, 328–32.

Kronauer, D. J. C., Schöning, C., Pedersen, J. S., Boomsma, J. J., and Gadau, J. (2004). Extreme queen-mating frequency and colony fission in African army ants. *Molecular Ecology*, **13**, 2381–88.

Kronauer, D. J. C., Schöning, C., and Boomsma, J. J. (2006a). Male parentage in army ants. *Molecular Ecology*, **15**, 1147–51.

Kronauer, D. J. C., Berghoff, S. M., Powell, S. et al. (2006b). A reassessment of the mating system characteristics of the army ant *Eciton burchellii*. *Naturwissenschaften*, **93**, 402–406.

Kronauer, D. J. C., Schöning, C., d'Ettorre, P., and Boomsma, J. J. (2010). Colony fusion and worker reproduction after queen loss in army ants. *Proceedings of the Royal Society B*, **277**, 755–63.

Kudô, K., Tsujita, S., Tsuchida, K. et al. (2005). Stable relatedness structure of the large-colony swarm-founding wasp *Polybia paulista*. *Behavioral Ecology and Sociobiology*, **58**, 27–35.

Kümmerli, R. and Keller, L. (2009). Patterns of split sex ratio in ants have multiple evolutionary causes based on different within-colony conflicts. *Biology Letters*, **5**, 713–16.

Kuzdzal-Fick, J. J., Foster, K. R., Queller, D. C., and Strassmann, J. E. (2007). Exploiting new terrain: an advantage to sociality in the slime mold *Dictyostelium discoideum*. *Behavioral Ecology*, **18**, 433–37.

Lachmann, M., Blackstone, N. W., Haig, D. et al. (2003). Cooperation and conflict in the evolution of genomes, cells and multicellular organisms. In P. Hammerstein, ed. *Genetic and cultural evolution of cooperation*, pp. 327–56. The MIT Press, Cambridge MA.

Langer, P., Hogendoorn, K., and Keller, L. (2004). Tug-of-war over reproduction in a social bee. *Nature*, **428**, 844–47.

Lattorff, H. M. G., Moritz, R. F. A., and Fuchs, S. (2005). A single locus determines thelytokous parthenogenesis of laying honeybee workers (*Apis mellifera capensis*). *Heredity*, **94**, 533–37.

Lattorff, H. M. G., Moritz, R. F. A., Crewe, R. M., and Solignac, M. (2007). Control of reproductive dominance by the *thelytoky* gene in honeybees. *Biology Letters*, **3**, 292–95.

Lehmann, L. and Keller, L. (2006a). The evolution of co-operation and altruism—a general framework and a classification of models. *Journal of Evolutionary Biology*, **19**, 1365–76.

Lehmann, L. and Keller, L. (2006b). Synergy, partner choice and frequency dependence: their integration into inclusive fitness theory and their interpretation in terms of direct and indirect fitness effects. *Journal of Evolutionary Biology*, **19**, 1426–36.

Lehmann, L., Keller, L., and Sumpter, D. J. T. (2007a). The evolution of helping and harming on graphs; the return of the inclusive fitness effect. *Journal of Evolutionary Biology*, **20**, 2284–95.

Lehmann, L., Keller, L., West, S., and Roze, D. (2007b). Group selection and kin selection: two concepts but one process. *Proceedings of the National Academy of Sciences, USA*, **104**, 6736–39.

Lehmann, L., Rousset, F., Roze, D., and Keller, L. (2007c). Strong reciprocity or strong ferocity? A population genetic view of the evolution of altruistic punishment. *American Naturalist*, **170**, 21–36.

Lehmann, L., Ravigné, V., and Keller, L. (2008). Population viscosity can promote the evolution of altruistic sterile helpers and eusociality. *Proceedings of the Royal Society B*, **275**, 1887–95.

Leigh, E. G. (1977). How does selection reconcile individual advantage with the good of the group? *Proceedings of the National Academy of Sciences, USA*, **74**, 4542–46.

Leigh, E. G. (1991). Genes, bees and ecosystems: the evolution of a common interest among individuals. *Trends in Ecology and Evolution*, **6**, 257–62.

Leigh, E. G. (1995). The major transitions of evolution. *Evolution*, **49**, 1302–306.

Leigh, E. G. (1999). *Tropical forest ecology: a view from Barro Colorado Island*. Oxford University Press, New York.

Lenoir, A., D'Ettorre, P., Errard, C., and Hefetz, A. (2001). Chemical ecology and social parasitism in ants. *Annual Review of Entomology*, **46**, 573–99.

Lessells, C. M., Snook, R. R., and Hosken, D. J. (2009). The evolutionary origin and maintenance of sperm: selection for a small, motile gamete mating type. In T. R. Birkhead, D. J. Hosken, and S. Pitnick, eds. *Sperm biology: an evolutionary perspective*, pp. 43–67. Academic Press, Burlington MA.

Ligon, J. D. and Burt, D. B. (2004). Evolutionary origins. In W. D. Koenig and J. L. Dickinson, eds. *Ecology and evolution of cooperative breeding in birds*, pp. 5–34. Cambridge University Press, Cambridge.

Lim, M. M., Wang, Z., Olazábal, D. E., Ren, X., Terwilliger, E. F., and Young, L. J. (2004). Enhanced partner preference in a promiscuous species by manipulating the expression of a single gene. *Nature*, **429**, 754–57.

Lin, N. and Michener, C. D. (1972). Evolution of sociality in insects. *Quarterly Review of Biology*, **47**, 131–59.

Linksvayer, T. A. and Wade, M. J. (2005). The evolutionary origin and elaboration of sociality in the aculeate Hymenoptera: maternal effects, sib-social effects, and heterochrony. *Quarterly Review of Biology*, **80**, 317–36.

Little, A. E. F. and Currie, C. R. (2009). Parasites may help stabilize cooperative relationships. *BMC Evolutionary Biology*, **9**, 124.

Lopez-Vaamonde, C., Koning, J. W., Jordan, W. C., and Bourke, A. F. G. (2003). No evidence that reproductive bumblebee workers reduce the production of new queens. *Animal Behaviour*, **66**, 577–84.

Lopez-Vaamonde, C., Koning, J. W., Brown, R. M., Jordan, W. C., and Bourke, A. F. G. (2004). Social parasitism by male-producing reproductive workers in a eusocial insect. *Nature*, **430**, 557–60.

Love, G. D., Grosjean, E., Stalvies, C. et al. (2009). Fossil steroids record the appearance of Demospongiae during the Cryogenian period. *Nature*, **457**, 718–21.

Lucas, E. R. (2009). Social structure and evolution of the apoid wasp *Microstigmus nigrophthalmus*, Ph.D. thesis, University of Sussex, U.K.

Lürling, M. and Van Donk, E. (2000). Grazer-induced colony formation in *Scenedesmus*: are there costs to being colonial? *Oikos*, **88**, 111–18.

Lutzoni, F. and Miadlikowska, J. (2009). Lichens. *Current Biology*, **19**, R502–503.

Lynch, M. (2007). The frailty of adaptive hypotheses for the origins of organismal complexity. *Proceedings of the National Academy of Sciences, USA*, **104**, 8597–604.

Maccoll, A. D. C. and Hatchwell, B. J. (2004). Determinants of lifetime fitness in a cooperative breeder, the long-tailed tit *Aegithalos caudatus*. *Journal of Animal Ecology*, **73**, 1137–48.

Mackie, G. O. (1964). Analysis of locomotion in a siphonophore colony. *Proceedings of the Royal Society of London B*, **159**, 366–91.

Mackie, G. O. (1986). From aggregates to integrates: physiological aspects of modularity in colonial animals. *Philosophical Transactions of the Royal Sociey of London B*, **313**, 175–96.

Magrath, R. D. and Heinsohn, R. G. (2000). Reproductive skew in birds: models, problems and prospects. *Journal of Avian Biology*, **31**, 247–58.

Marchetti, M., Capela, D., Glew, M. et al. (2010). Experimental evolution of a plant pathogen into a legume symbiont. *PLoS Biology*, **8**(1), e1000280.

Marcot, J. D. and McShea, D. W. (2007). Increasing hierarchical complexity throughout the history of life: phylogenetic tests of trend mechanisms. *Paleobiology*, **33**, 182–200.

Margulis, L. (1970). *Origin of eukaryotic cells*. Yale University Press, New Haven.

Martin, W. (2003). Gene transfer from organelles to the nucleus: frequent and in big chunks. *Proceedings of the National Academy of Sciences, USA*, **100**, 8612–14.

Martin, S. J., Beekman, M., Wossler, T. C., and Ratnieks, F. L. W. (2002). Parasitic Cape honeybee workers, *Apis mellifera capensis*, evade policing. *Nature*, **415**, 163–65.

Martin, S. J., Châline, N., Oldroyd, B. P., Jones, G. R., and Ratnieks, F. L. W. (2004). Egg marking pheromones of anarchistic worker honeybees (*Apis mellifera*). *Behavioral Ecology*, **15**, 839–44.

Martin, S. J., Vitikainen, E., Helanterä, H., and Drijfhout, F. P. (2008). Chemical basis of nestmate discrimination in the ant *Formica exsecta*. *Proceedings of the Royal Society B*, **275**, 1271–78.

Matsuura, M. and Yamane, S. (1990). *Biology of the vespine wasps*. Springer-Verlag, Berlin.

Matsuura, K., Vargo, E. L., Kawatsu, K. et al. (2009). Queen succession through asexual reproduction in termites. *Science*, **323**, 1687.

Mattila, H. R. and Seeley, T. D. (2007). Genetic diversity in honey bee colonies enhances productivity and fitness. *Science*, **317**, 362–64.

May, R. M. (1976). Models for two interacting populations. In R. M. May, ed. *Theoretical ecology: principles and applications*, pp. 49–70. Blackwell Scientific Publications, Oxford.

Maynard Smith, J. (1964). Group selection and kin selection. *Nature*, **201**, 1145–47.

Maynard Smith, J. (1982). *Evolution and the theory of games*. Cambridge University Press, Cambridge.

Maynard Smith, J. (1984). The ecology of sex. In J. R. Krebs and N. B. Davies, eds. *Behavioural ecology: an evolutionary approach*, 2nd edn, pp. 201–21. Blackwell Scientific Publications, Oxford.

Maynard Smith, J. (1988). Evolutionary progress and levels of selection. In M. H. Nitecki, ed. *Evolutionary progress*, pp. 219–30. University of Chicago Press, Chicago.

Maynard Smith, J. and Szathmáry, E. (1995). *The major transitions in evolution*. W. H. Freeman, Oxford.

Maynard Smith, J., Dowson, C. G., and Spratt, B. G. (1991). Localized sex in bacteria. *Nature*, **349**, 29–31.

Mayr, E. (1997). The objects of selection. *Proceedings of the National Academy of Sciences, USA*, **94**, 2091–94.

McCallum, H. (2008). Tasmanian devil facial tumour disease: lessons for conservation biology. *Trends in Ecology and Evolution*, **23**, 631–37.

McElreath, R. and Boyd, R. (2007). *Mathematical models of social evolution*. University of Chicago Press, Chicago.

McRae, S. B. (1997). A rise in nest predation enhances the frequency of intraspecific brood parasitism in a moorhen population. *Journal of Animal Ecology*, **66**, 143–53.

McShea, D. W. (1996). Metazoan complexity and evolution: is there a trend? *Evolution*, **50**, 477–92.

McShea, D. W. (2001). The minor transitions in hierarchical evolution and the question of a directional bias. *Journal of Evolutionary Biology*, **14**, 502–18.

McShea, D. W. and Changizi, M. A. (2003). Three puzzles in hierarchical evolution. *Integrative and Comparative Biology*, **43**, 74–81.

Mehdiabadi, N. J., Jack, C. N., Talley Farnham, T. et al. (2006). Kin preference in a social microbe. *Nature*, **442**, 881–82.

Metcalf, R. A. and Whitt, G. S. (1977). Relative inclusive fitness in the social wasp *Polistes metricus*. *Behavioral Ecology and Sociobiology*, **2**, 353–60.

Meunier, J., West, S. A., and Chapuisat, M. (2008). Split sex ratios in the social Hymenoptera: a meta-analysis. *Behavioral Ecology*, **19**, 382–90.

Michener, C. D. (1964). Reproductive efficiency in relation to colony size in Hymenopterous societies. *Insectes Sociaux*, **11**, 317–41.

Michener, C. D. (1969). Comparative social behavior of bees. *Annual Review of Entomology*, **14**, 299–342.

Michener, C. D. (1990). Reproduction and castes in social halictine bees. In W. Engels, ed. *Social insects: an evolutionary approach to castes and reproduction*, pp. 77–121. Springer-Verlag, Berlin.

Michod, R. E. (1983). Population biology of the first replicators: on the origin of the genotype, phenotype and organism. *American Zoologist*, **23**, 5–14.

Michod, R. E. (1996). Cooperation and conflict in the evolution of individuality. II. Conflict mediation. *Proceedings of the Royal Society of London B*, **263**, 813–22.

Michod, R. E. (1997a). Cooperation and conflict in the evolution of individuality. I. Multilevel selection of the organism. *American Naturalist*, **149**, 607–45.

Michod, R. E. (1997b). Evolution of the individual. *American Naturalist*, **150**, S5–21.

Michod, R. E. (1999). Individuality, immortality, and sex. In L. Keller, ed. *Levels of selection in evolution*, pp. 53–74. Princeton University Press, Princeton.

Michod, R. E. (2000). *Darwinian dynamics: evolutionary transitions in fitness and individuality*. Princeton University Press, Princeton.

Michod, R. E. (2003). Cooperation and conflict mediation during the origin of multicellularity. In P. Hammerstein, ed. *Genetic and cultural evolution of cooperation*, pp. 291–307. The MIT Press, Cambridge MA.

Michod, R. E. (2006). The group covariance effect and fitness trade-offs during evolutionary transitions in individuality. *Proceedings of the National Academy of Sciences, USA*, **103**, 9113–17.

Michod, R. E. (2007). Evolution of individuality during the transition from unicellular to multicellular life. *Proceedings of the National Academy of Sciences, USA*, **104**, 8613–18.

Michod, R. E. and Roze, D. (1997). Transitions in individuality. *Proceedings of the Royal Society of London B*, **264**, 853–57.

Michod, R. E. and Roze, D. (2001). Cooperation and conflict in the evolution of multicellularity. *Heredity*, **86**, 1–7.

Michod, R. E., Nedelcu, A. M., and Roze, D. (2003). Cooperation and conflict in the evolution of individuality IV. Conflict mediation and evolvability in *Volvox carteri*. *BioSystems*, **69**, 95–114.

Michod, R. E., Viossat, Y., Solari, C. A., Hurand, M., and Nedelcu, A. M. (2006). Life-history evolution and the origin of multicellularity. *Journal of Theoretical Biology*, **239**, 257–72.

Mikheyev, A. S., Bresson, S., and Conant, P. (2009). Single-queen introductions characterize regional and local invasions by the facultatively clonal little fire ant *Wasmannia auropunctata*. *Molecular Ecology*, **18**, 2937–44.

Mock, D. W. and Parker, G. A. (1997). *The evolution of sibling rivalry*. Oxford University Press, Oxford.

Mock, D. W. and Parker, G. A. (1998). Siblicide, family conflict and the evolutionary limits of selfishness. *Animal Behaviour*, **56**, 1–10.

Monnin, T. and Liebig, J. (2008). Understanding eusociality requires both proximate and ultimate thinking and due consideration of individual and colony-level interests. *Oikos*, **117**, 1441–43.

Monnin, T. and Ratnieks, F. L. W. (1999). Reproduction versus work in queenless ants: when to join a hierarchy of hopeful reproductives? *Behavioral Ecology and Sociobiology*, **46**, 413–22.

Monnin, T. and Ratnieks, F. L. W. (2001). Policing in queenless ponerine ants. *Behavioral Ecology and Sociobiology*, **50**, 97–108.

Monnin, T., Ratnieks, F. L. W., Jones, G. R., and Beard, R. (2002). Pretender punishment induced by chemical signalling in a queenless ant. *Nature*, **419**, 61–65.

Moore, T. and Haig, D. (1991). Genomic imprinting in mammalian development: a parental tug-of-war. *Trends in Genetics*, **7**, 45–49.

Moreau, C. S., Bell, C. D., Vila, R., Archibald, S. B., and Pierce, N. E. (2006). Phylogeny of the ants: diversification in the age of angiosperms. *Science*, **312**, 101–104.

Moritz, R. F. A., Pirk, C. W. W., Hepburn, H. R., and Neumann, P. (2008). Short-sighted evolution of virulence in parasitic honeybee workers (*Apis mellifera capensis* Esch.). *Naturwissenschaften*, **95**, 507–13.

Morrison, C. L., Ríos, R., and Duffy, J. E. (2004). Phylogenetic evidence for an ancient rapid radiation of Caribbean sponge-dwelling snapping shrimps (*Synalpheus*). *Molecular Phylogenetics and Evolution*, **30**, 563–81.

Mueller, U. G. (1991). Haplodiploidy and the evolution of facultative sex ratios in a primitively eusocial bee. *Science*, **254**, 442–44.

Mueller, U. G. (2002). Ant versus fungus versus mutualism: ant-cultivar conflict and the deconstruction of the attine ant-fungus symbiosis. *American Naturalist*, **160**, S67–98.

Mueller, U. G., Eickwort, G. C., and Aquadro, C. F. (1994). DNA fingerprinting analysis of parent-offspring conflict in a bee. *Proceedings of the National Academy of Sciences, USA*, **91**, 5143–47.

Murchison, E. P., Tovar, C., Hsu, A. et al. (2010). The Tasmanian Devil transcriptome reveals Schwann cell origins of a clonally transmissible cancer. *Science*, **327**, 84–87.

Murgia, C., Pritchard, J. K., Kim, S. Y., Fassati, A., and Weiss, R. A. (2006). Clonal origin and evolution of a transmissible cancer. *Cell*, **126**, 477–87.

Nadell, C. D., Xavier, J. B., and Foster, K. R. (2009). The sociobiology of biofilms. *FEMS Microbiology Reviews*, **33**, 206–24.

Nakayama, T. and Ishida, K. (2009). Another acquisition of a primary photosynthetic organelle is underway in *Paulinella chromatophora*. *Current Biology*, **19**, R284–85.

Nalepa, C. A. and Jones, S. C. (1991). Evolution of monogamy in termites. *Biological Reviews*, **66**, 83–97.

Nanork, P., Paar, J., Chapman, N. C., Wongsiri, S., and Oldroyd, B. P. (2005). Asian honeybees parasitize the future dead. *Nature*, **437**, 829.

Nanork, P., Chapman, N. C., Wongsiri, S., Lim, J., Gloag, R. S., and Oldroyd, B. P. (2007). Social parasitism by workers in queenless and queenright *Apis cerana* colonies. *Molecular Ecology*, **16**, 1107–14.

Narra, H. P. and Ochman, H. (2006). Of what use is sex to bacteria? *Current Biology*, **16**, R705–10.

Nedelcu, A. M. (2009). Environmentally induced responses co-opted for reproductive altruism. *Biology Letters*, **5**, 805–808.

Nedelcu, A. M. and Michod, R. E. (2006). The evolutionary origin of an altruistic gene. *Molecular Biology and Evolution*, **23**, 1460–64.

Neumann, P. and Hepburn, R. (2002). Behavioural basis for social parasitism of Cape honeybees (*Apis mellifera capensis*). *Apidologie*, **33**, 165–92.

Neumann, P. and Moritz, R. F. A. (2002). The Cape honeybee phenomenon: the sympatric evolution of a social parasite in real time? *Behavioral Ecology and Sociobiology*, **52**, 271–81.

Nishide, Y., Satoh, T., Hiraoka, T., Obara, Y., and Iwabuchi, K. (2007). Clonal structure affects the assembling behavior in the Japanese queenless ant *Pristomyrmex punctatus*. *Naturwissenschaften*, **94**, 865–69.

Noll, F. B. and Wenzel, J. W. (2008). Caste in the swarming wasps: 'queenless' societies in highly social insects. *Biological Journal of the Linnean Society*, **93**, 509–22.

Nonacs, P. (1986). Ant reproductive strategies and sex allocation theory. *Quarterly Review of Biology*, **61**, 1–21.

Nonacs, P. (1993). Male parentage and sexual deception in the social Hymenoptera. In D. L. Wrensch and M. A. Ebbert, eds. *Evolution and diversity of sex ratio in insects and mites*, pp. 384–401. Chapman and Hall, New York.

Nonacs, P. (2006). The rise and fall of transactional skew theory in the model genus *Polistes*. *Annales Zoologici Fennici*, **43**, 443–55.

Nonacs, P. (2007). Tug-of-war has no borders: it is the missing model in reproductive skew theory. *Evolution*, **61**, 1244–50.

Noonan, K. M. (1981). Individual strategies of inclusive-fitness-maximizing in *Polistes fuscatus* foundresses. In R. D. Alexander and D. W. Tinkle, eds. *Natural selection and social behavior*, pp. 18–44. Chiron Press, New York.

Nowak, M. A. (2006). Five rules for the evolution of cooperation. *Science*, **314**, 1560–63.

Nowak, M. A. and May, R. M. (1992). Evolutionary games and spatial chaos. *Nature*, **359**, 826–29.

Nowell, P. C. (1976). The clonal evolution of tumor cell populations. *Science*, **194**, 23–28.

Nunney, L. (1999a). Lineage selection: natural selection for long-term benefit. In L. Keller, ed. *Levels of selection in evolution*, pp. 238–52. Princeton University Press, Princeton.

Nunney, L. (1999b). Lineage selection and the evolution of multistage carcinogenesis. *Proceedings of the Royal Society of London B*, **266**, 493–98.

O'Donnell, S. (1998). Reproductive caste determination in eusocial wasps (Hymenoptera: Vespidae). *Annual Review of Entomology*, **43**, 323–46.

Ohkawara, K., Nakayama, M., Satoh, A., Trindl, A., and Heinze, J. (2006). Clonal reproduction and genetic caste differences in a queen-polymorphic ant, *Vollenhovia emeryi. Biology Letters*, **2**, 359–63.

Ohtsuki, H. and Tsuji, K. (2009). Adaptive reproduction schedule as a cause of worker policing in social Hymenoptera: a dynamic game analysis. *American Naturalist*, **173**, 747–58.

Ohtsuki, H., Hauert, C., Lieberman, E., and Nowak, M. A. (2006). A simple rule for the evolution of cooperation on graphs and social networks. *Nature*, **441**, 502–505.

Okasha, S. (2006). *Evolution and the levels of selection*. Clarendon Press, Oxford.

Oldroyd, B. P. (2002). The Cape honeybee: an example of a social cancer. *Trends in Ecology and Evolution*, **17**, 249–51.

Oldroyd, B. P. and Fewell, J. H. (2007). Genetic diversity promotes homeostasis in insect colonies. *Trends in Ecology and Evolution*, **22**, 408–13.

Oldroyd, B. P., Smolenski, A. J., Cornuet, J.-M., and Crozier, R. H. (1994). Anarchy in the beehive. *Nature*, **371**, 749.

O'Riain, M. J. and Faulkes, C. G. (2008). African mole-rats: eusociality, relatedness and ecological constraints. In J. Korb and J. Heinze, eds. *Ecology of social evolution*, pp. 207–23. Springer-Verlag, Berlin.

O'Riain, M. J. and Jarvis, J. U. M. (1997). Colony member recognition and xenophobia in the naked mole-rat. *Animal Behaviour*, **53**, 487–98.

O'Riain, M. J., Jarvis, J. U. M., Alexander, R., Buffenstein, R., and Peeters, C. (2000). Morphological castes in a vertebrate. *Proceedings of the National Academy of Sciences, USA*, **97**, 13194–97.

Osman, R. W. and Haugsness, J. A. (1981). Mutualism among sessile invertebrates: a mediator of competition and predation. *Science*, **211**, 846–48.

Oster, G. F. and Wilson, E. O. (1978). *Caste and ecology in the social insects*. Princeton University Press, Princeton.

Ostrowski, E. A., Katoh, M., Shaulsky, G., Queller, D. C., and Strassmann, J. E. (2008). Kin discrimination increases with genetic distance in a social amoeba. *PLoS Biology*, **6**(11), e287.

Pacala, S. W., Gordon, D. M., and Godfray, H. C. J. (1996). Effects of social group size on information transfer and task allocation. *Evolutionary Ecology*, **10**, 127–65.

Packer, L. (1990). Solitary and eusocial nests in a population of *Augochlorella striata* (Provancher) (Hymenoptera; Halictidae) at the northern edge of its range. *Behavioral Ecology and Sociobiology*, **27**, 339–44.

Packer, L. and Owen, R. E. (1994). Relatedness and sex ratio in a primitively eusocial halictine bee. *Behavioral Ecology and Sociobiology*, **34**, 1–10.

Pál, C. and Szathmáry, E. (2000). The concept of fitness and individuality revisited. *Journal of Evolutionary Biology*, **13**, 352–55.

Pamilo, P. (1989). Estimating relatedness in social groups. *Trends in Ecology and Evolution*, **4**, 353–55.

Pamilo, P. (1990). Sex allocation and queen-worker conflict in polygynous ants. *Behavioral Ecology and Sociobiology*, **27**, 31–36.

Pamilo, P. (1991). Evolution of the sterile caste. *Journal of Theoretical Biology*, **149**, 75–95.

Pamilo, P. and Crozier, R. H. (1982). Measuring genetic relatedness in natural populations: methodology. *Theoretical Population Biology*, **21**, 171–93.

Pancer, Z. and Cooper, M. D. (2006). The evolution of adaptive immunity. *Annual Review of Immunology*, **24**, 497–518.

Parker, G. A., Baker, R. R., and Smith, V. G. F. (1972). The origin and evolution of gamete dimorphism and the male-female phenomenon. *Journal of Theoretical Biology*, **36**, 529–53.

Passera, L. (1994). Characteristics of tramp species. In D. F. Williams, ed. *Exotic ants: biology, impact, and control of introduced species*, pp. 23–43. Westview Press, Boulder.

Payne, J. L., Boyer, A. G., Brown, J. H. et al. (2009). Two-phase increase in the maximum size of life over 3.5 billion years reflects biological innovation and environmental opportunity. *Proceedings of the National Academy of Sciences, USA*, **106**, 24–27.

Pearcy, M., Aron, S., Doums, C., and Keller, L. (2004). Conditional use of sex and parthenogenesis for worker and queen production in ants. *Science*, **306**, 1780–83.

Pearse, A.-M. and Swift, K. (2006). Transmission of devil facial-tumour disease. *Nature*, **439**, 549.

Peeters, C. (1993). Monogyny and polygyny in ponerine ants with or without queens. In L. Keller, ed. *Queen number and sociality in insects*, pp. 234–61. Oxford University Press, Oxford.

Pen, I. and Weissing, F. J. (2000). Towards a unified theory of cooperative breeding: the role of ecology and life history re-examined. *Proceedings of the Royal Society of London B*, **267**, 2411–18.

Pennisi, E. (2009). Agreeing to disagree. *Science*, **323**, 706–708.

Pfeiffer, T. and Bonhoeffer, S. (2003). An evolutionary scenario for the transition to undifferentiated multicellularity. *Proceedings of the National Academy of Sciences, USA*, **100**, 1095–98.

Pike, N. and Foster, W. A. (2008). The ecology of altruism in a clonal insect. In J. Korb and J. Heinze, eds. *Ecology of social evolution*, pp. 37–56. Springer-Verlag, Berlin.

Pineda-Krch, M. and Lehtilä, K. (2004a). Costs and benefits of genetic heterogeneity within organisms. *Journal of Evolutionary Biology*, **17**, 1167–77.

Pineda-Krch, M. and Lehtilä, K. (2004b). Challenging the genetically homogeneous individual. *Journal of Evolutionary Biology*, **17**, 1192–94.

Pirk, C. W. W., Neumann, P., and Ratnieks, F. L. W. (2003). Cape honeybees, *Apis mellifera capensis*, police worker-laid eggs despite the absence of relatedness benefits. *Behavioral Ecology*, **14**, 347–52.

Platt, T. G. and Bever, J. D. (2009). Kin competition and the evolution of cooperation. *Trends in Ecology and Evolution*, **24**, 370–77.

Pomiankowski, A. (1999). Intragenomic conflict. In L. Keller, ed. *Levels of selection in evolution*, pp. 121–52. Princeton University Press, Princeton.

Queller, D. C. (1989). The evolution of eusociality: reproductive head starts of workers. *Proceedings of the National Academy of Sciences, USA*, **86**, 3224–26.

Queller, D. C. (1992). Quantitative genetics, inclusive fitness, and group selection. *American Naturalist*, **139**, 540–58.

Queller, D. C. (1994). Genetic relatedness in viscous populations. *Evolutionary Ecology*, **8**, 70–73.

Queller, D. C. (1995). The spaniels of St. Marx and the Panglossian paradox: a critique of a rhetorical programme. *Quarterly Review of Biology*, **70**, 485–89.

Queller, D. C. (1996). The measurement and meaning of inclusive fitness. *Animal Behaviour*, **51**, 229–32.

Queller, D. C. (1997). Cooperators since life began. *Quarterly Review of Biology*, **72**, 184–88.

Queller, D. C. (2000). Relatedness and the fraternal major transitions. *Philosophical Transactions of the Royal Society of London B*, **355**, 1647–55.

Queller, D. C. (2003). Theory of genomic imprinting conflict in social insects. *BMC Evolutionary Biology*, **3**, 15.

Queller, D. C. (2006). To work or not to work. *Nature*, **444**, 42–43.

Queller, D. C. and Strassmann, J. E. (1998). Kin selection and social insects. *BioScience*, **48**, 165–75.

Queller, D. C. and Strassmann, J. E. (2002). The many selves of social insects. *Science*, **296**, 311–13.

Queller, D. C. and Strassmann, J. E. (2009). Beyond society: the evolution of organismality. *Philosophical Transactions of the Royal Society B*, **364**, 3143–55.

Queller, D. C., Strassmann, J. E., Solís, C. R., Hughes, C. R., and DeLoach, D. M. (1993). A selfish strategy of social insect workers that promotes social cohesion. *Nature*, **365**, 639–41.

Queller, D. C., Zacchi, F., Cervo, R. et al. (2000). Unrelated helpers in a social insect. *Nature*, **405**, 784–87.

Queller, D. C., Ponte, E., Bozzaro, S., and Strassmann, J. E. (2003). Single-gene greenbeard effects in the social amoeba *Dictyostelium discoideum. Science*, **299**, 105–106.

Rabeling, C., Lino-Neto, J., Cappellari, S. C., Dos-Santos, I. A., Mueller, U. G., and Bacci, M. (2009). Thelytokous parthenogenesis in the fungus-gardening ant *Mycocepurus smithii* (Hymenoptera: Formicidae). *PLoS One*, **4**(8), e6781.

Raff, R. A. (1988). The selfish cell lineage. *Cell*, **54**, 445–46.

Rainey, P. B. (2007). Unity from conflict. *Nature*, **446**, 616.

Rainey, P. B. and Rainey, K. (2003). Evolution of cooperation and conflict in experimental bacterial populations. *Nature*, **425**, 72–74.

Ramesh, M. A., Malik, S.-B., and Logsdon, J. M. (2005). A phylogenomic inventory of meiotic genes: evidence for sex in *Giardia* and an early eukaryotic origin of meiosis. *Current Biology*, **15**, 185–91.

Rand, D. M., Haney, R. A., and Fry, A. J. (2004). Cytonuclear coevolution: the genomics of cooperation. *Trends in Ecology and Evolution*, **19**, 645–53.

Rankin, D. J., Bargum, K., and Kokko, H. (2007a). The tragedy of the commons in evolutionary biology. *Trends in Ecology and Evolution*, **22**, 643–51.

Rankin, D. J., López-Sepulcre, A., Foster, K. R., and Kokko, H. (2007b). Species-level selection reduces selfishness through competitive exclusion. *Journal of Evolutionary Biology*, **20**, 1459–68.

Rasmussen, B., Fletcher, I. R., Brocks, J. J., and Kilburn, M. R. (2008a). Reassessing the first appearance of eukaryotes and cyanobacteria. *Nature*, **455**, 1101–104.

Rasmussen, G. S. A., Gusset, M., Courchamp, F., and Macdonald, D. W. (2008b). Achilles' heel of sociality revealed by energetic poverty trap in cursorial hunters. *American Naturalist*, **172**, 508–18.

Ratnieks, F. L. W. (1988). Reproductive harmony via mutual policing by workers in eusocial Hymenoptera. *American Naturalist*, **132**, 217–36.

Ratnieks, F. L. W. (1991). The evolution of genetic odor-cue diversity in social Hymenoptera. *American Naturalist*, **137**, 202–26.

Ratnieks, F. L. W. (1993). Egg-laying, egg-removal, and ovary development by workers in queenright honey bee colonies. *Behavioral Ecology and Sociobiology*, **32**, 191–98.

Ratnieks, F. L. W. (1995). Evidence for a queen-produced egg-marking pheromone and its use in worker policing in the honey bee. *Journal of Apicultural Research*, **34**, 31–37.

Ratnieks, F. L. W. (2001). Heirs and spares: caste conflict and excess queen production in *Melipona* bees. *Behavioral Ecology and Sociobiology*, **50**, 467–73.

Ratnieks, F. L. W. and Helanterä, H. (2009). The evolution of extreme altruism and inequality in insect societies. *Philosophical Transactions of the Royal Society B*, **364**, 3169–79.

Ratnieks, F. L. W. and Reeve, H. K. (1992). Conflict in single-queen hymenopteran societies: the structure of conflict and processes that reduce conflict in advanced eusocial species. *Journal of Theoretical Biology*, **158**, 33–65.

Ratnieks, F. L. W. and Visscher, P. K. (1989). Worker policing in the honeybee. *Nature*, **342**, 796–97.

Ratnieks, F. L. W. and Wenseleers, T. (2005). Policing insect societies. *Science*, **307**, 54–56.

Ratnieks, F. L. W. and Wenseleers, T. (2008). Altruism in insect societies and beyond: voluntary or enforced? *Trends in Ecology and Evolution*, **23**, 45–52.

Ratnieks, F. L. W., Monnin, T., and Foster, K. R. (2001). Inclusive fitness theory: novel predictions and tests in eusocial Hymenoptera. *Annales Zoologici Fennici*, **38**, 201–14.

Ratnieks, F. L. W., Foster, K. R., and Wenseleers, T. (2006). Conflict resolution in insect societies. *Annual Review of Entomology*, **51**, 581–608.

Read, A. F. and Harvey, P. H. (1993). The evolution of virulence. *Nature*, **362**, 500–501.

Rebbeck, C. A., Thomas, R., Breen, M., Leroi, A. M., and Burt, A. (2009). Origins and evolution of a transmissible cancer. *Evolution*, **63**, 2340–49.

Reber, A., Castella, G., Christe, P., and Chapuisat, M. (2008). Experimentally increased group diversity improves disease resistance in an ant species. *Ecology Letters*, **11**, 682–89.

Reeve, H. K. (1989). The evolution of conspecific acceptance thresholds. *American Naturalist*, **133**, 407–35.

Reeve, H. K. (1991). *Polistes*. In K. G. Ross and R. W. Matthews, eds. *The social biology of wasps*, pp. 99–148. Comstock Publishing Associates, Ithaca.

Reeve, H. K. (1998a). Acting for the good of others: kinship and reciprocity with some new twists. In C. Crawford and D. Krebs, eds. *Handbook of evolutionary psychology*, pp. 43–85. Lawrence Erlbaum, Hillsdale NJ.

Reeve, H. K. (1998b). Game theory, reproductive skew, and nepotism. In L. A. Dugatkin and H. K. Reeve, eds. *Game theory and animal behavior*, pp. 118–45. Oxford University Press, New York.

Reeve, H. K. (2000a). Multi-level selection and human cooperation. *Evolution and Human Behavior*, **21**, 65–72.

Reeve, H. K. (2000b). A transactional theory of within-group conflict. *American Naturalist*, **155**, 365–82.

Reeve, H. K. (2001). In search of unified theories in sociobiology: help from social wasps. In L. A. Dugatkin, ed. *Model systems in behavioral ecology*, pp. 57–71. Princeton University Press, Princeton.

Reeve, H. K. and Hölldobler, B. (2007). The emergence of a superorganism through intergroup competition. *Proceedings of the National Academy of Sciences, USA*, **104**, 9736–40.

Reeve, H. K. and Jeanne, R. L. (2003). From individual control to majority rule: extending transactional models of reproductive skew in animal societies. *Proceedings of the Royal Society of London B*, **270**, 1041–45.

Reeve, H. K. and Keller, L. (2001). Tests of reproductive-skew models in social insects. *Annual Review of Entomology*, **46**, 347–85.

Reeve, H. K. and Ratnieks, F. L. W. (1993). Queen-queen conflicts in polygynous societies: mutual tolerance and reproductive skew. In L. Keller, ed. *Queen number and sociality in insects*, pp. 45–85. Oxford University Press, Oxford.

Reeve, H. K. and Shen, S.-F. (2006). A missing model in reproductive skew theory: the bordered tug-of-war. *Proceedings of the National Academy of Sciences, USA*, **103**, 8430–34.

Reeve, H. K., Westneat, D. F., Noon, W. A., Sherman, P. W., and Aquadro, C. F. (1990). DNA 'fingerprinting' reveals high levels of inbreeding in colonies of the eusocial naked mole-rat. *Proceedings of the National Academy of Sciences, USA*, **87**, 2496–500.

Reeve, H. K., Emlen, S. T., and Keller, L. (1998). Reproductive sharing in animal societies: reproductive incentives or incomplete control by dominant breeders? *Behavioral Ecology*, **9**, 267–78.

Reeve, H. K., Starks, P. T., Peters, J. M., and Nonacs, P. (2000). Genetic support for the evolutionary theory of reproductive transactions in social wasps. *Proceedings of the Royal Society of London B*, **267**, 75–79.

Ribeiro, M. de F., Wenseleers, T., Santos Filho, P. de S., and Alves, D. de A. (2006). Miniature queens in stingless bees: basic facts and evolutionary hypotheses. *Apidologie*, **37**, 191–206.

Richardson, D. H. S. (1999). War in the world of lichens: parasitism and symbiosis as exemplified by lichens and lichenicolous fungi. *Mycological Research*, **103**, 641–50.

Richardson, D. S., Burke, T., and Komdeur, J. (2002). Direct benefits and the evolution of female-biased cooperative breeding in Seychelles warblers. *Evolution*, **56**, 2313–21.

Richardson, D. S., Komdeur, J., and Burke, T. (2003a). Altruism and infidelity among warblers. *Nature*, **422**, 580.

Richardson, D. S., Burke, T., and Komdeur, J. (2003b). Sex-specific associative learning cues and inclusive fitness benefits in the Seychelles warbler. *Journal of Evolutionary Biology*, **16**, 854–61.

Richardson, D. S., Burke, T., and Komdeur, J. (2007). Grandparent helpers: the adaptive significance of older, postdominant helpers in the Seychelles warbler. *Evolution*, **61**, 2790–800.

Richerson, P. J., Boyd, R. T., and Henrich, J. (2003). Cultural evolution of human cooperation. In P. Hammerstein, ed. *Genetic and cultural evolution of cooperation*, pp. 357–88. The MIT Press, Cambridge MA.

Ridley, M. (2000). *Mendel's demon: gene justice and the complexity of life*. Weidenfeld & Nicolson, London.

Ridley, M. and Grafen, A. (1981). Are green beard genes outlaws? *Animal Behaviour*, **29**, 954–55.

Robinson, G. E., Grozinger, C. M., and Whitfield, C. W. (2005). Sociogenomics: social life in molecular terms. *Nature Review Genetics*, **6**, 257–70.

Robinson, G. E., Fernald, R. D., and Clayton, D. F. (2008). Genes and social behavior. *Science*, **322**, 896–900.

Rodríguez-Ezpeleta, N. and Philippe, H. (2006). Plastid origin: replaying the tape. *Current Biology*, **16**, R53–56.

Rodríguez-Ezpeleta, N., Brinkmann, H., Burey, S. C. et al. (2005). Monophyly of primary photosynthetic eukaryotes: green plants, red algae, and glaucophytes. *Current Biology*, **15**, 1325–30.

Rokas, A. (2008). The origins of multicellularity and the early history of the genetic toolkit for animal development. *Annual Review of Genetics*, **42**, 235–51.

Rolff, J. and Reynolds, S. E., eds. (2009). *Insect infection and immunity: evolution, ecology, and mechanisms*. Oxford University Press, Oxford.

Ross, K. G. and Keller, L. (1995). Ecology and evolution of social organization: insights from fire ants and other highly eusocial insects. *Annual Review of Ecology and Systematics*, **26**, 631–56.

Ross, K. G. and Matthews, R. W. (1989). Population genetic structure and social evolution in the sphecid wasp *Microstigmus comes*. *American Naturalist*, **134**, 574–98.

Ross-Gillespie, A., Gardner, A., West, S. A., and Griffin, A. S. (2007). Frequency dependence and cooperation: theory and a test with bacteria. *American Naturalist*, **170**, 331–42.

Rousset, F. (2004). *Genetic structure and selection in subdivided populations*. Princeton University Press, Princeton.

Roze, D. and Michod, R. E. (2001). Mutation, multilevel selection, and the evolution of propagule size during the origin of multicellularity. *American Naturalist*, **158**, 638–54.

Rubenstein, D. R. and Lovette, I. J. (2007). Temporal environmental variability drives the evolution of cooperative breeding in birds. *Current Biology*, **17**, 1414–19.

Ruiz-Trillo, I., Burger, G., Holland, P. W. H. et al. (2007). The origins of multicellularity: a multi-taxon genome initiative. *Trends in Genetics*, **23**, 113–18.

Russell, A. F. (2004). Mammals: comparisons and contrasts. In W. D. Koenig and J. L. Dickinson, eds. *Ecology and evolution of cooperative breeding in birds*, pp. 210–27. Cambridge University Press, Cambridge.

Russell, A. F. and Hatchwell, B. J. (2001). Experimental evidence for kin-biased helping in a cooperatively breeding vertebrate. *Proceedings of the Royal Society of London B*, **268**, 2169–74.

Russell, A. F. and Lummaa, V. (2009). Maternal effects in cooperative breeders: from hymenopterans to humans. *Philosophical Transactions of the Royal Society B*, **364**, 1143–67.

Ruvinsky, A. (1997). Sex, meiosis and multicellularity. *Acta Biotheoretica*, **45**, 127–41.

Sachs, J. L. (2008). Resolving the first steps to multicellularity. *Trends in Ecology and Evolution*, **23**, 245–48.

Sachs, J. L. and Bull, J. J. (2005). Experimental evolution of conflict mediation between genomes. *Proceedings of the National Academy of Sciences, USA*, **102**, 390–95.

Sachs, J. L. and Wilcox, T. P. (2006). A shift to parasitism in the jellyfish symbiont *Symbiodinium microadriaticum*. *Proceedings of the Royal Society B*, **273**, 425–29.

Sachs, J. L., Mueller, U. G., Wilcox, T. P., and Bull, J. J. (2004). The evolution of cooperation. *Quarterly Review of Biology*, **79**, 135–60.

Santorelli, L. A., Thompson, C. R. L., Villegas, E. et al. (2008). Facultative cheater mutants reveal the genetic complexity of cooperation in social amoebae. *Nature*, **451**, 1107–10.

Schaap, P., Winckler, T., Nelson, M. et al. (2006). Molecular phylogeny and evolution of morphology in the social amoebas. *Science*, **314**, 661–63.

Schmid-Hempel, P. (1998). *Parasites in social insects*. Princeton University Press, Princeton.

Schnable, P. S. and Wise, R. P. (1998). The molecular basis of cytoplasmic male sterility and fertility restoration. *Trends in Plant Science*, **3**, 175–80.

Schopf, J. W. (1994). The early evolution of life: solution to Darwin's dilemma. *Trends in Ecology and Evolution*, **9**, 375–77.

Schopf, J. W. (2006). Fossil evidence of Archaean life. *Philosophical Transactions of the Royal Society B*, **361**, 869–85.

Schurko, A. M., Neiman, M., and Logsdon, J. M. (2009). Signs of sex: what we know and how we know it. *Trends in Ecology and Evolution*, **24**, 208–17.

Schwander, T., Humbert, J.-Y., Brent, C. S. et al. (2008). Maternal effect on female caste determination in a social insect. *Current Biology*, **18**, 265–69.

Schwarz, M. P., Richards, M. H., and Danforth, B. N. (2007). Changing paradigms in insect social evolution: insights from halictine and allodapine bees. *Annual Review of Entomology*, **52**, 127–50.

Seeley, T. D. (1995). *The wisdom of the hive: the social physiology of honey bee colonies.* Harvard University Press, Cambridge MA.

Seeley, T. D. (1997). Honey bee colonies are group-level adaptive units. *American Naturalist*, **150**, S22–41.

Seeley, T. D. and Tarpy, D. R. (2007). Queen promiscuity lowers disease within honeybee colonies. *Proceedings of the Royal Society B*, **274**, 67–72.

Sessions, A. L., Doughty, D. M., Welander, P. V., Summons, R. E., and Newman, D. K. (2009). The continuing puzzle of the great oxidation event. *Current Biology*, **19**, R567–74.

Sherman, P. W., Seeley, T. D., and Reeve, H. K. (1988). Parasites, pathogens, and polyandry in social Hymenoptera. *American Naturalist*, **131**, 602–10.

Sherman, P. W., Lacey, E. A., Reeve, H. K., and Keller, L. (1995). The eusociality continuum. *Behavioral Ecology*, **6**, 102–108.

Sherman, P. W., Reeve, H. K., and Pfennig, D. W. (1997). Recognition systems. In J. R. Krebs and N. B. Davies, eds. *Behavioural ecology: an evolutionary approach*, 4th edn, pp. 69–96. Blackwell Science Ltd, Oxford.

Shik, J. Z. (2008). Ant colony size and the scaling of reproductive effort. *Functional Ecology*, **22**, 674–81.

Shimkets, L. J. (1990). Social and developmental biology of the Myxobacteria. *Microbiological Reviews*, **54**, 473–501.

Shimkets, L. J. (1999). Intercellular signaling during fruiting-body development of *Myxococcus xanthus*. *Annual Review of Microbiology*, **53**, 525–49.

Siddle, H. V., Kreiss, A., Eldridge, M. D. B. et al. (2007). Transmission of a fatal clonal tumor by biting occurs due to depleted MHC diversity in a threatened carnivorous marsupial. *Proceedings of the National Academy of Sciences, USA*, **104**, 16221–26.

Sigmund, K. (1992). On prisoners and cells. *Nature*, **359**, 774.

Sigmund, K. (2007). Punish or perish? Retaliation and collaboration among humans. *Trends in Ecology and Evolution*, **22**, 593–600.

Silk, J. B. (2009). Nepotistic cooperation in non-human primate groups. *Philosophical Transactions of the Royal Society B*, **364**, 3243–54.

Sledge, M. F., Dani, F. R., Cervo, R., Dapporto, L., and Turillazzi, S. (2001). Recognition of social parasites as nest-mates: adoption of colony-specific host cuticular odours by the paper wasp parasite *Polistes sulcifer*. *Proceedings of the Royal Society of London B*, **268**, 2253–60.

Smith, C. R., Toth, A. L., Suarez, A. V., and Robinson, G. E. (2008). Genetic and genomic analyses of the division of labour in insect societies. *Nature Reviews Genetics*, **9**, 735–48.

Smith, S. M., Beattie, A. J., Kent, D. S., and Stow, A. J. (2009). Ploidy of the eusocial beetle *Austroplatypus incompertus* (Schedl) (Coleoptera, Curculionidae) and implications for the evolution of eusociality. *Insectes Sociaux*, **56**, 285–88.

Smukalla, S., Caldara, M., Pochet, N. et al. (2008). *FLO1* is a variable green beard gene that drives biofilm-like cooperation in budding yeast. *Cell*, **135**, 726–37.

Sober, E. and Wilson, D. S. (1998). *Unto others: the evolution and psychology of unselfish behavior.* Harvard University Press, Cambridge MA.

Solari, C. A., Michod, R. E., and Goldstein, R. E. (2008). *Volvox barberi*, the fastest swimmer of the Volvocales (Chlorophyceae). *Journal of Phycology*, **44**, 1395–98.

Solomon, N. G. and French, J. A., eds. (1997). *Cooperative breeding in mammals.* Cambridge University Press, Cambridge.

Solomon, N. G., Richmond, A. R., Harding, P. A. et al. (2009). Polymorphism at the *avpr1a* locus in male prairie voles correlated with genetic but not social monogamy in field populations. *Molecular Ecology*, **18**, 4680–95.

Stacey, P. B. and Ligon, J. D. (1991). The benefits-of-philopatry hypothesis for the evolution of cooperative breeding: variation in territory quality and group size effects. *American Naturalist*, **137**, 831–46.

Stajich, J. E., Berbee, M. L., Blackwell, M. et al. (2009). The Fungi. *Current Biology*, **19**, R840–45.

Stander, P. E. (1992). Cooperative hunting in lions: the role of the individual. *Behavioral Ecology and Sociobiology*, **29**, 445–54.

Stanley, S. M. (1973). An ecological theory for the sudden origin of multicellular life in the Late Precambrian. *Proceedings of the National Academy of Sciences, USA*, **70**, 1486–89.

Stark, R. E. (1992). Cooperative nesting in the multivoltine large carpenter bee *Xylocopa sulcatipes* Maa (Apoidea: Anthophoridae): do helpers gain or lose to solitary females? *Ethology*, **91**, 301–10.

Starr, C. K. (1984). Sperm competition, kinship, and sociality in the Aculeate Hymenoptera. In R. L. Smith, ed. *Sperm competition and the evolution of animal mating systems*, pp. 427–64. Academic Press, Orlando.

Stearns, S. C. (2007). Are we stalled part way through a major evolutionary transition from individual to group? *Evolution*, **61**, 2275–80.

Stern, D. L. (1994). A phylogenetic analysis of soldier evolution in the aphid family Hormaphididae. *Proceedings of the Royal Society of London B*, **256**, 203–209.

Stern, D. L. (1998). Phylogeny of the tribe Cerataphidini (Homoptera) and the evolution of the horned soldier aphids. *Evolution*, **52**, 155–65.

Stern, D. L. and Foster, W. A. (1996). The evolution of soldiers in aphids. *Biological Reviews*, **71**, 27–79.

Stevens, M. I., Hogendoorn, K., and Schwarz, M. P. (2007). Evolution of sociality by natural selection on variances in reproductive fitness: evidence from a social bee. *BMC Evolutionary Biology*, **7**, 153.

Strassmann, J. E. and Queller, D. C. (2004). Genetic conflicts and intercellular heterogeneity. *Journal of Evolutionary Biology*, **17**, 1189–91.

Strassmann, J. E. and Queller, D. C. (2007). Insect societies as divided organisms: the complexities of purpose and cross-purpose. *Proceedings of the National Academy of Sciences, USA*, **104**, 8619–26.

Strassmann, J. E., Gastreich, K. R., Queller, D. C., and Hughes, C. R. (1992). Demographic and genetic evidence for cyclical changes in queen number in a neotropical wasp, *Polybia emaciata*. *American Naturalist*, **140**, 363–72.

Sudd, J. H. and Franks, N. R. (1987). *The behavioural ecology of ants*. Blackie, Glasgow.

Sundström, L. (1994). Sex ratio bias, relatedness asymmetry and queen mating frequency in ants. *Nature*, **367**, 266–68.

Sundström, L., Chapuisat, M., and Keller, L. (1996). Conditional manipulation of sex ratios by ant workers: a test of kin selection theory. *Science*, **274**, 993–95.

Sundström, L., Seppä, P., and Pamilo, P. (2005). Genetic population structure and dispersal patterns in *Formica* ants - a review. *Annales Zoologici Fennici*, **42**, 163–77.

Sutovsky, P., Moreno, R. D., Ramalho-Santos, J., Dominko, T., Simerly, C., and Schatten, G. (2000). Ubiquitinated sperm mitochondria, selective proteolysis, and the regulation of mitochondrial inheritance in mammalian embryos. *Biology of Reproduction*, **63**, 582–90.

Szathmáry, E. (2006). The origin of replicators and reproducers. *Philosophical Transactions of the Royal Society B*, **361**, 1761–76.

Szathmáry, E. and Demeter, L. (1987). Group selection of early replicators and the origin of life. *Journal of Theoretical Biology*, **128**, 463–86.

Szathmáry, E. and Wolpert, L. (2003). The transition from single cells to multicellularity. In P. Hammerstein, ed. *Genetic and cultural evolution of cooperation*, pp. 271–90. The MIT Press, Cambridge MA.

Szilágyi, A., Scheuring, I., Edwards, D. P., Orivel, J., and Yu, D. W. (2009). The evolution of intermediate castration virulence and ant coexistence in a spatially structured environment. *Ecology Letters*, **12**, 1306–16.

Takahashi, J., Martin, S. J., Ono, M., and Shimizu, I. (2010). Male production by non-natal workers in the bumblebee, *Bombus deuteronymus* (Hymenoptera: Apidae). *Journal of Ethology*, **28**, 61–66.

Tao, Y., Masly, J. P., Araripe, L., Ke, Y., and Hartl, D. L. (2007). A *sex-ratio* meiotic drive system in *Drosophila simulans*. I: An autosomal suppressor. *PLoS Biology*, **5**(11), e292.

Taylor, P. D. (1992). Altruism in viscous populations—an inclusive fitness model. *Evolutionary Ecology*, **6**, 352–56.

Taylor, P. D. and Frank, S. A. (1996). How to make a kin selection model. *Journal of Theoretical Biology*, **180**, 27–37.

Taylor, C. and Nowak, M. A. (2007). Transforming the dilemma. *Evolution*, **61**, 2281–92.

Taylor, T. N., Hass, H., Remy, W., and Kerp, H. (1995). The oldest fossil lichen. *Nature*, **378**, 244.

Taylor, P. D., Wild, G., and Gardner, A. (2007). Direct fitness or inclusive fitness: how shall we model kin selection? *Journal of Evolutionary Biology*, **20**, 301–309.

Thompson, J. N. (2005). *The geographic mosaic of coevolution*. University of Chicago Press, Chicago.

Thompson, G. J. (2006). Kin selection in disguise? *Insectes Sociaux*, **53**, 496–97.

Thompson, J. N. and Cunningham, B. M. (2002). Geographic structure and dynamics of coevolutionary selection. *Nature*, **417**, 735–38.

Thompson, G. J. and Oldroyd, B. P. (2004). Evaluating alternative hypotheses for the origin of eusociality in corbiculate bees. *Molecular Phylogenetics and Evolution*, **33**, 452–56.

Thompson, G. J., Kucharski, R., Maleszka, R., and Oldroyd, B. P. (2008). Genome-wide analysis of genes related to ovary activation in worker honey bees. *Insect Molecular Biology*, **17**, 657–65.

Thorne, B. L. (1997). Evolution of eusociality in termites. *Annual Review of Ecology and Systematics*, **28**, 27–54.

Tice, M. M. (2008). Modern life in ancient mats. *Nature*, **452**, 40–41.

Tofilski, A., Couvillon, M. J., Evison, S. E. F., Helanterä, H., Robinson, E. J. H., and Ratnieks, F. L. W. (2008). Preemptive defensive self-sacrifice by ant workers. *American Naturalist*, **172**, E239–43.

Toth, A. L., Varala, K., Newman, T. C. et al. (2007). Wasp gene expression supports an evolutionary link between maternal behavior and eusociality. *Science*, **318**, 441–44.

Tóth, E. and Duffy, J. E. (2008). Influence of sociality on allometric growth and morphological differentiation in sponge-dwelling alpheid shrimp. *Biological Journal of the Linnean Society*, **94**, 527–40.

Traulsen, A. and Nowak, M. A. (2006). Evolution of cooperation by multilevel selection. *Proceedings of the National Academy of Sciences, USA*, **103**, 10952–55.

Travis, J. M. J. (2004). The evolution of programmed death in a spatially structured population. *Journal of Gerontology: Biological Sciences*, **59A**, 301–305.

Travisano, M. and Velicer, G. J. (2004). Strategies of microbial cheater control. *Trends in Microbiology*, **12**, 72–78.

Trivers, R. L. (1971). The evolution of reciprocal altruism. *Quarterly Review of Biology*, **46**, 35–57.

Trivers, R. L. (1974). Parent-offspring conflict. *American Zoologist*, **14**, 249–64.

Trivers, R. (1985). *Social evolution*. Benjamin/Cummings, Menlo Park CA.

Trivers, R. L. and Hare, H. (1976). Haplodiploidy and the evolution of the social insects. *Science*, **191**, 249–63.

Tsuji, K. (1994). Inter-colonial selection for the maintenance of cooperative breeding in the ant *Pristomyrmex pungens*: a laboratory experiment. *Behavioral Ecology and Sociobiology*, **35**, 109–13.

Tudge, C. (2000). *The variety of life: a survey and celebration of all the creatures that have ever lived*. Oxford University Press, Oxford.

Turillazzi, S., Sledge, M. F., Dani, F. R., Cervo, R., Massolo, A., and Fondelli, L. (2000). Social hackers: integration in the host chemical recognition system by a paper wasp social parasite. *Naturwissenschaften*, **87**, 172–76.

Turner, P. E. and Chao, L. (1999). Prisoner's dilemma in an RNA virus. *Nature*, **398**, 441–43.

Van Zweden, J. S., Dreier, S., and d'Ettorre, P. (2009). Disentangling environmental and heritable nestmate recognition cues in a carpenter ant. *Journal of Insect Physiology*, **55**, 159–64.

Vehrencamp, S. L. (1979). The roles of individual, kin, and group selection in the evolution of sociality. In P. Marler and J. G. Vandenbergh, eds. *Handbook of behavioral neurobiology. Vol. 3. Social behavior and communication*, pp. 351–94. Plenum Press, New York.

Velicer, G. J. and Vos, M. (2009). Sociobiology of the Myxobacteria. *Annual Review of Microbiology*, **63**, 599–623.

Velicer, G. J. and Yu, Y. N. (2003). Evolution of novel cooperative swarming in the bacterium *Myxococcus xanthus*. *Nature*, **425**, 75–78.

Vermeij, G. J. (2006). Historical contingency and the purported uniqueness of evolutionary innovations. *Proceedings of the National Academy of Sciences, USA*, **103**, 1804–809.

Visscher, P. K. (1989). A quantitative study of worker reproduction in honey bee colonies. *Behavioral Ecology and Sociobiology*, **25**, 247–54.

Vos, M. and Velicer, G. J. (2009). Social conflicts in centimeter- and global-scale populations of the bacterium *Myxococcus xanthus*. *Current Biology*, **19**, 1763–67.

Wade, M. J. (1980). Kin selection: its components. *Science*, **210**, 665–67.

Wahaj, S. A., Van Horn, R. C., Van Horn, T. L. et al. (2004). Kin discrimination in the spotted hyena (*Crocuta crocuta*): nepotism among siblings. *Behavioral Ecology and Sociobiology*, **56**, 237–47.

Wahaj, S. A., Place, N. J., Weldele, M. L., Glickman, S. E., and Holekamp, K. E. (2007). Siblicide in the spotted hyena: analysis with ultrasonic examination of wild and captive individuals. *Behavioral Ecology*, **18**, 974–84.

Wai, T., Teoli, D., and Shoubridge, E. A. (2008). The mitochondrial DNA genetic bottleneck results from replication of a subpopulation of genomes. *Nature Genetics*, **40**, 1484–88.

Walboomers, J. M. M., Jacobs, M. V., Manos, M. M. et al. (1999). Human papillomavirus is a necessary cause of invasive cervical cancer worldwide. *Journal of Pathology*, **189**, 12–19.

Wang, J., Ross, K. G., and Keller, L. (2008). Genome-wide expression patterns and the genetic architecture of a fundamental social trait. *PLoS Genetics*, **4**(7), e1000127.

Wattanachaiyingcharoen, W., Oldroyd, B. P., Wongsiri, S., Palmer, K., and Paar, R. (2003). A scientific note on the mating frequency of *Apis dorsata*. *Apidologie*, **34**, 85–86.

Wcislo, W. T. and Danforth, B. N. (1997). Secondarily solitary: the evolutionary loss of social behavior. *Trends in Ecology and Evolution*, **12**, 468–74.

Wcislo, W. T. and Tierney, S. M. (2009). The evolution of communal behavior in bees and wasps: an alternative to eusociality. In J. Gadau and J. Fewell, eds. *Organization of insect societies: from genome to sociocomplexity*, pp. 148–69. Harvard University Press, Cambridge MA.

Wenseleers, T. and Ratnieks, F. L. W. (2001). Towards a general theory of conflict: the sociobiology of Mendelian segregation. In T. Wenseleers, *Conflict from cell to colony*, Ph.D. thesis, pp. 164–73, Katholieke Universiteit Leuven, Belgium.

Wenseleers, T. and Ratnieks, F. L. W. (2004). Tragedy of the commons in *Melipona* bees. *Proceedings of the Royal Society of London B (Supplement)*, **271**, S310–12.

Wenseleers, T. and Ratnieks, F. L. W. (2006a). Comparative analysis of worker reproduction and policing in eusocial Hymenoptera supports relatedness theory. *American Naturalist*, **168**, E163–79.

Wenseleers, T. and Ratnieks, F. L. W. (2006b). Enforced altruism in insect societies. *Nature*, **444**, 50.

Wenseleers, T., Ratnieks, F. L. W., and Billen, J. (2003). Caste fate conflict in swarm-founding social Hymenoptera: an inclusive fitness analysis. *Journal of Evolutionary Biology*, **16**, 647–58.

Wenseleers, T., Helanterä, H., Hart, A., and Ratnieks, F. L. W. (2004a). Worker reproduction and policing in insect societies: an ESS analysis. *Journal of Evolutionary Biology*, **17**, 1035–47.

Wenseleers, T., Hart, A. G., and Ratnieks, F. L. W. (2004b). When resistance is useless: policing and the evolution of reproductive acquiescence in insect societies. *American Naturalist*, **164**, E154–67.

Wenseleers, T., Gardner, A., and Foster, K. R. (2010). Social evolution theory: a review of methods and approaches. In T. Székely, J. Komdeur, and A. J. Moore, eds. *Social behaviour: genes, ecology and evolution*, pp. 132–158. Cambridge University Press, Cambridge.

Wenzel, J. W. and Pickering, J. (1991). Cooperative foraging, productivity, and the central limit theorem. *Proceedings of the National Academy of Sciences, USA*, **88**, 36–38.

Werren, J. H., Skinner, S. W., and Huger, A. M. (1986). Male-killing bacteria in a parasitic wasp. *Science*, **231**, 990–92.

Werren, J. H., Nur, U., and Wu, C.-I. (1988). Selfish genetic elements. *Trends in Ecology and Evolution*, **3**, 297–302.

West, S. A. (2009). *Sex allocation*. Princeton University Press, Princeton.

West, S. A. and Kiers, E. T. (2009). Evolution: what is an organism? *Current Biology*, **19**, R1080–82.

West, S. A., Lively, C. M., and Read, A. F. (1999). A pluralist approach to sex and recombination. *Journal of Evolutionary Biology*, **12**, 1003–12.

West, S. A., Murray, M. G., Machado, C. A., Griffin, A. S., and Herre, E. A. (2001). Testing Hamilton's rule with competition between relatives. *Nature*, **409**, 510–13.

West, S. A., Pen, I., and Griffin, A. S. (2002a). Cooperation and competition between relatives. *Science*, **296**, 72–75.

West, S. A., Kiers, E. T., Pen, I., and Denison, R. F. (2002b). Sanctions and mutualism stability: when should less beneficial mutualists be tolerated? *Journal of Evolutionary Biology*, **15**, 830–37.

West, S. A., Griffin, A. S., and Gardner, A. (2007a). Evolutionary explanations for cooperation. *Current Biology*, **17**, R661–72.

West, S. A., Griffin, A. S., and Gardner, A. (2007b). Social semantics: altruism, cooperation, mutualism, strong reciprocity and group selection. *Journal of Evolutionary Biology*, **20**, 415–32.

West, S. A., Diggle, S. P., Buckling, A., Gardner, A., and Griffin, A. S. (2007c). The social lives of microbes. *Annual Review of Ecology, Evolution, and Systematics*, **38**, 53–77.

West-Eberhard, M. J. (1975). The evolution of social behavior by kin selection. *Quarterly Review of Biology*, **50**, 1–33.

West-Eberhard, M. J. (1978). Polygyny and the evolution of social behavior in wasps. *Journal of the Kansas Entomological Society*, **51**, 832–56.

West-Eberhard, M. J. (1981). Intragroup selection and the evolution of insect societies. In R. D. Alexander and D. W. Tinkle, eds. *Natural selection and social behavior*, pp. 3–17. Chiron Press, New York.

West-Eberhard, M. J. (1987). Flexible strategy and social evolution. In Y. Itô, J. L. Brown, and J. Kikkawa, eds. *Animal societies. theories and facts*, pp. 35–51. Japan Scientific Societies Press, Tokyo.

West-Eberhard, M. J. (1988). Phenotypic plasticity and 'genetic' theories of insect sociality. In G. Greenberg and E. Tobach, eds. *Evolution of social behavior and integrative levels*, pp. 123–33. Lawrence Erlbaum, Hillsdale NJ.

West-Eberhard, M. J. (1992). Genetics, epigenetics, and flexibility: a reply to Crozier. *American Naturalist*, **139**, 224–26.

West-Eberhard, M. J. (2003). *Developmental plasticity and evolution*. Oxford University Press, New York.

Wheeler, W. M. (1910). *Ants. Their structure, development and behavior*. Columbia University Press, New York.

Wheeler, W. M. (1911). The ant-colony as an organism. *Journal of Morphology*, **22**, 307–25.

Wheeler, D. E. (1986). Developmental and physiological determinants of caste in social Hymenoptera: evolutionary implications. *American Naturalist*, **128**, 13–34.

Wheeler, D. E. (1991). The developmental basis of worker caste polymorphism in ants. *American Naturalist*, **138**, 1218–38.

White, D. J., Wolff, J. N., Pierson, M., and Gemmell, N. J. (2008). Revealing the hidden complexities of mtDNA inheritance. *Molecular Ecology*, **17**, 4925–42.

Whitehouse, M. E. A. and Lubin, Y. (2005). The functions of societies and the evolution of group living: spider societies as a test case. *Biological Reviews*, **80**, 347–61.

Wieser, W. (1997). A major transition in darwinism. *Trends in Ecology and Evolution*, **12**, 367–70.

Wilfert, L., Gadau, J., and Schmid-Hempel, P. (2007). Variation in genomic recombination rates among animal taxa and the case of social insects. *Heredity*, **98**, 189–97.

Willensdorfer, M. (2008). Organism size promotes the evolution of specialized cells in multicellular digital organisms. *Journal of Evolutionary Biology*, **21**, 104–10.

Willensdorfer, M. (2009). On the evolution of differentiated multicellularity. *Evolution*, **63**, 306–23.

Williams, G. C. (1966). *Adaptation and natural selection*. Princeton University Press, Princeton.

Williams, K. P., Sobral, B. W., and Dickerman, A. W. (2007). A robust species tree for the *Alphaproteobacteria*. *Journal of Bacteriology*, **189**, 4578–86.

Wilson, E. O. (1966). Behaviour of social insects. In P. T. Haskell, ed. *Insect behaviour*. RES Symposium No. 3, pp. 81–96. Royal Entomological Society, London.

Wilson, E. O. (1971). *The insect societies*. Belknap Press of Harvard University Press, Cambridge MA.

Wilson, E. O. (1975). *Sociobiology: the new synthesis*. Belknap Press of Harvard University Press, Cambridge MA.

Wilson, E. O. (1985). The sociogenesis of insect colonies. *Science*, **228**, 1489–95.

Wilson, E. O. (1992). *The diversity of life*. Belknap Press of Harvard University Press, Cambridge MA.

Wilson, D. S. (1997). Altruism and organism: disentangling the themes of multilevel selection theory. *American Naturalist*, **150**, S122–34.

Wilson, E. O. (2005). Kin selection as the key to altruism: its rise and fall. *Social Research*, **72**, 159–66.

Wilson, E. O. (2008). One giant leap: how insects achieved altruism and colonial life. *BioScience*, **58**, 17–25.

Wilson, E. O. and Hölldobler, B. (2005). Eusociality: origin and consequences. *Proceedings of the National Academy of Sciences, USA*, **102**, 13367–71.

Wilson, D. S. and Sober, E. (1989). Reviving the superorganism. *Journal of Theoretical Biology*, **136**, 337–56.

Wilson, D. S. and Wilson, E. O. (2007). Rethinking the theoretical foundation of sociobiology. *Quarterly Review of Biology*, **82**, 327–48.

Wilson, D. S. and Wilson, E. O. (2008). Evolution 'for the good of the group'. *American Scientist*, **96**, 380–89.

Wilson, D. S., Pollock, G. B., and Dugatkin, L. A. (1992). Can altruism evolve in purely viscous populations? *Evolutionary Ecology*, **6**, 331–41.

Wilson-Rich, N., Spivak, M., Fefferman, N. H., and Starks, P. T. (2009). Genetic, individual, and group facilitation of disease resistance in insect societies. *Annual Review of Entomology*, **54**, 405–23.

Winston, M. L. (1987). *The biology of the honey bee*. Harvard University Press, Cambridge MA.

Woelfing, B., Traulsen, A., Milinski, M., and Boehm, T. (2009). Does intra-individual major histocompatibility complex diversity keep a golden mean? *Philosophical Transactions of the Royal Society B*, **364**, 117–28.

Wolff, J. N. and Gemmell, N. J. (2008). Lost in the zygote: the dilution of paternal mtDNA upon fertilization. *Heredity*, **101**, 429–34.

Wolpert, L. (1990). The evolution of development. *Biological Journal of the Linnean Society*, **39**, 109–24.

Wolpert, L. and Szathmáry, E. (2002). Evolution and the egg. *Nature*, **420**, 745.

Wood, K. J., Bushell, A. R., and Jones, N. D. (2010). The discovery of immunological tolerance: now more than just a laboratory solution. *Journal of Immunology*, **184**, 3–4.

Woyciechowski, M. and Lomnicki, A. (1987). Multiple mating of queens and the sterility of workers among eusocial Hymenoptera. *Journal of Theoretical Biology*, **128**, 317–27.

Xu, J. (2005). The inheritance of organelle genes and genomes: patterns and mechanisms. *Genome*, **48**, 951–58.

Yamamura, N. and Higashi, M. (1992). An evolutionary theory of conflict resolution between relatives: altruism, manipulation, compromise. *Evolution*, **46**, 1236–39.

Yoon, H. S., Hackett, J. D., Ciniglia, C., Pinto, G., and Bhattacharya, D. (2004). A molecular timeline for the origin of photosynthetic eukaryotcs. *Molecular Biology and Evolution*, **21**, 809–18.

Young, L. J. and Hammock, E. A. D. (2007). On switches and knobs, microsatellites and monogamy. *Trends in Genetics*, **23**, 209–12.

Yu, D. W. (2001). Parasites of mutualisms. *Biological Journal of the Linnean Society*, **72**, 529–46.

Yuan, X., Xiao, S., and Taylor, T. N. (2005). Lichen-like symbiosis 600 million years ago. *Science*, **308**, 1017–20.

Zanette, L. R. S. and Field, J. (2008). Genetic relatedness in early associations of *Polistes dominulus*: from related to unrelated helpers. *Molecular Ecology*, **17**, 2590–97.

Zimmer, C. (2009a). On the origin of eukaryotes. *Science*, **325**, 666–68.

Zimmer, C. (2009b). On the origin of sexual reproduction. *Science*, **324**, 1254–56.

Author Index

Index to first authors and to second authors in papers with two authors.

Aanen, D. K. 78, 96, 133, 142, 158
Abbot, P. 102, 146
Abedin, M. 101
Abouheif, E. 171, 188–89
Adams, E. S. 177
Addicott, J. F. 112
Agnarsson, I. 14
Agrawal, A. F. 111, 122
Akino, T. 133
Alexander, R. D. 3, 21, 45, 50, 60, 66, 101–102, 113, 119, 126, 130, 153, 159, 163–64, 179, 185, 192
Alonso, W. J. 63, 68
Amdam, G. V. 69
Amoah-Buahin, E. 100
Anderson, C. 164, 168, 172, 193
Araten, D. J. 182
Archibald, J. M. 12, 75, 149
Arnold, G. 160
Austad, S. N. 142
Avilés, L. 14
Axelrod, R. 22, 105, 141
Ayre, D. J. 15

Baer, B. 136
Baglione, V. 115
Baldauf, S. L. 11
Banfield, W. G. 143
Bantinaki, E. 100
Bargum, K. 55, 63, 65–66, 101–102
Barr, C. M. 149
Baudry, E. 146

Beekman, M. 43, 132, 146, 153, 159–60, 193
Bell, G. 12, 89, 111, 115, 118, 137, 164, 169–70
Bennett, B. 132
Bennett, P. M. 14, 113, 121
Bentolila, S. 152
Bergmüller, R. 30, 116
Bergstrom, C. T. 76, 105, 111
Bever, J. D. 47–48
Bhattacharya, D. 12
Bijma, P. 78, 202
Birky, C. W. 149–50
Birmingham, A. L. 146
Blackstone, N. W. 105
Bonabeau, E. 193
Bone, Q. 126
Bonhoeffer, S. 127
Bonner, J. T. 3–4, 11, 13, 24, 52, 87–88, 92, 99, 109, 113–14, 122, 124–26, 162–64, 168–70, 172–73, 175–77, 179, 184
Boomsma, J. J. 5, 14, 21, 24–25, 42, 46, 60–61, 66–68, 86, 90, 101–103, 108, 110, 134, 136, 146, 163–64, 171, 176, 178, 192–93
Boraas, M. E. 98–100, 118
Borgia, G. 3, 45, 50, 130
Boswell, G. P. 177
Boucher, D. H. 111
Bourke, A. F. G. 3, 6, 10, 20–22, 25, 28, 31–33, 35, 42–44, 46, 50, 52, 56–62, 65–68, 70, 75, 85, 87, 90–92, 102, 107–110, 125, 130, 145, 152, 155, 159, 163–65, 168, 170–71, 173, 176–79, 181, 185–86, 189–93, 202–204
Boyd, R. 25, 31, 35, 47–48, 81
Boyle, R. A. 25
Brady, S. G. 17–18, 174
Branda, S. S. 13
Brandt, M. 135

Breed, M. D. 132
Brockmann, H. J. 113, 119
Brocks, J. J. 12
Bronstein, J. L. 112
Bshary, R. 30, 158
Bull, J. J. 105–106, 158
Burki, F. 12
Burland, T. M. 104
Burt, A. 5, 43–44, 50, 89, 106, 132–33, 136, 138–39, 142, 144, 148–52, 182
Burt, D. B. 14
Buss, L. W. 2–4, 8, 24, 52, 59, 86–90, 125, 130, 133, 141, 168–69, 173, 179–81, 183
Buston, P. M. 80
Butterfield, N. J. 12–14
Buttner, M. J. 13

Cairns-Smith, A. G. 11
Camazine, S. 25, 193
Cameron, S. A. 17
Canestrari, D. 117, 133
Cant, M. A. 80, 202
Cao, L. 150
Carlin, N. F. 165
Carr, M. 13
Carroll, S. B. 13, 26
Cavalier-Smith, T. 11
Châline, N. 43
Changizi, M. A. 4, 15, 21, 204
Chao, L. 59
Chapman, R. E. 204
Chapman, N. C. 147
Chapman, T. W. 17, 103–104, 119
Charlat, S. 153
Charlesworth, B. 52
Charnov, E. L. 60, 110
Chenoweth, L. B. 17, 178
Choe, J. C. 85, 101, 113, 163, 185, 187–88, 191
Clutton-Brock, T. H. 22, 25, 30, 59, 74, 108, 148, 154, 202
Cockburn, A. 14
Coggin, J. H. 143
Cole, B. J. 145–46, 176
Colman, A. M. 24
Cooper, M. D. 132
Cornwallis, C. K. 62
Cosmides, L. M. 3, 43, 149, 159
Costa, J. T. 201

Cousens, R. 91
Couzin, I. D. 142
Craig, R. 191–92
Creel, S. 126, 153
Cremer, S. 132–33
Crespi, B. J. 14, 45, 59, 92, 113, 119, 141
Crozier, R. H. 16, 20, 31, 33, 42, 59, 63, 65–67, 107, 163, 171–72
Cruz, Y. P. 17
Cunningham, B. M. 112
Currie, C. R. 131, 148

Dacks, J. 12, 111, 175
Danforth, B. N. 178
Darwin, C. 77
Davies, N. B. 135
Dawkins, R. 3–4, 8, 11, 13, 21, 29, 32, 40, 45, 50, 57–59, 65, 67, 79, 86–88, 105–106, 110, 130, 135, 152, 159, 177
Deisboeck, T. S. 142
Demeter, L. 24
Dendy, A. 2
Denison, R. F. 158
D'Ettorre, P. 132–33
Dickinson, J. L. 59, 108
Diggle, S. P. 61
Dijkstra, M. B. 146, 171
Dingli, D. 144
Dobata, S. 147
Dodgson, J. 100
Doebeli, M. 76–77, 112
Donaldson, Z. R. 52
Douglas, A. E. 158
Douzery, E. J. P. 12–13
Duffy, J. E. 17, 103, 119, 127, 171
Dugatkin, L. A. 24, 30, 65, 74
Duncan, L. 53
Dunn, C. W. 15, 108

Eberhard, W. G. 3, 43, 149
Edwards, D. P. 158
Edwards, S. V. 136
Eickwort, G. C. 118
Ekman, J. 121

Ellis, L. 153
Embley, T. M. 12, 25, 149
Emlen, S. T. 14, 42, 59, 67, 80, 108, 116, 119
Endler, A. 156–57, 160
Endler, J. A. 177
Engel, M. S. 16–17, 175, 177
Engelstädter, J. 43, 153

Farnworth, E. G. 111
Faulkes, C. G. 17, 104, 114, 127
Fehr, E. 25, 85, 155
Fewell, J. H. 136
Field, J. 93, 118–19, 124, 132, 202
Fink, S. 54
Fjerdingstad, E. J. 163, 171–72
Flärdh, K. 13
Fletcher, J. A. 76–77
Floreano, D. 60
Fischbacher, U. 25, 85
Foitzik, S. 133
Foster, K. R. 28, 35, 63, 65–68, 76–77, 80–81,
    83, 112, 126, 156
Foster, W. A. 16, 119–20
Fournier, D. 136
Frank, S. A. 5–6, 22, 24, 28, 35, 42, 57, 60–62,
    76, 80–81, 83–86, 89, 112, 141, 148, 155,
    158–59, 182
Frank, U. 144
Franks, N. R. 3, 6, 20, 22, 25, 28, 31–33, 35,
    42–43, 50, 52, 57, 58–59, 65–67, 70, 87, 91,
    102, 107–10, 130, 152, 159, 184, 189–90, 193,
    202–204
French, J. A. 14, 108
Fuchs, J. 15

Gächter, S. 155
Gadagkar, R. 60, 69–70, 119, 122, 124, 178
Gadau, J. 52, 202
Galtier, N. 149
Gardner, A. 6, 20, 28, 30, 35, 41–42, 50–51, 56,
    58–59, 66, 76, 80–81, 85, 103
Gaston, K. J. 111
Gautrais, J. 193
Gemmell, N. J. 150

Genoud, M. 192
Getz, W. M. 134
Gibbons, D. W. 133
Gilbert, O. M. 53, 55, 61, 93, 108–109
Gilley, D. C. 58
Giraud, T. 203
Giron, D. 17, 103
Gobin, B. 146
Golden, J. W. 13, 125, 169
Golley, F. B. 111
Goodnight, C. J. 57
Gordon, D. M. 70
Gotzek, D. 54–55
Gould, S. J. 57
Grafen, A. 6, 20, 22–23, 28, 31–34, 36–37, 41–42,
    48, 50–51, 57–61, 65, 68, 80, 152, 160, 202
Gray, M. W. 12
Grbić, M. 103
Griffin, A. S. 61–63, 106, 127, 202
Grimaldi, D. 16–17, 175, 177
Grogan, K. E. 146
Grosberg, R. K. 13, 15, 21, 24, 75, 86–89, 92, 97,
    99, 113–14, 125, 127, 130, 133, 142, 164, 173,
    179, 202
Gunnels, C. W. 116

Haig, D. 43, 55, 60, 152, 160, 202
Haine, E. R. 131
Hamilton, W. D. 3, 5, 21–23, 28, 30, 32, 34–35,
    41–42, 50, 57, 62, 65–66, 88, 105, 108, 130,
    136, 149–50, 175, 194, 198, 202
Hammerstein, P. 22, 24–25, 84–85
Hammock, E. A. D. 54
Hammond, R. L. 60, 62, 171
Hardin, G. 46, 79, 84
Hare, H. 42, 60–62, 68, 145, 153
Hart, A. G. 190
Härtel, S. 146
Hartmann, A. 156
Harvell, C. D. 15
Harvey, P. H. 105
Hatchwell, B. J. 59, 67, 69, 108, 113, 121
Haugsness, J. A. 112
Hawlena, H. 30, 41
Hedges, S. B. 12–13
Hedrick, P. W. 136
Heg, D. 118
Heinsohn, R. G. 60, 133

Heinze, J. 5, 24, 58, 69, 113, 119, 133, 202
Helanterä, H. 43, 55, 63, 65–66, 69, 86, 101–102, 171, 203–204
Helmkampf, M. 176
Henshaw, M. T. 93
Hepburn, H. R. 146
Herre, E. A. 158
Herron, M. D. 6, 13, 97, 106, 164, 168, 173–75
Higashi, M. 42
Hillesheim, E. 140
Himler, A. G. 90, 137
Hines, H. M. 17
Hochberg, M. E. 90, 127
Hoekstra, R. F. 43
Hoffman, P. F. 25
Hogendoorn, K. 68
Hölldobler, B. 9, 14, 58, 63–66, 69–70, 132, 136, 172, 177–78, 187, 190, 204
Holway, D. A. 204
Honeybee Genome Sequencing Consortium 26
Hornett, E. A. 153
Hou, C. 190
Hughes, W. O. H. 16–17, 46, 63, 65–66, 69, 101–103, 107, 136–37, 176, 193
Hunt, G. J. 52
Hunt, J. H. 17, 63–70, 127
Hurst, G. D. D. 43, 153
Hurst, L. D. 43–45, 149–50, 153, 159–60, 183

Inward, D. J. G. 16–17, 107, 174
Ishida, K. 12
Isoda, T. 143–44
Itô, Y. 70

Jaenike, J. 150–51
Jaffé, R. 176
Janeway, C. A. 132
Janzen, D. H. 112
Jarvis, J. U. M. 114, 132
Jeanne, R. L. 87, 163, 180–82, 185–86, 192
Jeanson, R. 193
Jemielity, S. 165
Jeon, J. 101, 163, 185, 187–88, 191
Johns, P. M. 133

Johnson, L. J. 183
Johnson, P. C. D. 102
Johnstone, R. A. 80
Jones, M. E. 143, 145
Jones, S. C. 102

Kaiser, D. 13, 26, 113, 118, 124–25, 169
Karatygin, I. V. 15
Keeling, P. J. 12
Keim, C. N. 13
Keller, L. 5–6, 22–24, 28, 30, 38, 41, 43, 51–52, 54–56, 60–62, 67, 69, 75, 78, 160, 165, 171, 189, 192, 202, 204
Kent, D. S. 16, 102
Kenyon, C. 89
Kerszberg, M. 115
Kessin, R. H. 118
Khila, A. 171, 188–89
Kiers, E. T. 9, 158
Kikuta, N. 156
Killingback, T. 22
King, N. 13, 16, 20, 26, 101, 113, 118
Kirk, D. L. 53, 88, 97–98, 106, 115, 118, 168
Kirkwood, T. B. L. 142
Kleiman, M. 122
Knoll, A. H. 2, 11–12, 14, 25, 121, 149
Knowlton, N. 112
Kock, D. 104
Koenig, W. D. 108, 113, 121
Kokko, H. 121
Komdeur, J. 59, 115–16, 133, 202
Kopp, R. E. 25
Korb, J. 5, 24, 53, 58, 69, 106, 113, 119, 202
Koufopanou, V. 90, 115, 125
Krause, J. 20, 118
Krebs, J. R. 135, 177
Krieger, M. J. B. 54
Kronauer, D. J. C. 171, 176
Kudô, K. 93
Kümmerli, R. 61
Kuzdzal-Fick, J. J. 126

Lachmann, M. 5, 86, 113–14, 126, 173
Langer, P. 60–61, 102, 116

Lattorff, H. M. G. 54
Lehmann, L. 6, 22–24, 28, 30, 38, 41, 49, 51, 78, 85
Lehtilä, K. 136
Leigh, E. G. 3, 5–6, 9, 44, 50, 95, 106, 112, 130, 148, 151, 160
Lenoir, A. 132–33, 135
Lenton, T. M. 25
Lessells, C. M. 150
Leys, R. 68
Liebig, J. 63, 65, 69
Ligon, J. D. 14, 115, 119
Lim, M. M. 54
Lin, N. 106, 117–18
Linksvayer, T. A. 202
Little, A. E. F. 148
Lomnicki, A. 145
Lopez-Vaamonde, C. 146
Love, G. D. 14
Lovette, I. J. 114, 127
Lubin, Y. 14
Lucas, E. R. 101–103
Lummaa, V. 202
Lürling, M. 97
Lutzoni, F. 15, 76, 122
Lynch, M. 26

Maccoll, A. D. C. 69
Macdonald, D. 126, 153
Macdonald, K. S. 17, 103, 127
Mackie, G. O. 2, 15, 19, 126
Magrath, R. D. 60
Marchetti, M. 96
Marcot, J. D. 4
Mardulyn, P. 17
Margulis, L. 11
Martin, S. J. 56, 133, 146–47
Martin, W. 12, 25, 148–49
Matsuura, K. 136
Matsuura, M. 70
Matthews, R. W. 101, 103
Mattila, H. R. 136
May, R. M. 112
Maynard Smith, J. 3–12, 24–25, 31, 60, 81, 86–87, 90, 92, 95, 106, 122, 137, 149–50, 152, 168, 178–79, 183, 202
Mayr, E. 4
McCallum, H. 143–45

McElreath, R. 31, 35, 47–48, 81
McRae, S. B. 133
McShea, D. W. 4, 15, 21, 164, 168, 172, 193, 204
Medzhitov, R. 132
Mehdiabadi, N. J. 108, 132
Metcalf, R. A. 60
Meunier, J. 61
Miadlikowska, J. 15, 76, 122
Michener, C. D. 96, 106, 117–18, 177–78
Michod, R. E. 5–6, 8, 13, 21, 24, 53, 56–57, 59, 86, 88–90, 97–98, 106, 127, 141, 163–64, 168, 173–75, 179–80, 182–83, 194
Mikheyev, A. S. 90
Mock, D. W. 40, 42
Monnin, T. 63, 65, 68–69, 153–55, 160
Mooers, A. O. 89, 164, 169–70
Moore, T. 60
Moreau, C. S. 17, 174
Moritz, R. F. A. 146–47
Morrison, C. L. 17
Mueller, U. G. 61, 77, 96, 158
Murchison, E. P. 143
Murgia, C. 132, 143–45

Nadell, C. D. 13
Nakayama, T. 12
Nalepa, C. A. 102
Nanork, P. 146
Narra, H. P. 9, 12
Nedelcu, A. M. 53, 56
Neumann, P. 146–47
Nishide, Y. 140
Noll, F. B. 174
Nonacs, P. 60–61, 80, 146, 156, 189
Noonan, K. M. 60
Nowak, M. A. 22, 112, 141, 144
Nowell, P. C. 141
Nunney, L. 129, 141, 182

Ochman, H. 9, 12
O'Donnell, S. 172
Ohkawara, K. 136
Ohtsuki, H. 22, 156, 187
Okasha, S. 4, 57, 59

Oldroyd, B. P. 17, 56, 132, 136, 146
O'Riain, M. J. 17, 104, 114, 132, 167, 170
Osman, R. W. 112
Oster, G. F. 122, 125, 165, 174
Ostrowski, E. A. 108, 132
Owen, R. E. 68
Owens, I. P. F. 14, 113, 121

Pacala, S. W. 20
Packer, L. 56, 68
Pál, C. 89–90, 168, 179, 183
Pamilo, P. 20, 31, 33, 42, 59, 61, 68, 107
Pancer, Z. 132
Parker, G. A. 22, 40, 42, 148, 150, 154
Passera, L. 204
Payne, J. L. 25
Pearcy, M. 136
Pearse, A.-M. 143
Peeters, C. 178
Pen, I. 121
Pennisi, E. 64
Pfeiffer, T. 127
Philippe, H. 12
Pickering, J. 126
Pike, N. 119
Pineda-Krch, M. 136
Pirk, C. W. W. 156
Platt, T. G. 47, 48
Pomiankowski, A. 43–45, 160

Queller, D. C. 5–9, 21, 23–24, 48–49, 52–53, 55,
    57–58, 60–61, 65, 67, 74–75, 77, 85–87, 89–93,
    95, 99, 107, 113, 119, 121–22, 130, 136,
    141–42, 149, 164

Rabeling, C. 91
Raff, R. A. 90
Rainey, K. 46, 100, 106, 126–27, 138–39
Rainey, P. B. 46, 90, 100, 106, 126–27, 138–39
Ramesh, M. A. 12

Rand, D. M. 11, 148
Rankin, D. J. 46–47, 79–80, 145
Rasmussen, B. 12
Rasmussen, G. S. A. 126
Ratnieks, F. L. W. 5–6, 22, 42–43, 46, 58–60, 62,
    64, 67–68, 79–80, 84–86, 90, 92, 130, 145–46,
    148, 153–56, 159–60, 165, 171, 176, 181,
    185–86, 189–91
Read, A. F. 105
Rebbeck, C. A. 143
Reber, A. 136
Reeve, H. K. 41–43, 57–58, 60–61, 64–65, 75,
    80, 104, 130, 134–35, 145–46, 153–56, 159–60,
    163, 180–82, 185–87, 192
Reynolds, S. E. 132
Ribeiro, M. de F. 191
Rice, W. R. 105, 158
Richardson, D. H. S. 131
Richardson, D. S. 67, 69
Richerson, P. J. 25
Ridley, M. 11, 13, 20, 24, 26, 51, 88–89, 106,
    148–50, 152, 160, 164, 175
Robinson, G. E. 26, 52, 55, 202
Rodríguez-Ezpeleta, N. 12
Roger, A. J. 12, 111, 175
Rokas, A. 26
Rolff, J. 132
Ross, K. G. 54–56, 101, 103, 204
Ross-Gillespie, A. 137
Rousset, F. 6, 28
Roze, D. 86, 90, 141, 179, 194
Rubenstein, D. R. 114, 127
Ruiz-Trillo, I. 11
Russell, A. F. 14, 67, 202
Ruvinsky, A. 106, 175
Ruxton, G. D. 20, 118

Sachs, J. L. 15, 24, 30, 76, 79, 95–96, 105–106,
    111–12, 122, 147, 158, 174
Santorelli, L. A. 92
Schaap, P. 13, 92, 169
Schmid-Hempel, P. 132, 136
Schnable, P. S. 44, 152
Schopf, J. W. 11–13
Schuck-Paim, C. 63, 68
Schurko, A. M. 111
Schwander, T. 189
Schwarz, M. P. 17, 101–102, 118

Seeley, T. D. 25, 58, 136
Sessions, A. L. 25
Shen, S.-F. 80
Sherman, P. W. 14, 133–34, 136
Shik, J. Z. 190
Shimkets, L. J. 13, 87
Siddle, H. V. 143–44
Sigmund, K. 24, 112, 155
Silk, J. B. 67
Simpson, J. A. 16, 102
Sixt, M. 132–33
Sledge, M. F. 135
Smith, A. 184
Smith, C. R. 52
Smith, S. M. 16, 102
Smukalla, S. 53, 100, 114
Sober, E. 57–58
Solari, C. A. 97
Solomon, N. G. 14, 54, 108
Stacey, P. B. 115, 119
Stajich, J. E. 174
Stander, P. E. 125
Stanley, S. M. 118
Stark, R. E. 60, 68, 117, 125
Starr, C. K. 145
Stearns, S. C. 24, 201
Stern, D. L. 16, 120
Stevens, M. I. 126
Strassmann, J. E. 6, 9, 24, 52, 58, 60–61, 67, 93,
    95, 113, 119, 130, 136, 142, 164
Strathmann, R. R. 13, 21, 24, 75, 86–89, 92, 97,
    99, 113–14, 125, 127, 130, 142, 164, 173,
    179, 202
Stuart, A. E. 131
Sudd, J. H. 184, 190
Summers, K. 141
Sundström, L. 40, 60, 93, 171
Sutovsky, P. 149
Swift, K. 143
Szathmáry, E. 3–11, 24–26, 60, 86–90, 92, 95,
    106, 113, 115, 122, 149–50, 152, 168, 173,
    178–79, 183, 202
Szilágyi, A. 112

Takahashi, J. 146
Tannenbaum, E. 122
Tao, Y. 151
Tarpy, D. R. 58, 136

Taylor, C. 22
Taylor, P. D. 48, 81, 83
Taylor, T. N. 15
Thompson, G. J. 17, 56, 63, 65–66
Thompson, J. N. 96, 112
Thorne, B. L. 16, 101–102, 174
Tice, M. M. 13
Tierney, S. M. 108
Tofilski, A. 185
Tooby, J. 3, 43, 149, 159
Toth, A. L. 26, 69
Tóth, E. 171
Traulsen, A. 22
Travis, J. M. J. 60
Travisano, M. 24, 80, 148
Trivers, R. L. 5, 22, 24, 42–44, 50, 60–62, 68, 76,
    89, 105–106, 132–33, 136, 138–39, 142,
    144–45, 148–53, 182
Trueman, E. R. 126
Tschinkel, W. R. 177
Tsuji, K. 140, 156, 187
Tudge, C. 11–13
Turillazzi, S. 135
Turner, P. E. 59

Van Donk, E. 97
Van Zweden, J. S. 133
Vehrencamp, S. L. 80
Velicer, G. J. 13, 24, 80, 87, 92, 99–100,
    108–109, 126, 148
Vermeij, G. J. 11
Visscher, P. K. 60, 85, 155
Vos, M. 13, 87, 92, 99–100, 108–109,
    126

Wade, M. J. 65, 70, 202
Wagner, G. P. 15
Wahaj, S. A. 40, 67
Wai, T. 150
Walboomers, J. M. M. 142
Wang, J. 54–55
Wattanachaiyingcharoen, W. 107
Wcislo, W. T. 108, 178
Weissing, F. J. 121

Wenseleers, T. 5–6, 22, 28, 35, 43, 46, 60, 62, 64–65, 68, 76–77, 81, 83–86, 112, 145, 148, 153–56, 159, 163, 185, 187–89
Wenzel, J. W. 126, 174
Werren, J. H. 43, 130, 153
West, S. A. 9, 22–24, 28–30, 38, 41, 47–51, 56–57, 59–60, 62, 65–66, 85, 111, 122, 158, 202
West-Eberhard, M. J. 26, 52, 66–67, 90, 106, 191–92, 194
Wheeler, D. E. 52, 172, 189–90
Wheeler, W. M. 3, 166
White, D. J. 149–50
Whitehouse, M. E. A. 14
Whitt, G. S. 60
Wieser, W. 4
Wilcox, T. P. 147
Wilfert, L. 202
Willensdorfer, M. 127, 169, 184
Williams, G. C. 3
Williams, K. P. 11
Wilson, D. S. 20, 48, 57–59, 63, 66–67
Wilson, E. O. 3, 5, 9, 11, 14–15, 20, 57–59, 63–67, 69–70, 122, 125, 132, 136, 153, 159, 163–65, 168, 172, 174, 177–78, 185, 190, 201, 204
Wilson-Rich, N. 132
Winston, M. L. 165, 181
Wise, R. P. 44, 152
Woelfing, B. 136
Wolff, J. N. 150
Wolpert, L. 26, 86–88, 90, 92, 113, 115, 173

Wood, K. J. 133
Woyciechowski, M. 145

Xu, J. 149–50

Yamamura, N. 42
Yamane, S. 70
Yanega, D. 14
Yoon, H. S. 12
Yoon, H.-S. 13, 125, 169
Young, L. J. 52, 54
Yu, D. W. 130
Yu, Y. N. 100
Yuan, X. 15

Zanette, L. R. S. 93
Zimmer, C. 12
Zink, A. G. 80
Zwick, M. 76

Subject Index

Acquiescence *see* Self-restraint
Actor
  Definition of 28–29
Adaptation 57, 200
  Group-level 58–59, 194
Aerobic respiration 111, 121
Ageing 60
  Evolutionary theory of 142
Aggregation development 86–88,
    92, 99
  Number of origins of 92
  Selfish mutants and 92
Alate 92
  Definition of 47
ALL/lymphoma 143
Allorecognition 131, 133
Altruism 5, 8, 14, 21, 28–30, 57, 65, 74, 86, 90,
    106–108, 188, 203
  Between- and within-species 74–78
  Definition of 28, 30
  Dispersal and 48–49, 92, 109
  Green-beard 50–51
  Hamilton's rule for 36–40, 48
  Problem of 21, 23
  Reciprocal 76
Anarchic phenotype (in Honey bee) 56
Ancient asexual 111
Anisogamy 150
Arrhenotoky 54
  Definition of 33
Arms race 134–36, 177
Asexual reproduction 11–12, 31, 54,
    105, 111, 136–37,
    140, 146
  See also Arrhenotoky; Thelytoky
Assured fitness return 122, 124
Autoimmunity 134

Autosome 151
Avpr1a (gene) 54

Balancing selection 137
Beneficiary 28
Benefits of philopatry 119
Biofilm 13
Biological hierarchy 2–4, 43, 57, 59, 178, 200
Body size 25, 89, 113, 121, 176–77, 184
  See also Multicellular organism, Size of (cell
    number)
Bottleneck 86–88, 91–93, 97, 134, 144
  Definition of 86
Broad-sense cooperation 21, 153, 155–56
  Definition of 30
Brood parasitism 30, 113, 117, 133–35
Budding-off 87, 91

Cadherin 99
Cancer 10, 88–89, 141–45, 182
  Non-transmissible 141–42
  Relatedness and 141
  Transmissible 132, 142–45, 147
  See also Cell lineage, Selfish
Canine transmissible venereal tumour 142–45
Cannibalism 30, 115
Caste 7, 14, 52, 137, 165
  See also Queen-worker dimorphism; Worker
    polymorphism

Caste determination 165, 172, 203
  Queen effect on 188–89
  Timing of in development 172,
    189–90
Caste dimorphism 10, 167–68, 170–71
  See also Queen-worker dimorphism
Caste fate 155, 185, 191, 194
  Conflict over 62, 155, 191
  Definition of 155
Cell *see* Eukaryotic cell; Prokaryotic cell
Cell adhesion 53, 100–101
Cell adhesion molecule 55, 99
Cell lineage
  Selfish 10, 88, 90, 132, 141, 183
  See also Cancer
Cell number *see* Multicellular organism, Size of
  (cell number)
Cell wall retaining daughter cells 97, 99–100
Chain rule 82
Cheating 10, 22, 90, 129, 137, 148, 155, 158–60
  Costs to group and 140–48
  Frequency-dependence and 137–40
Chemical mimicry 135
Chemical profile 156–57
Chimera 108, 136
Chloroplast 1, 11–12, 95, 121–22, 148–50
  Date of origin of 12
  Genome of 148
  Origin of 11–12
  See also Organelle
Chromosome 7, 152
  See also Sex chromosome; X chromosome;
    Y chromosome
Clonal society 90, 137, 140
Clone 31, 37, 42, 58, 66, 80, 88, 107–109, 127,
  143, 146, 179
Coalition of interest 160
Coercion 6, 22, 75, 80, 129, 148–58, 198
  Evolution of 22, 46, 84–86
  Relatedness and 84–86
Coincidence of fitness interests 42, 59, 75, 96,
  101, 109, 131, 181, 195, 200
Colony cycle 90–91
Colony foundation 30, 86–87, 90
Colony size *see* Eusocial society, Size of
Communal nesting 108
Community of interest 159
Complex social group
  Definition of 164–65, 167–68
  Delayed evolution of 174–75
  Number of origins of 172–74
  Sexual reproduction and 136, 175

  See also Eusocial society, Complex;
    Multicellular organism, Complex
Complexity 1–4, 25, 175–76, 200
Competition
  Between-group 70, 177
  Between-kin *see* Kin competition
Component of selection 65, 70
Conditionality in social evolution *see* Genes
  influencing social behaviour, Conditionality
  of
Conflict 3, 8, 10, 41–47, 58, 68–70, 79, 136–37
  Actual 43, 80
  Information and 43, 160
  Potential 42–43, 80, 87, 91–93, 203
  Power and 43, 153, 159–60, 185, 191
  See also Intergenomic conflict; Intragenomic
    conflict; Parent-offspring conflict;
    Queen-worker conflict
Conflict resolution 6, 46, 58, 79–86, 129, 191, 194
Constraint 43, 79–80
Contagious reticulum cell sarcoma 143
Contingent irreversibility 178
Cooperation 28–30, 106
  Evolution of 4, 21
  See also Broad-sense cooperation; Narrow-
    sense cooperation
Cooperative breeding 14, 106, 108, 114, 119,
  124, 168, 202
  Number of origins of 14–15
  See also Eusocial society, Simple
Cooperator association 112
  Definition of 112
Costs to group productivity 140–48
csA (gene) 53, 55
Cue-scrambling 160
Cultural inheritance 25
Cytoplasmic element 43
  See also Selfish genetic element
Cytoplasmic male sterility 44–45, 138–39, 149,
  152–53, 194
  Inclusive fitness theory and 44–45

Dance language 58
Defence 69
  Cooperative 130–31, 133
Development 25–26, 88, 101, 168
  See also Aggregative development; Unitary
    development

Devil facial tumour disease 142–45
Diploidy 31, 33, 67, 122, 181, 186
    Relatedness values under 31–33, 181
Direct benefit 21, 28, 51, 74, 78, 85, 123, 155
Direct fitness 34, 65, 69, 202
Direct fitness modelling approach 81
Dispersal 47–49, 67, 114, 122, 124
    Alternation with non-dispersal of 47, 49, 91–93
    See also Altruism, Dispersal and; Population
        viscosity
Division of labour 7, 122, 125, 136, 184
    Social group size and 184, 190
    See also Non-reproductive division of labour;
        Reproductive division of labour
DNA 7, 11, 149
    Selfish 10
    See also Mitochondrial DNA
Dog leukocyte antigen 144
Domatia 158
Dominance 10, 84, 140, 153, 156, 180, 185
    Definition of 153
    Social group size and 153, 159, 185–86
Dominance hierarchy 68
Drifter worker 146
Dufour's gland 154

Ecological constraint 119
Ecological factors
    Eusociality and 66, 113–19, 121
    Interspecific mutualisms and 111–13, 202
    Multicellularity and 113–19, 121
    Sexual reproduction and 111, 202
    Size of social groups and 176–77
    Social group formation and 110–21
Egalitarian major transition 7–9, 74–79, 86, 194,
    198, 201
    Definition of 7
Egg 32, 52, 92, 104, 135, 150, 152, 156, 165, 189
    Marking of 156–57, 160
Egg-eating 84–85, 155–56
    See also Worker policing
Embryogenesis 169
    See also Germline
Endosymbiosis 11, 121
    Number of occurrences of 12
    Primary 12
    Secondary 12
Environmental stress 113–114

Epigenetic theory of eusocial evolution 67
Ergonomic phase 187
ESS *see* Evolutionarily stable strategy
Eukaryote
    Definition of 1, 11
    Phylogeny of 16
Eukaryotic cell 5, 8, 11–12, 195
    Date of origin of 9, 12
    Number of origins of 12, 16
    Origin of 1, 3, 10–12, 75, 78, 95, 110, 121
Eusocial society 8, 10, 14, 58, 78, 90, 93, 106,
    147, 153–58, 178, 203
    Complex 14, 21, 110, 164–68, 170–72, 176, 192
    Relatedness levels within 102–104, 107
    Simple 14, 21, 101, 164–65, 167–68, 170–72
    Size of 91, 153, 162–65, 170–72, 176–77, 184–93
    See also Eusociality; Social group
Eusociality 5, 9, 14–15, 17, 19, 75
    Date of origin of 9, 14–18
    Definition of 14
    Evolution of 6, 20
    Number of origins of 9, 14–18
    Origin of 3, 9–10, 14–15, 20, 66–68, 78, 96,
        105, 113–19, 121–23
    Pathway in origin of 101–104
    Relatedness levels at origin of 102–104, 107
    See also Eusocial society
Evolutionarily stable strategy 80–81
    Definition of 81
Extracellular matrix 98–100

Facultative sociality 56, 60, 68, 113, 118
Fever
    Communal 132
Fitness trade-off model 179–80
Flight 2
*FLO*1 (gene) 53, 100
Flocculation 53
Food supply 113, 115
    Variance in 126
Foraging 13, 30, 58, 69, 100, 122–23, 125–26,
    177, 185, 193
    Central-place 122
Fortress defender 113, 119
Fossil 11–15, 18
Fossil molecule 12, 14
Fossil record 4, 11–12, 14–15, 175
Foundress association 86–87, 93

Fraternal major transition 7–9, 74–79, 194, 198, 201, 204
   Definition of 7
Frequency-dependence *see* Negative frequency-dependence
Fruiting body 13, 53, 55, 87–88, 108–109, 114, 124, 169
Functional monogyny 102–104
   Definition of 107
Fusion of insect colonies 133

Game theory 123
Gamete 12, 20, 32, 43–44, 87, 106, 112, 136, 141, 150, 152, 182–83
   Size of 150
Gender 150
Gene expression 26, 55
   Facultative *see* Genes influencing social behaviour, Conditionality of
Genes influencing social behaviour 26
   Conditionality of 45, 52, 67–68, 75, 89–90
   Examples of 52–56
Gene's-eye view *see* Selfish gene theory
Genetic code 7, 11
Genetic variation within social groups 69, 136–37, 175–76, 203
   Resistance to parasites and 136
Genome 5, 8–9, 24
   Evolution of 9–10, 202
   Nuclear 43, 75–76, 88, 105, 139, 148, 150, 152
   Organellar 11, 43, 75–76, 88, 148–50
   Origin of 10, 24
   Size of 175
   See also Organelle, Transfer of genes from
Genomic imprinting 60
Geographic mosaic theory of coevolution 112
Germline 10, 30, 89–90, 115, 164, 180, 182
   Cell number and 168–69
   Definition of 20
   Divergence time of 10, 89, 164, 168–69, 179, 182
   Number of cell divisions in 182
   Segregation of 10, 89–90, 164, 179, 182–83, 190–91
'Good for the species' view 41
gp80 (protein) 53–54
Gp-9 (gene) 54
Green beard gene 50–51, 65–66
   Definition of 50

Dynamics of 50–51
Examples of 53–56, 100
Outlaw gene concept and 51
Polymorphism in 56
Group selection 64–65
   Intrademic 57, 70
Group size *see* Eusocial society, Size of
Groupmate recognition *see* Nestmate recognition

Habitat saturation 116, 119, 204
Hamilton's rule 32, 34, 65–66, 79, 101, 106, 123, 191–92, 202
   Derivation of 35–36
   Dispersal and 40, 47–49
   Locus for social behaviour and 49–50
   Marginal 83
   Relatedness values and 34, 36–41
   Tests of 60, 69
Haplodiploidy 32–33, 66, 90, 102, 145, 181, 186, 189
   Definition of 32
   Relatedness values under 32–33, 181
Haplodiploidy hypothesis 66, 202
Haploidy 53
Helper 14, 60, 62, 67, 75, 102, 107, 115–18, 134, 147, 165, 171
Hermaphrodite 44, 138–39, 152
Heterocyst 125, 169
Heteroplasmy 149–50
Hierarchical organization *see* Biological hierarchy
Horizontal transmission 96, 105, 131, 143–44, 147–48, 158
Host–parasite relationship 60
Human leukocyte antigen 144
Hunting 125–26
Hydrocarbon 157
Hydrogenosome 149

Immune system 130–34, 144
Inbreeding 104
Inclusive fitness 34, 57–58
Inclusive fitness theory 3, 5, 21–23, 28–41, 43
   Assumptions of 47, 49–52, 55–56

Conflict and 42
Critiques of 22–23, 63–64, 68–70
Defence of 65–68
Evidence for 59–64
Falsifiability of 70–71
Gene expression and 52, 67
General theory of social evolution and 22, 41
Group selection and 65
Major transitions and 74
Multilevel selection theory and 57–59
Value of 57–71
See also Hamilton's rule
Indirect fitness 34, 65, 69, 202
Indirect genetic effect 202
Individual 20
Definition of 8
See also Individuality
Individuality 4, 9, 45, 194
Evolution of 2–3, 5, 7–8, 10–11, 20, 162, 195
See also Individual
Individual-level selection 4, 41
Infanticide 30, 156
Information *see* Conflict, Information and
Intergenomic conflict 58, 150, 152–53
Definition of 43
Interspecific mutualism 5, 9, 15, 75–78, 86, 147–48, 194–95, 201
Coercion in 158
Date of origin of 15
Evolution of 9
Number of origins of 15
Open 95–96, 147
Origin of 10, 15, 78, 111–12, 121–22
Intragenomic conflict 50, 58, 150–53
Definition of 43
Invasive species 204
Isogamy 150

Kin competition 40, 47–49, 63, 91–92
Kin discrimination 43, 62, 108, 132–33
Within-group 43, 65, 67, 202
Kin recognition *see* Kin discrimination
Kin selection 5, 28, 31
See also Inclusive fitness theory
Kin-selected conflict *see* Conflict
King (in social insects) 21, 30

Language 7, 25
Levels of selection *see* Multilevel selection
Life cycle 47
Classification of 87
Major transitions and 86–93
Life insurer 113, 119
Lifespan 182, 192
Limitation of exploitation 129, 159–60
Local population regulation 48
Locomotion 122, 126
Locus 31
Longevity *see* Lifespan

Major histocompatibility complex 132, 136, 144–45
Major transition in evolution 2–3, 7, 10–11, 70, 198, 200–201, 204
Classification of 78
Definition of 6–8
Evolution of individuality and 9–11
Genetic changes in 26
Number of 7, 9–11
Stages of 15, 20, 23
See also Egalitarian major transition; Fraternal major transition
Major transitions view of evolution 4, 24, 30, 60
Male-killer 45, 153
See also Selfish genetic element
Male parentage (in eusocial Hymenoptera) 42, 62, 64, 145, 156
See also Worker reproduction
Maternal effect 188, 202
Matriline 160
Meiosis 12
Fair 43, 86, 106, 149–52
Origin of 106, 202
Meiotic drive 10, 43, 50, 86, 150–52, 160
Recombination and 152, 160
Suppressors of 44, 86
Metastasis 141
Mitochondria 1, 11, 44, 95, 121, 138, 148–50
Altruism in 45
Clonality of 45
Date of origin of 12
Genome of 148
Origin of 11–12, 96
See also Organelle

Mitochondrial DNA 150
Mitosis 88, 105–106
Mitosome 149
Modifier gene 44, 49–51, 151–53
Monandry 101–104, 107
   Definition of 33
Monogamy 54
   Lifetime 110, 134, 192
Monogyny 101
   Definition of 33
   See also Functional monogyny
Multicellular organism 2–4, 8–10, 13–15, 20, 45,
   55, 58, 78, 87–88, 92–93, 106, 203
   Complex 13, 20, 25, 55, 92, 164–65, 167–68,
   175
   Simple 13, 20, 52, 106, 164–65, 168–69
   Size of (cell number) 89–91, 97–99, 162–65,
   168–70, 176–77, 179–84, 190–91
   See also Multicellularity; Social group
Multicellularity 4–5, 9, 13–14, 75
   Aquatic environment and 99, 109
   Date of origin of 9, 13–14
   Evolution of 6
   Experimental evolution of 98, 100
   Hamilton's rule and 101, 127
   Number of origins of 9, 13
   Origin of 1, 3, 9–10, 13–14, 78, 96–97, 105,
   113–19, 121–23, 202
   Oxygen levels and 25
   Pathway in origin of 97–99, 101
   Terrestrial environment and 99, 109
   See also Multicellular organism
Multicoloniality 171, 203
   Definition of 171–72
Multilevel selection 3, 57–59
Multiple mating *see* Polyandry
Mutation 92, 99, 101, 182–83
   See also Somatic mutation
Mutual benefit 29
Mutualism *see* Interspecific mutualism

Narrow-sense cooperation 5, 8, 21–22, 28, 30, 51,
   74–76, 78
   Definition of 28, 30
   Hamilton's rule for 36–38
Natural selection 3–4, 21–23, 25, 30, 32, 34,
   40–41, 57, 159, 200

Negative frequency-dependence 137–40
   Definition of 137
   Examples of 138–40
Neighbour-modulated modelling approach 81
Neofem2 (gene) 53
Nepotism 67, 160
   See also Kin discrimination
Nest 67, 93, 96, 116–119, 165, 171, 190, 203–204
   Building of 69
   Inheritance of 106, 187
Nest-site limitation 116, 118–19, 121
Nestmate recognition 130–33
   Cues used for 133
Nitrogen fixation 112, 125, 169
Non-reproductive division of labour 164–65,
   183–84, 190, 193, 204
   Definition of 164–65
Nucleus 11, 148
   See also Genome, Nuclear

Oocyte 188
Organelle 1, 8, 11, 76, 148–50
   Elimination of 86, 150
   Transfer of genes from 10, 75–76, 148–49
   Uniparental inheritance of 105, 148–50, 152
   See also Chloroplast; Genome, Organellar;
   Mitochondria
Organismality 9
Origin of life 11
Outbreeding 31–33, 43, 105, 149
Outlaw gene 50–51
   Definition of 50
Ovary 155, 171, 186, 188–89
Ovule 44–45, 138–39, 152
Oxygen 25, 184
   Rise in level of 25, 110

Parasitism 96, 117, 129–32, 136, 148, 175, 203
   See also Brood parasitism; Social parasitism
Parent-offspring conflict 42
Parental care 67, 107
Parental manipulation 189
Parliament of genes 152, 160

Partial differentiation 82
Partner choice 158
Partner fidelity 105, 112, 158
Pathway of social group formation 95–99, 101
    See also Semisociality; Subsociality
Patriline 43, 160
Pheromonal trail 193
Photosynthesis 2, 98, 121, 125
Physical worker caste *see* Worker polymorphism
Plastid 11
    See also Chloroplast
Point of no return 178
Policing 10, 22, 56, 81, 83, 155–56, 160, 203
    Definition of 84, 155
    Evolution of 84–86
    Social group size and 159, 186–89
    See also Worker policing; Worker reproduction
Pollen 44–45, 112, 138–39, 152
Polyandry 33, 103–104, 107, 136–37, 176, 186, 193
    Definition of 107
Polydomy 171
    Definition of 171
Polyembryony 104
Polygamy 54
Polygyny 107, 136–37
    Definition of 107
Polyp 15, 87, 108, 122
Population genetics 28, 34, 57
Population viscosity 47–48, 93, 108–109, 122
    Definition of 47
Positive feedback *see* Social evolution, Self-
    reinforcing
Power *see* Conflict, Power and
Predation 99, 110, 113, 118–19
Progress in evolution 4, 178
Prokaryote
    Definition of 1
    Multicellularity in 13
Prokaryotic cell 11
    Date of origin of 9
    Inclusive fitness theory and origin of 24
    Number of origins of 11
    Origin of 3, 8, 11, 24, 105, 202
Propagule 86
    Group 86–87, 91–93
    Unitary 86–88, 91–93, 181
Proximate factors in evolution 25, 69
Public good 45–46, 106
Punishment 10, 22, 24, 84, 154–56
    Definition of 153–54
    Social group size and 159

Queen (in social insects) 21, 30, 53–54, 86,
    156–57, 160, 165, 185
    Asexual production of 136
    Cull of (in stingless bees) 60, 155, 191
    Cycles in number of 93
    Dwarf 155, 191
    Frequency of larvae developing as 62, 68
    Replacement of 185, 192
Queen-worker conflict
    Male parentage and 42
    Sex ratio and 42–43
    See also Conflict
Queen-worker dimorphism 137, 165–66, 172,
    185–86, 189, 191
    See also Caste dimorphism

Recipient
    Definition of 28–29
Reciprocity 22, 25
    Strong 85
    See also Altruism, Reciprocal
Recognition systems 130–37, 147, 160
    Imperfection of 134–35, 156
Recombination 152, 160, 202
regA (gene) 53
Relatedness 22, 31–34, 66, 79–80, 180–81, 202
    Between- and within-species 75–78
    Bottleneck and 87–93
    Definition of 31
    Genetic valuation of others and 34, 40
    Evidence for effects on social behaviour
        of 61–62
    Hamilton's rule and 34–41
    Kinship and 49–50
    Levels at origin of eusociality of 102–104, 107
    Life-for-life 33
    Negative 36, 38, 41
    Regression 31, 33, 36, 83
    Secondary decrease in 69, 137
    Tragedy of the commons and 80–83
    See also Diploidy, Relatedness values under;
        Haplodiploidy, Relatedness values under
Relatedness asymmetry 44, 60–61, 76, 151,
    189
    Definition of 42
Replicator 7, 10–11, 57, 59, 78, 152, 200
Replicator-vehicle distinction 59

Reproductive (form in social insects) 21, 77, 90–91
Reproductive division of labour 14–15, 89, 164–65, 179–90, 204
  Definition of 14
Reproductive skew 61, 80, 123, 180, 187
  Definition of 60
Reproductive value 33
Restorer gene 152
  See also Modifier gene
Reversals in social evolution *see* Social evolution, Reversals in
RNA 7, 11
  Messenger 188
RNA interference 151
Robot 59
Russian doll analogy 59

Sanctions 10, 86, 158
Sedentariness 112, 122–23, 126
Seed 44–45, 91–92, 112
Self-limitation 6, 80, 84, 129, 159
Self-organization 25–26, 193
Self-replicating molecule 11
Self-restraint 85, 187–89, 203
Self versus non-self
  Interspecific mutualisms and 130–31
  Recognition of 130–37
Selfish gene theory 3–4, 21, 29, 40, 57, 59
Selfish genetic element 5, 129, 183
  See also Cytoplasmic element; Male-killer; Sex-ratio distorter
Selfishness 21, 28–30
  Definition of 28–30
  ESS level of 80–83
  Hamilton's rule for 36–37, 40, 80
Semisociality 99, 101, 108–109
  Definition of 96
  Relatedness levels and 108
Sessile habit *see* Sedentariness
Sex *see* Sexual reproduction
Sex allocation 60
  See also Sex ratio
Sex chromosome 150
Sex ratio 42, 60–61, 68, 151, 153
  See also Split sex ratio

Sex-ratio distorter 152–53
  See also Selfish genetic element
Sexual reproduction 12–13, 105
  Complexity and 175
  Date of origin of 13
  Evolution of 7, 10, 74, 86, 136, 175, 202
  Facultative 12
  Number of origins of 12
  Origin of 7, 12, 78, 95–96, 105, 111, 122, 202
Sexual selection 176
Shared genes 78–79, 101, 105, 121, 195
Shared reproductive fate 75, 78–79, 101, 105–106, 121, 148, 152, 158, 195
  Experimental evolution of 106
Siblicide 40
Siderophore 63
Simple social group
  Definition of 164–65, 167–68
  See also Eusocial society, Simple; Multicellular organism, Simple
Size of social group *see* Eusocial society, Size of; Multicellular organism, Size of (cell number); Social group, Size of
Size-complexity hypothesis 163, 168, 176, 179, 193–95, 198, 203
  Associations predicted by 164–72
  Interspecific mutualisms and 194–95
  Theoretical basis of 179–90
Size-complexity rule 162
Slug (of cellular slime mould) 87–88, 126, 169
Snowball Earth hypothesis 25
Social action 34
  Between- and within-species 74–78
  Definition of 28–30
  Evolvable types of 34, 36–41
  Gene for 32, 34
Social evolution
  Expanded view of 5, 200
  Molecular-genetic basis of 26, 202
  Open questions in 201–203
  Principles of 5–6, 23, 198–200
  Reversals in 18, 178
  Self-reinforcing 190–94
Social group
  Definition of 15, 20
  Size of 26, 159

See also Complex social group; Eusocial
   society; Multicellular organism; Simple
   social group
Social group formation 10, 15, 20–21, 23, 95, 198,
   204
   Definition of 20
Social group maintenance 10,15, 20–23, 79, 129,
   198
   Classification of processes of 129
   Definition of 20
Social group transformation 6, 10, 15, 20–21, 23,
   162, 198, 204
   Definition of 20
Social immunity 132
Social parasitism 30, 113, 117, 132, 134–35, 146–47
Social polymorphism 54
   See also Facultative sociality
Social traits
   Antecedents of 69
Soldier caste 119–20
Soma
   Definition of 20
Somatic cell 30, 53, 75, 89–90, 115, 164, 179
   Number of types of 20, 164, 168–70, 183–84, 190
   Polymorphism in 26
Somatic fusion 133, 136
Somatic mutation 88–90, 180
   Rate of 89–90, 182
   Number of cells and 89–90, 182
Sperm 87, 149–52
Sperm receptacle 87, 171, 186
Spermatogenesis 151
Spite 21, 28–30
   Definition of 28, 30
   Hamilton's rule for 36–37, 41
Split sex ratio 61
   Definition of 60
Spore 12, 20, 55, 87–89, 92, 109, 124, 183
Sterility in helper caste 39, 75, 134, 165
   See also Worker, Total sterility in
Sting 154
Subsociality 97–99, 101–104, 107, 109, 204
   Altruism and 110
   Definition of 96
   Relatedness levels and 107
Supercolony 203–204
Superorganism 9, 58
Suppressor gene *see* Modifier gene
Surface area to volume ratio 184
Symbiosis *see* Endosymbiosis; Interspecific mutualism

Synergistic factors
   Eukaryotic cell and 121
   Eusociality and 122–26
   Interspecific mutualisms and 121–22
   Multicellularity and 122–26
   Sexual reproduction and 122
   Social group formation and 121–26

Temporal facilitation 118, 121
Terminal differentiation 106, 164, 169, 171,
   173–74, 179
Territory 30, 115–16, 177
Thelytoky 55, 136, 140, 147
   Definition of 54
thelytoky (gene) 54
Thelytoky phenotype (in Honey bee) 56
Three-quarters relatedness hypothesis *see*
   Haplodiploidy hypothesis
Totipotency 89, 164–65, 183, 193, 204
   Definition of 89
Tragedy of the commons 45–47, 79–80, 84–85
   Model of 81–83
Tree 1, 177
Tropical environment 111

Ubiquitin 149
Ultimate factors in evolution 25, 69
Unicoloniality 75, 107, 133, 201, 203–204
   Definition of 203
   Ecological success of 204
   Number of origins of 203
Uniparental inheritance *see* Organelle,
   Uniparental inheritance of
Unitary development 86–89, 127, 136

V1aR (receptor) 54
Vehicle
   Definition of 59

Vertical transmission 96, 105, 131, 143–44, 148, 158
Virtual dominant 180–82, 185–86, 188, 192
 Definition of 182
Virulence 143–45, 147, 183
 Evolutionary theory of 105, 144, 148

Waste disposal 190
Wing 171
Worker (in social insects) 8, 14, 21, 30, 47, 53–54, 67, 75, 77, 90, 107, 134, 145, 156–57, 160, 165, 171, 203
 Frequency of reproductive 62, 68, 81, 85–86, 187–88
 Reproductive potential of 165, 171–72, 185–89, 191–92, 194
 Total sterility in 171, 174, 187–89, 203–204
 See also Worker reproduction
Worker policing 60, 62, 68, 84–86, 145–46, 155–56, 186–89, 194
 Theory of 145, 155
 See also Policing; Worker reproduction
Worker polymorphism 166, 174, 185, 193
 Colony size and 172–73, 190
 Definition of 165
 Degree of 165, 173
 Genetic component of 176
Worker reproduction 53, 56, 85–86, 145–47, 155–56, 181, 191–92, 203
 Colony size and 185–89
 Costs of 145–46, 156, 188, 203
 Intraspecific social parasitism and 146–47
 See also Male parentage (in eusocial Hymenoptera); Worker, Frequency of reproductive; Worker, Reproductive potential of
Wrinkly spreader mutant (in bacteria) 100, 138–39
wspf (gene) 100

X chromosome 150–51

Y chromosome 150–51

Zooid 15, 19, 30, 108, 122, 126, 171
Zygote 9, 26, 52, 86–90, 95, 105, 107, 136, 149–50, 182
 Evolution of 9
 Origin of 10, 12
 See also Sexual reproduction

Taxonomic Index

Acrocephalus sechellensis (Seychelles warbler) 115–16
Actinobacteria 13
Alga 12, 15, 53, 56, 76, 88, 97, 122, 131, 147–48, 150, 170, 195
  Brown 13, 174
  Cryptophyte 12
  Green 13, 97–98, 100
  Red 12–14, 174
  Volvocine 6, 13, 97–98, 106, 115, 118, 125, 165, 168, 173–74, 180
Amoebozoa 13
Anabaena (bacterium) 13, 125, 169
Anelosimus (spider) 14
Animal (Kingdom Animalia) 1, 7, 13, 99, 148, 152, 167–70, 174, 183
Animalia *see* Animal
Annelida 169
Ant 16–17, 30, 32, 40, 61–62, 85–87, 93, 102, 107, 133, 136–37, 147, 154, 156–58, 160, 165, 168, 171–75, 177, 189, 193, 195, 201, 203–204
  Argentine 203
  Army 166, 171, 177
  Fungus-growing 77, 96, 130–31, 148, 158, 190
  Leaf-cutter 171
  Ponerine 178
  Red imported fire 54–56, 177
Ant-plant 158, 195
Aphid 16, 102, 104, 119–20, 146
Apini 17, 102
Apis (bee) 107
Apis dorsata (Giant honey bee) 107
Apis mellifera (Honey bee) 43
Apis mellifera capensis (Cape honey bee) 54–55, 140, 146–147
Apis mellifera scutellata (bee) 146

Arabidopsis (plant) 149
Archaea 1, 11–12
Arthropod 119
Arthropoda 169
Ascomycota 15, 174
Atta (ant) 171–72, 176
Attini 130
Augochlorella striata (bee) 61
Austroplatypus incompertus (beetle) 16, 102, 104

Baboon 5
Bacillus subtilis (bacterium) 13
Bacteria 1, 11–12, 25, 30, 41, 45, 61, 63, 100, 106, 111, 126–27, 131–32, 138–39, 148, 153, 198
  Rhizobial 112, 121–22
Bangiomorpha (red alga) 12, 14
Basidiomycota 174
Bathyergidae 17, 104, 114
Bee 5, 17, 32, 60–62, 69, 85–86, 93, 102, 108, 171
  Allodapine 17, 61, 101–102, 104, 108–109, 116, 126, 168, 178
  Bumble 136, 146
  Cape honey 54–55, 140, 146
  Carpenter 117, 125
  Corbiculate 17, 102
  Giant honey 107
  Halictid 18, 56, 61, 118, 168, 178
  Honey 43, 56, 58, 84–85, 132, 136, 146–47, 155, 160, 162, 165, 181, 191
  Stingless 60, 62, 155, 165, 171, 191
  Trigonine stingless 191

Beetle
   Ambrosia 16, 102, 119
Bird 14, 40, 42, 135
   Social (cooperatively-breeding) 62, 67, 69,
     108, 133, 198
Blattodea 16–17, 102
Bombini 17, 102
Bombus terrestris (bee) 136
Bryozoa 15, 112
Butterfly
   Blue Moon 153

Caenorhabditis elegans (nematode) 89
Camponotus floridanus (ant) 156–57
Canis familiaris (Domestic dog) 143
Carnivore 153
Cassiopea xamachana (Upside-down jellyfish) 147
Cattle 46
Cherry 183
Chimpanzee 154
Chlamydomonas reinhardtii (alga) 97
Chlorella vulgaris (alga) 98–100, 118, 127
Chlorophyceae 13
Choanoflagellate 13, 99
Chordata 169
Chough
   White-winged 133
Clamator glandarius (Great spotted cuckoo) 117,
   133
Cnidaria 15, 169
Cockroach 17
Coleoptera 16, 102
Colonial marine invertebrate *see* Invertebrate,
   Colonial marine
Colophina monstrifica (aphid) 120
Coral 15, 112
Corcorax melanorhamphos (White-winged
   chough) 133
Cordia nodosa (plant) 158
Corvus corone (Carrion crow) 115, 117, 133
Crow
   Carrion 115, 117, 133
Crustacea 14
Cryptomys damarensis see Fukomys damarensis
Cryptotermes secundus (termite) 53
Ctenophora 169
Cuckoo
   Great spotted 117, 133

Cyanobacteria 11–13, 15, 76, 121–22, 125, 149,
   169

Devil
   Tasmanian 142–45
Dictyostelium (cellular slime mould) 13
Dictyostelium discoideum (cellular slime
   mould) 53, 55, 61, 87–88, 93, 108,
   118, 126
Dinoponera quadriceps (ant) 154, 160
Dog
   African wild 126
   Domestic 89, 142–44
Dorylus (ant) 171–72, 176–77
Dorylus helvolus (ant) 166
Drosophila (fly) 151
Drosophila simulans (fly) 151

Eciton (ant) 171–72, 176
Encyrtidae 17, 103
Entoprocta 169
Epiponini 174
Escovopsis (fungus) 130–31, 148
Eudorina (alga) 106
Eudorina elegans (alga) 97
Eukaryote 1, 3, 5, 7–13, 16, 25, 75, 92, 95, 105,
   110–11, 121, 148–50, 168, 170, 174–75,
   194–95
Euprymna scolopes (Bobtail squid) 95
Exoneura nigrescens (bee) 61, 104,
   116, 126
Exoneura robusta (bee) 104

Firmicutes 13
Fish 2, 89, 137
   Cichlid 116, 118
   Cleaner 158
Formica (ant) 171
Formicidae 17, 102
Fruitfly 151

Fungi (Kingdom Fungi) 1, 7, 13, 15, 76–77, 96, 122, 130–31, 148, 158, 167, 170, 174, 183, 195
Fukomys damarensis (Damaraland mole-rat) 104

Gallinula chloropus (Moorhen) 133
Gastrotricha 169
Gonium pectorale (alga) 97

Halictidae 17, 103
Halictus quadricinctus (bee) 18
Hamster
    Syrian 143
Hemiptera 16, 102
Heterocephalus glaber (Naked mole-rat) 104
Homo sapiens (human) 143
Hormaphididae 16, 102
Human 7, 24–25, 46, 88–89, 95, 130, 142–44, 155, 182, 184, 201
Hydra (hydrozoan) 87
Hydroid 112
Hymenoptera 17, 32, 66–69, 102, 107, 145
    Eusocial 24, 33, 42, 46, 52, 60–62, 64, 67–69, 81, 85, 87, 90–91, 102, 107, 119, 137, 140, 145–46, 153, 155–56, 160, 165–66, 171, 176, 181, 186–89, 193, 198
Hypolimnas bolina (butterfly) 153

Insect 14, 16, 68, 101, 107, 135, 137, 188
    Social 8–9, 20–21, 30, 47, 59–60, 63–64, 67–69, 70, 77, 87, 91–92, 95–96, 108, 132–34, 136, 138, 140, 165, 170–72, 181, 198, 202
Invertebrate 112, 122, 131–32, 170–71, 201
    Colonial marine 15, 19, 30, 108, 122–23, 126, 133, 168, 171, 190, 198
Isoptera 17

Jellyfish
    Upside-down 147–48

Labroides dimidiatus (fish) 158
Lasioglossum calceatum (bee) 18
Lasioglossum lustrans (bee) 18
Lepiotaceae 130
Lichen 15, 76, 122, 131, 195, 201
Linepithema humile (ant) 203
Lion
    African 125
Liostenogaster flavolineata (wasp) 118, 124
Lizard 137
Lycaon pictus (African wild dog) 126

Maize 44
Mammal 2, 14, 40, 42, 60, 142, 144, 149–50
    Social (cooperatively-breeding) 62, 67, 108, 198
Melipona (bee) 62, 155, 165, 191
Meliponini 17, 102
Mesocricetus auratus (Syrian hamster) 143
Mesozoa 169
Micro-organism 59, 134
Microstigmus (wasp) 17, 101–103, 108–109
Microstigmus comes (wasp) 101, 103–104
Microstigmus nigrophthalmus (wasp) 101, 103–104
Microtus (vole) 54
Mischocyttarus mexicanus (wasp) 116
Mole-rat 14, 17, 104, 114, 119
    Damaraland 104
    Naked 104, 123, 132, 167–68, 170, 198
Mollusca 169
Moorhen 133
Mosquito 143
Mouse 89
Myxobacteria 13, 87, 92, 99–100, 108–109, 126
Myxococcus xanthus (bacterium) 87, 100

Nanomia cara (siphonophore) 19
Nematoda 169
Nematode 89, 118
Neolamprologus pulcher (fish) 116, 118

Reptile 2
Rhodophyta 13
Rice 44
Rotifer
   Bdelloid 111
Rotifera 169

Ochromonas vallescia (alga) 98, 100

Pan troglodytes (Chimpanzee) 154
Panthera leo (African lion) 125
Papillomavirus 142
Paulinella chromatophora (cercozoan
   amoeba) 12
Pelican 30
Pemphigidae 16, 102
Pemphigus bursarius (aphid) 104
Pemphigus obesinymphae (aphid) 104
Pemphigus spyrothecae (aphid) 104, 120
Pemphredoninae 17, 103
Phaeophyta 13
Phlaeothripinae 17, 103
Placozoa 169
Plant (Kingdom Plantae) 1, 7, 13, 60, 91, 99, 112,
   119–120, 137, 141, 149, 158, 167, 170, 174,
   177, 183, 195, 198
   Flowering 44–45, 138–39, 152
   Leguminous 121
Plantae *see* Plant
Platyhelminthes 169
Pleodorina californica (alga) 97
Polistes dominulus (wasp) 93, 107
Polistinae 17, 103
Porifera 169
Primate 7, 153
Pristomyrmex punctatus (ant) 140, 147
*Pristomyrmex pungens see Pristomyrmex
   punctatus*
Prokaryote 1, 7, 9, 11–13, 105, 121, 169, 174–75,
   201
Proteobacteria 11, 13, 121
Pseudomonas aeruginosa (bacterium) 61, 63
Pseudomonas fluorescens (bacterium) 45, 100,
   126, 138–39
Pseudonocardia (bacterium) 131
Pseudoregma bambucicola (aphid) 120

Saccharomyces cerevisiae (Budding yeast) 53,
   55, 100, 114
Salmon 150
Salps 15
Salpa fusiformis (tunicate) 126
Sarcophilus harrisii (Tasmanian devil) 143
Scenedesmus acutus (alga) 97, 99
Schizosaccharomyces pombe (Fission
   yeast) 100
Sea Anemone 15
Shrimp 14, 17, 103, 119, 171
Siphonophore 15, 19, 30
Slime mould
   Cellular 13, 53, 55, 61, 87–88, 92– 93, 99,
   108–109, 114, 118–19, 126, 132, 165,
   169–70
Solenopsis invicta (Red imported fire ant) 54–56,
   177
Spider 14
Sponge 14
Squid
   Bobtail 95, 121
Starling
   African 114
Stenogastrinae 17, 103
Streptomyces (bacterium) 13
Sturnidae 114
Symbiodinium microadriaticum (alga) 147
Synalpheus (shrimp) 17, 103

Termite 16–17, 53, 87, 102, 106–107, 119, 136,
   168, 171, 174–75, 177, 198
   Fungus-growing 96, 158
Thrips 17, 103–104, 119
Thysanoptera 17, 103
Tiger 1
Tunicate 15, 126

Urochordata 15

Vertebrate 89, 91, 132–33, 136,
    170, 201
  Social (cooperatively-breeding) 14, 59, 70,
    119, 133, 153, 168, 170, 202
Vespidae 17, 103
Vespinae 17, 103
Vibrio fischeri (bacterium)
    95, 121
Viroid 143
Virus 59, 142–43
  Bacteriophage 106
Vole 54
Volvocine alga *see* Alga,
    Volvocine
Volvox (alga) 88
Volvox aureus (alga) 97
Volvox barberi (alga) 97
Volvox carteri (alga) 53, 56,
    97, 115

Warbler, Seychelles 115–16
Wasp 17, 32, 60–62, 69, 85–86, 93, 101, 103,
    108–109, 171
  Fig 47
  Paper 93, 107, 116, 153, 162
  Parasitoid 17, 103
  Polyembryonic 17, 102–104
  Stenogastrine 118, 124
  Swarm-founding 87, 93, 174
Wolbachia (bacterium) 45, 153
Wolf 92, 143

Xylocopa sulcatipes (bee) 117, 125

Yeast 55
  Budding 53, 55, 100, 114
  Fission 100